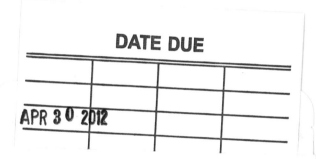

# NONLINEAR
# CONTROL SYSTEMS

# NONLINEAR CONTROL SYSTEMS
## ANALYSIS AND DESIGN

**HORACIO J. MARQUEZ**

*Department of Electrical and Computer Engineering,*
*University of Alberta, Canada*

A JOHN WILEY & SONS, INC., PUBLICATION

*Library of Congress Cataloging-in-Publication Data is available.*

ISBN 0-471-42799-3

Printed in the United States of America.

10 9 8 7 6 5 4 3 2

*To my wife, Goody (Christina);*
*son, Francisco;*
*and daughter, Madison*

# Contents

## 10 Feedback Linearization                                                    255

## 11 Nonlinear Observers                                                        291

# Preface

I began writing this textbook several years ago. At that time my intention was to write a research monograph with focus on the input–output theory of systems and its connection with robust control, including a thorough discussion of passivity and dissipativity of systems. In the middle of that venture I began teaching a first-year graduate-level course in nonlinear control, and my interests quickly shifted into writing something more useful to my students. The result of this effort is the present book, which doesn't even resemble the original plan. I have tried to write the kind of textbook that I would have enjoyed myself as a student. My goal was to write something that is *thorough*, yet *readable*.

The first chapter discusses linear and nonlinear systems and introduces phase plane analysis. Chapter 2 introduces the notation used throughout the book and briefly summarizes the basic mathematical notions needed to understand the rest of the book. This material is intended as a reference source and not a full coverage of these topics. Chapters 3 and 4 contain the essentials of the Lyapunov stability theory. Autonomous systems are discussed in Chapter 3 and nonautonomous systems in Chapter 4. I have chosen this separation because I am convinced that the subject is better understood by developing the main ideas and theorems for the simpler case of autonomous systems, leaving the more subtle technicalities for later. Chapter 5 briefly discusses feedback stabilization based on backstepping. I find that introducing this technique right after the main stability concepts greatly increases students' interest in the subject. Chapter 6 considers input–output systems. The chapter begins with the basic notions of extended spaces, causality, and system gains and introduces the concept of input-output stability. The same chapter also discusses the stability of feedback interconnections via the celebrated small gain theorem. The approach in this chapter is classical; input–output systems are considered without assuming the existence of an internal (i.e. state space) description. As such, Chapters 3-5 and 6 present two complementary views of the notion of stability: Lyapunov, where the focus is on the stability of equilibrium points of *unforced* systems (i.e. without external excitations); and the input–output theory, where systems are assumed to be relaxed (i.e. with zero initial conditions) and subject to an external input. Chapter 7 focuses on the important concept of input-to-state stability and thus starts to bridge across the two alternative views of stability. In Chapters 8 and 9 we pursue a rather complete discussion of dissipative systems, an active area of research, including its importance in the so-called nonlinear $\mathcal{L}_2$ gain control problem. Passive systems are studied first in Chapter

8, along with some of the most important results that derive from this concept. Chapter 9 generalizes these ideas and introduces the notion of dissipative system. I have chosen this presentation for historical reasons and also because it makes the presentation easier and enhances the student's understanding of the subject. Finally, Chapters 10 and 11 provide a brief introduction to feedback linearization and nonlinear observers, respectively.

Although some aspects of control design are covered in Chapters 5, 9, and 10, the emphasis of the book is on analysis and covers the fundamentals of the theory of nonlinear control. I have restrained myself from falling into the temptation of writing an encyclopedia of everything ever written on nonlinear control, and focused on those parts of the theory that seem more fundamental. In fact, I would argue that most of the material in this book is essential enough that it should be taught to every graduate student majoring in control systems.

There are many examples scattered throughout the book. Most of them are not meant to be real-life applications, but have been designed to be pedagogical. My philosophy is that real physical examples tend to be complex, require elaboration, and often distract the reader's attention from the main point of the book, which is the explanation of a particular technique or a discussion of its limitations.

I have tried my best to clean up all the typographical errors as well as the more embarrassing mistakes that I found in my early writing. However, like many before me, and the many that will come after, I am sure that I have failed! I would very much appreciate to hear of any error found by the readers. Please email your comments to

marquez@ee.ualberta.ca

I will keep an up-to-date errata list on my website:

http://www.ee.ualberta.ca/~marquez

Like most authors, I owe much to many people who directly or indirectly had an influence in the writing of this textbook. I will not provide a list because I do not want to forget anyone, but I would like to acknowledge four people to whom I feel specially indebted: Panajotis Agathoklis (University of Victoria), Chris Damaren (University of Toronto), Chris Diduch, and Rajamani Doraiswami (both of the University of New Brunswick). Each one of them had a profound impact in my career, and without their example this book would have never been written. I would also like to thank the many researchers in the field, most of whom I never had the pleasure to meet in person, for the beautiful things that they have published. It was through their writings that I became interested in the subject. I have not attempted to list every article by every author who has made a contribution to nonlinear control, simply because this would be impossible. I have tried to acknowledge those references that have drawn my attention during the preparation of my lectures and later during the several stages of the writing of this book. I sincerely apologize to every author who may feel that his or her work has not been properly acknowledged here and encourage them to write to me.

I am deeply grateful to the University of Alberta for providing me with an excellent working environment and to the Natural Sciences and Engineering Research Council of Canada (NSERC) for supporting my research. I am also thankful to John Wiley and Son's representatives: John Telecki, Kristin Cooke Fasano, Kirsten Rohstedt and Brendan Cody, for their professionalism and assistance.

I would like to thank my wife Goody for her encouragement during the writing of this book, as well as my son Francisco and my daughter Madison. To all three of them I owe many hours of quality time. Guess what guys? It's over (until the next project). Tonight I'll be home early.

Horacio J. Marquez

Edmonton, Alberta

# Chapter 1

# Introduction

This first chapter serves as an introduction to the rest of the book. We present several simple examples of dynamical systems, showing the evolution from linear time-invariant to nonlinear. We also define the several classes of systems to be considered throughout the rest of the book. Phase-plane analysis is used to show some of the elements that characterize nonlinear behavior.

## 1.1 Linear Time-Invariant Systems

In this book we are interested in nonlinear dynamical systems. The reader is assumed to be familiar with the basic concepts of state space analysis for linear time-invariant (LTI) systems. We recall that a state space realization of a finite-dimensional LTI system has the following form

$$\dot{x} = Ax + Bu \qquad (1.1)$$
$$y = Cx + Du \qquad (1.2)$$

where $A$, $B$, $C$, and $D$ are (real) constant matrices of appropriate dimensions. Equation (1.1) determines the dynamics of the response. Equation (1.2) is often called the read out equation and gives the desired output as a linear combination of the states.

**Example 1.1** *Consider the mass–spring system shown in the Figure 1.1. Using Newton's second law, we obtain the equilibrium equation*

1

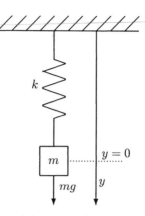

Figure 1.1: mass–spring system.

$$my'' = \sum forces$$
$$= f(t) - f_k - f_\beta$$

where $y$ is the displacement from the reference position, $f_\beta$ is the viscous friction force, and $f_k$ represents the restoring force of the spring. Assuming linear properties, we have that $f_\beta = \beta \dot{y}$, and $f_k = ky$. Thus,

$$m\ddot{y} + \beta \dot{y} + ky = mg.$$

Defining states $x_1 = y, x_2 = \dot{y}$, we obtain the following state space realization

$$\begin{cases} \dot{x}_1 = x_2 \\ \dot{x}_2 = -\frac{k}{m}x_1 - \frac{\beta}{m}x_2 + g \end{cases}$$

or

$$\begin{bmatrix} \dot{x}_1 \\ \dot{x}_2 \end{bmatrix} = \begin{bmatrix} 0 & 1 \\ -\frac{k}{m} & -\frac{\beta}{m} \end{bmatrix} \begin{bmatrix} x_1 \\ x_2 \end{bmatrix} + \begin{bmatrix} 0 \\ 1 \end{bmatrix} g$$

If our interest is in the displacement $y$, then

$$y = x_1 = \begin{bmatrix} 1 & 0 \end{bmatrix} \begin{bmatrix} x_1 \\ x_2 \end{bmatrix}$$

thus, a state space realization for the mass–spring systems is given by

$$\dot{x} = Ax + Bu$$
$$y = Cx + Du$$

*with*

$$A = \begin{bmatrix} 0 & 1 \\ -\frac{k}{m} & -\frac{\beta}{m} \end{bmatrix} \quad B = \begin{bmatrix} 0 \\ 1 \end{bmatrix} \quad C = \begin{bmatrix} 1 & 0 \end{bmatrix} \quad D = [0]$$

□

## 1.2 Nonlinear Systems

Most of the book focuses on nonlinear systems that can be modeled by a finite number of first-order ordinary differential equations:

$$\dot{x}_1 = f_1(x_1, \cdots, x_n, t, u_1, \cdots, u_p)$$
$$\vdots \tag{1.3}$$
$$\dot{x}_n = f_n(x_1, \cdots, x_n, t, u_1, \cdots, u_p)$$

Defining vectors

$$x = \begin{bmatrix} x_1 \\ \vdots \\ x_n \end{bmatrix}, \quad u = \begin{bmatrix} u_1 \\ \vdots \\ u_p \end{bmatrix}, \quad f(t, x, u) = \begin{bmatrix} f_1(x, t, u) \\ \vdots \\ f_n(x, t, u) \end{bmatrix}$$

we can re write equation (1.3) as follows:

$$\dot{x} = f(x, t, u). \tag{1.4}$$

Equation (1.4) is a generalization of equation (1.1) to nonlinear systems. The vector $x$ is called the *state vector* of the system, and the function $u$ is the input. Similarly, the system output is obtained via the so-called *read out equation*

$$y = h(x, t, u). \tag{1.5}$$

Equations (1.4) and (1.5) are referred to as the *state space realization* of the nonlinear system.

**Special Cases:**

- An important special case of equation (1.4) is when the input $u$ is identically zero. In this case, the equation takes the form

$$\dot{x} = f(x, t, 0) = f(x, t). \tag{1.6}$$

This equation is referred to as the *unforced* state equation. Notice that, in general, there is no difference between the unforced system with $u = 0$ or any other given function $u(x,t)$ (i.e., $u$ is not an arbitrary variable). Substituting $u = \gamma(x,t)$ in equation (1.4) eliminates $u$ and yields the unforced state equation.

- The second special case occurs when $f(x,t)$ *is not* a function of time. In this case we can write

$$\dot{x} = f(x) \qquad\qquad (1.7)$$

  in which case the system is said to be *autonomous*. Autonomous systems are invariant to shifts in the time origin in the sense that changing the time variable from $t$ to $\tau = t - \alpha$ does not change the right-hand side of the state equation.

Throughout the rest of this chapter we will restrict our attention to *autonomous systems*.

**Example 1.2** *Consider again the mass spring system of Figure 1.1. According to Newton's law*

$$
\begin{aligned}
m\ddot{y} &= \sum forces \\
&= f(t) - f_k - f_\beta.
\end{aligned}
$$

*In Example 1.1 we assumed linear properties for the spring. We now consider the more realistic case of a hardening spring in which the force strengthens as $y$ increases. We can approximate this model by taking*

$$f_k = ky(1 + a^2 y^2).$$

*With this constant, the differential equation results in the following:*

$$m\ddot{y} + \beta\dot{y} + ky + ka^2 y^3 = f(t).$$

*Defining state variables $x_1 = y, x_2 = \dot{y}$ results in the following state space realization*

$$
\begin{cases}
\dot{x}_1 = x_2 \\
\dot{x}_2 = -\frac{k}{m}x_1 - \frac{k}{m}a^2 x_1^3 - \frac{\beta}{m}x_2 + \frac{f(t)}{m}
\end{cases}
$$

*which is of the form $\dot{x} = f(x,u)$. In particular, if $u = 0$, then*

$$
\begin{cases}
\dot{x}_1 = x_2 \\
\dot{x}_2 = -\frac{k}{m}x_1 - \frac{k}{m}a^2 x_1^3 - \frac{\beta}{m}x_2
\end{cases}
$$

*or $\dot{x} = f(x)$.*                                                                                        □

## 1.3   Equilibrium Points

An important concept when dealing with the state equation is that of *equilibrium point*.

**Definition 1.1** *A point $x = x_e$ in the state space is said to be an equilibrium point of the autonomous system*

$$\dot{x} = f(x)$$

*if it has the property that whenever the state of the system starts at $x_e$, it remains at $x_e$ for all future time.*

According to this definition, the equilibrium points of (1.6) are the real roots of the equation $f(x_e) = 0$. This is clear from equation (1.6). Indeed, if

$$\dot{x} = \frac{dx}{dt} = f(x_e) = 0$$

it follows that $x_e$ is constant and, by definition, it is an equilibrium point. Equilibrium point for unforced nonautonomous systems can be defined similarly, although the time dependence brings some subtleties into this concept. See Chapter 4 for further details.

**Example 1.3** *Consider the following first-order system*

$$\dot{x} = r + x^2$$

*where $r$ is a parameter. To find the equilibrium points of this system, we solve the equation $r + x^2 = 0$ and immediately obtain that*

(i) *If $r < 0$, the system has two equilibrium points, namely $x = \pm\sqrt{r}$.*

(ii) *If $r = 0$, both of the equilibrium points in (i) collapse into one and the same, and the unique equilibrium point is $x = 0$.*

(iii) *Finally, if $r > 0$, then the system has no equilibrium points.* □

## 1.4   First-Order Autonomous Nonlinear Systems

It is often important and illustrative to compare linear and nonlinear systems. It will become apparent that the differences between linear and nonlinear behavior accentuate as the order of the state space realization increases. In this section we consider the simplest case, which is that of first order (linear and nonlinear) *autonomous* systems. In other words, we consider a system of the form

$$\dot{x} = f(x) \tag{1.8}$$

where $x(t)$ is a real-valued function of time. We also assume that $f(\cdot)$ is a continuous function of $x$. A very special case of (1.8) is that of a first-order *linear* system. In this case, $f(x) = ax$, and (1.8) takes the form

$$\dot{x} = ax. \tag{1.9}$$

It is immediately evident from (1.9) that the only equilibrium point of the first-order linear system is the origin $x = 0$. The simplicity associated with the linear case originates in the simple form of the differential equation (1.9). Indeed, the solution of this equation with an arbitrary initial condition $x_0 \neq 0$ is given by

$$x(t) = e^{at} x_0 \tag{1.10}$$

A solution of the differential equation (1.8) or (1.9) starting at $x_0$ is called a *trajectory*. According to (1.10), the trajectories of a first order linear system behave in one of two possible ways:

- Case (1), $a < 0$: Starting at $x_0$, $x(t)$ exponentially converges to the origin.

- Case (2), $a > 0$: Starting at $x_0$, $x(t)$ diverges to infinity as $t$ tends to infinity.

Thus, the equilibrium point of a first-order linear system can be either attractive or repelling. Attractive equilibrium points are called *stable*[1], while repellers are called *unstable*.

Consider now the nonlinear system (1.8). Our analysis in the linear case was guided by the luxury of knowing the solution of the differential equation (1.9). Unfortunately, most nonlinear equations cannot be solved analytically[2]. Short of a solution, we look for a qualitative understanding of the behavior of the trajectories. One way to do this is to acknowledge the fact that the differential equation

$$\dot{x} = f(x)$$

represents a *vector field* on the line; that is, at each $x$, $f(x)$ dictates the "velocity vector" $\dot{x}$ which determines how "fast" $x$ is changing. Thus, representing $\dot{x} = f(x)$ in a two-dimensional plane with axis $x$ and $\dot{x}$, the sign of $\dot{x}$ indicates the direction of the motion of the trajectory $x(t)$.

---

[1]See Chapter 3 for a more precise definition of the several notions of stability.
[2]This point is discussed in some detail in Chapter 2.

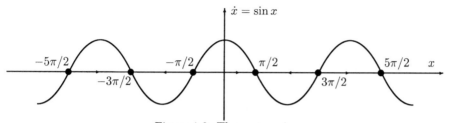

Figure 1.2: The system $\dot{x} = \cos x$.

**Example 1.4** *Consider the system*

$$\dot{x} = \cos x \tag{1.11}$$

*To analyze the trajectories of this system, we plot $\dot{x}$ versus $x$ as shown in Figure 1.2. From the figure we notice the following:*

- *The points where $\dot{x} = 0$, that is where $\cos x$ intersects the real axis, are the equilibrium points of (1.11). Thus, all points of the form $x = (1+2k)\pi/2, \quad k = 0, \pm 1, \pm 2, \cdots$ are equilibrium points of (1.11).*

- *Whenever $\dot{x} > 0$, the trajectories move to the right hand side, and vice versa. The arrows on the horizontal axis indicate the direction of the motion.*

*From this analysis we conclude the following:*

1. *The system (1.11) has an infinite number of equilibrium points.*

2. *Exactly half of these equilibrium points are attractive or stable, and the other half are unstable, or repellers.* □

The behavior described in example 1.4 is typical of first-order autonomous nonlinear systems. Indeed, a bit of thinking will reveal that the dynamics of these systems is dominated by the equilibrium points, in the sense that the only events that can occur to the trajectories is that either (1) they approach an equilibrium point or (2) they diverge to infinity. In all cases, trajectories are forced to either converge or diverge from an equilibrium point *monotonically*. To see this, notice that $\dot{x}$ can be either positive or negative, but it cannot change sign without passing through an equilibrium point. Thus, oscillations around an equilibrium point can <u>never</u> exist in first order systems. Recall from (1.10) that a similar behavior was found in the case of *linear* first-order systems.

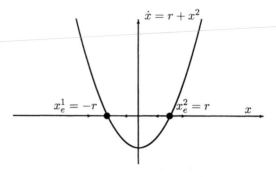

Figure 1.3: The system $\dot{x} = r + x^2$, $r < 0$.

**Example 1.5** *Consider again the system of example 1.3, that is*

$$\dot{x} = r + x^2$$

*Plotting $\dot{x}$ versus $x$ under the assumption that $r < 0$, we obtain the diagram shown in Figure 1.3. The two equilibrium points, namely $x_e^1$, and $x_e^2$, are shown in the Figure. From the analysis of the sign of $\dot{x}$, we see that $x_e^1$ is attractive, while $x_e^2$ is a repeller. Any trajectory starting in the interval $x_0 = (-\infty, -r) \cup (-r, r)$ monotonically converge to $x_e^1$. Trajectories initiating in the interval $x_0 = (r, \infty)$, on the other hand, diverge toward infinity.*  □

## 1.5   Second-Order Systems: Phase-Plane Analysis

In this section we consider second-order systems. This class of systems is useful in the study of nonlinear systems because they are easy to understand and unlike first-order systems, they can be used to explain interesting features encountered in the nonlinear world. Consider a second-order autonomous system of the form:

$$\dot{x}_1 = f_1(x_1, x_2) \tag{1.12}$$
$$\dot{x}_2 = f_2(x_1, x_2) \tag{1.13}$$

or

$$\dot{x} = f(x). \tag{1.14}$$

Throughout this section we assume that the differential equation (1.14) with initial condition $x(0) = x_0 = [x_{10}, x_{20}]^T$ has a unique solution of the form $x(t) = [x_1(t), x_2(t)]^T$. As in the case of first-order systems, this solution is called a *trajectory* from $x_0$ and can be represented graphically in the $x_1 - x_2$ plane. Very often, when dealing with second-order systems, it is useful to visualize the trajectories corresponding to various initial conditions in the $x_1 - x_2$ plane. The technique is known as phase-plane analysis, and the $x_1 - x_2$ plane is usually referred to as the phase-plane.

From equation (1.14), we have that

$$\dot{x} = \begin{bmatrix} \dot{x}_1 \\ \dot{x}_2 \end{bmatrix} = \begin{bmatrix} f_1(x) \\ f_2(x) \end{bmatrix} = f(x).$$

The function $f(x)$ is called a *vector field* on the state plane. This means that to each point $x^*$ in the plane we can assign a vector with the amplitude and direction of $f(x^*)$. For easy visualization we can represent $f(x)$ as a vector based at $x$; that is, we assign to $x$ the directed line segment from $x$ to $x + f(x)$. Repeating this operation at every point in the plane, we obtain a *vector field diagram*. Notice that if

$$x(t) = \begin{bmatrix} x_1(t) \\ x_2(t) \end{bmatrix}$$

is a solution of the differential equation $\dot{x} = f(t)$ starting at a certain initial state $x_0$, then $\dot{x} = f(x)$ represents the *tangent vector* to the curve. Thus it is possible to construct the trajectory starting at an arbitrary point $x_0$ from the vector field diagram.

**Example 1.6** *Consider the second-order system*

$$\begin{aligned} \dot{x}_1 &= x_2 \\ \dot{x}_2 &= -x_1^2 - x_2. \end{aligned}$$

*Figure 1.4 shows a phase-plane diagram of trajectories of this system, along with the vector field diagram. Given any initial condition $x_0$ on the plane, from the phase diagram it is easy to sketch the trajectories from $x_0$. In this book we do not emphasize the manual construction of these diagrams. Several computer packages can be used for this purpose. This plot, as well as many similar ones presented throughout this book, was obtained using MAPLE 7.* $\square$

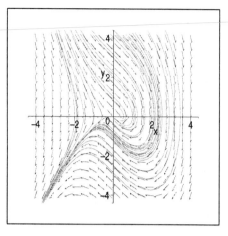

Figure 1.4: Vector field diagram form the system of Example 1.6.

## 1.6   Phase-Plane Analysis of Linear Time-Invariant Systems

Now consider a linear time-invariant system of the form

$$\dot{x} = Ax, \qquad A \in \mathbb{R}^{2 \times 2} \tag{1.15}$$

where the symbol $\mathbb{R}^{2 \times 2}$ indicates the set of $2 \times 2$ matrices with real entries. These systems are well understood, and the solution of this differential equation starting at $t = 0$ with an initial condition $x_0$ has the following well-established form:

$$x(t) = e^{At} x_0.$$

We are interested in a qualitative understanding of the form of the trajectories. To this end, we consider several cases, depending on the properties of the matrix $A$. Throughout this section we denote by $\lambda_1, \lambda_2$ the eigenvalues of the matrix $A$, and $v_1, v_2$, the corresponding eigenvectors.

### CASE 1: Diagonalizable Systems

Consider the system (1.15). Assume that the eigenvalues of the matrix $A$ are real, and define the following coordinate transformation:

$$x = Ty, \quad T \in \mathbb{R}^{2 \times 2}, \ T \text{ nonsingular}. \tag{1.16}$$

Given that $T$ is nonsingular, its inverse, denoted $T^{-1}$ exists, and we can write

$$\dot{y} = T^{-1}ATy = Dy \qquad (1.17)$$

Transformation of the form $D = T^{-1}AT$ are very well known in linear algebra, and the matrices $A$ and $D$ share several interesting properties:

**Property 1**: The matrices $A$ and $D$ share the same eigenvalues $\lambda_1$ and $\lambda_2$. For this reason the matrices $A$ and $D$ are said to be *similar*, and transformations of the form $T^{-1}AT$ are called *similarity transformations*.

**Property 2**: Assume that the eigenvectors $v_1, v_2$ associated with the real eigenvalues $\lambda_1, \lambda_2$ are linearly independent. In this case the matrix $T$ defined in (1.17) can be formed by placing the eigenvectors $v_1$ and $v_2$ as its columns. In this case we have that

$$D = \begin{bmatrix} \lambda_1 & 0 \\ 0 & \lambda_2 \end{bmatrix}$$

that is, in this case the matrix $A$ is similar to the diagonal matrix $D$. The importance of this transformation is that in the new coordinates $y = [y_1 \ y_2]^T$ the system is uncoupled, i.e.,

$$\begin{bmatrix} \dot{y}_1 \\ \dot{y}_2 \end{bmatrix} = \begin{bmatrix} \lambda_1 & 0 \\ 0 & \lambda_2 \end{bmatrix} \begin{bmatrix} y_1 \\ y_2 \end{bmatrix}$$

or

$$\dot{y}_1 = \lambda_1 y_1 \qquad (1.18)$$
$$\dot{y}_2 = \lambda_2 y_2 \qquad (1.19)$$

Both equations can be solved independently, and the general solution is given by (1.10). This means that the trajectories along each of the coordinate axes $y_1$ and $y_2$ are independent of one another.

Several interesting cases can be distinguished, depending the sign of the eigenvalues $\lambda_1$ and $\lambda_2$. The following examples clarify this point.

**Example 1.7** *Consider the system*

$$\begin{bmatrix} \dot{x}_1 \\ \dot{x}_2 \end{bmatrix} = \begin{bmatrix} -1 & 3 \\ 0 & -2 \end{bmatrix} \begin{bmatrix} x_1 \\ x_2 \end{bmatrix}.$$

*The eigenvalues in this case are $\lambda_1 = -1, \lambda_2 = -2$. The equilibrium point of a system where both eigenvalues have the same sign is called a* **node**. *$A$ is diagonalizable and $D = T^{-1}AT$ with*

$$T = \begin{bmatrix} 1 & -3 \\ 0 & 1 \end{bmatrix}, \quad D = \begin{bmatrix} -1 & 0 \\ 0 & -2 \end{bmatrix}.$$

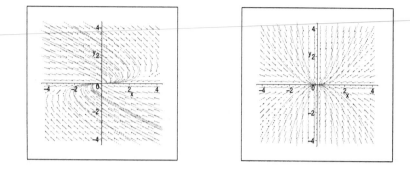

Figure 1.5: System trajectories of Example 1.7: (a) original system, (b) uncoupled system.

*In the new coordinates $y = Tx$, the modified system is*

$$\dot{y}_1 = -y_1$$
$$\dot{y}_2 = -2y_2$$

*or $\dot{y} = Dy$, which is uncoupled. Figure 1.5 shows the trajectories of both the original system [part (a)] and the uncoupled system after the coordinate transformation [part (b)]. It is clear from part (b) that the origin is attractive in both directions, as expected given that both eigenvalues are negative. The equilibrium point is thus said to be a* stable node. *Part (a) retains this property, only with a distortion of the coordinate axis. It is in fact worth nothing that Figure 1.5 (a) can be obtained from Figure 1.5 (b) by applying the linear transformation of coordinates $x = Ty$.*                                                                            □

**Example 1.8** *Consider the system*

$$\begin{bmatrix} \dot{x}_1 \\ \dot{x}_2 \end{bmatrix} = \begin{bmatrix} 1 & 3 \\ 0 & 2 \end{bmatrix} \begin{bmatrix} x_1 \\ x_2 \end{bmatrix}.$$

*The eigenvalues in this case are $\lambda_1 = 1, \lambda_2 = 2$. Applying the linear coordinate transformation $x = Ty$, we obtain*

$$\dot{y}_1 = y_1$$
$$\dot{y}_2 = 2y_2$$

*Figure 1.6 shows the trajectories of both the original and the uncoupled systems after the coordinate transformation. It is clear from the figures that the origin is repelling on both*

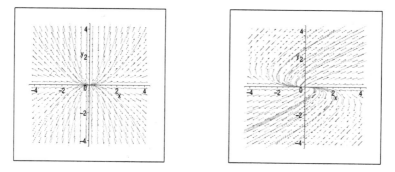

Figure 1.6: System trajectories of Example 1.8 (a) uncoupled system, (b) original system

*directions, as expected given that both eigenvalues are positive. The equilibrium point in this case is said to be an* unstable node. ☐

**Example 1.9** *Finally, consider the system*

$$\left[\begin{array}{c} \dot{x}_1 \\ \dot{x}_2 \end{array}\right] = \left[\begin{array}{cc} -1 & 3 \\ 0 & 2 \end{array}\right] \left[\begin{array}{c} x_1 \\ x_2 \end{array}\right].$$

*The eigenvalues in this case are $\lambda_1 = -1, \lambda_2 = 2$. Applying the linear coordinate transformation $x = Ty$, we obtain*

$$\begin{array}{rcl} \dot{y}_1 & = & -y_1 \\ \dot{y}_2 & = & 2y_2 \end{array}$$

*Figure 1.7 shows the trajectories of both the original and the uncoupled systems after the coordinate transformation. Given the different sign of the eigenvalues, the equilibrium point is attractive in one direction but repelling in the other. The equilibrium point in this case is said to be a* saddle. ☐

## CASE 2: Nondiagonalizable Systems

Assume that the eigenvalues of the matrix $A$ are *real* and *identical* (i.e., $\lambda_1 = \lambda_2 = \lambda$). In this case it may or may not be possible to associate two linearly independent eigenvalues $v_1$ and $v_2$ with the sole eigenvalue $\lambda$. If this is possible the matrix $A$ is diagonalizable and the trajectories can be analyzed by the previous method. If, on the other hand, only one linearly independent eigenvector $v$ can be associated with $\lambda$, then the matrix $A$ is not

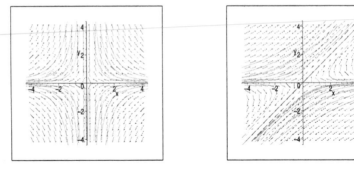

Figure 1.7: System trajectories of Example 1.9 (a) uncoupled system, (b) original system

diagonalizable. In this case, there always exist a similarity transformation $P$ such that

$$P^{-1}AP = J = \begin{bmatrix} \lambda_1 & 1 \\ 0 & \lambda_2 \end{bmatrix}.$$

The matrix $J$ is in the so-called *Jordan canonical form*. In this case the transformed system
is

$$\dot{y} = P^{-1}APy$$

or

$$\begin{aligned} \dot{y}_1 &= \lambda y_1 + y_2 \\ \dot{y}_2 &= \lambda y_2 \end{aligned}$$

the solution of this system of equations with initial condition $y_0 = [y_{10}, y_{20}]^T$ is as follows

$$\begin{aligned} y_1 &= y_{10}e^{\lambda t} + y_{20}te^{\lambda t} \\ y_2 &= y_{20}e^{\lambda t}. \end{aligned}$$

The shape of the solution is a somewhat distorted form of those encountered for diagonalizable systems. The equilibrium point is called a *stable node* if $\lambda < 0$ and *unstable node* if
$\lambda > 0$.

**Example 1.10** *Consider the system*

$$\begin{bmatrix} \dot{x}_1 \\ \dot{x}_2 \end{bmatrix} = \begin{bmatrix} -2 & 1 \\ 0 & -2 \end{bmatrix} \begin{bmatrix} x_1 \\ x_2 \end{bmatrix}.$$

*The eigenvalues in this case are $\lambda_1 = \lambda_2 = \lambda = -2$ and the matrix $A$ is not diagonalizable.
Figure 1.8 shows the trajectories of the system. In this example, the eigenvalue $\lambda < 0$ and
thus the equilibrium point $[0,0]$ is a stable node.*                                    □

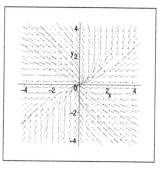

Figure 1.8: System trajectories for the system of Example 1.10.

## CASE 3: Systems with Complex Conjugate Eigenvalues

The most interesting case occurs when the eigenvalues of the matrix $A$ are complex conjugate, $\lambda_{1,2} = \alpha \pm \jmath\beta$. It can be shown that in this case a similarity transformation $M$ can be found that renders the following similar matrix:

$$M^{-1}AM = Q = \begin{bmatrix} \alpha & -\beta \\ \beta & \alpha \end{bmatrix}.$$

Thus the transform system has the form

$$\dot{y}_1 = \alpha y_1 - \beta y_2 \tag{1.20}$$
$$\dot{y}_2 = \beta y_1 + \alpha y_2 \tag{1.21}$$

The solution of this system of differential equations can be greatly simplified by introducing polar coordinates:

$$\rho = \sqrt{y_1^2 + y_2^2}$$
$$\theta = \tan^{-1}\left(\frac{y_2}{y_1}\right).$$

Converting (1.20) and (1.21) to polar coordinates, we obtain

$$\dot{\rho} = \alpha\rho$$
$$\dot{\theta} = \beta$$

which has the following solution:

$$\rho = \rho_0 e^{\alpha t} \tag{1.22}$$
$$\theta = \theta_0 + \beta t. \tag{1.23}$$

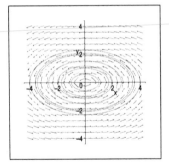

Figure 1.9: Trajectories for the system of Example 1.11

From here we conclude that

- In the polar coordinate system $\rho$ either increases exponentially, decreases exponentially, or stays constant depending whether the real part $\alpha$ of the eigenvalues $\lambda_{1,2}$ is positive, negative, or zero.

- The phase angle increases linearly with a "velocity" that depends on the imaginary part $\beta$ of the eigenvalues $\lambda_{1,2}$.

- In the $y_1 - y_2$ coordinate system (1.22)–(1.23) represent an exponential spiral. If $\alpha > 0$, the trajectories diverge from the origin as $t$ increases. If $\alpha < 0$ on the other hand, the trajectories converge toward the origin. The equilibrium $[0, 0]$ in this case is said to be a *stable focus* (if $\alpha < 0$) or *unstable focus* (if $\alpha > 0$).

- If $\alpha = 0$ the trajectories are closed ellipses. In this case the equilibrium $[0, 0]$ is said to be a *center*.

**Example 1.11**  *Consider the following system:*

$$\begin{bmatrix} \dot{x}_1 \\ \dot{x}_2 \end{bmatrix} = \begin{bmatrix} 0 & -3 \\ 1 & 0 \end{bmatrix} \begin{bmatrix} x_1 \\ x_2 \end{bmatrix}. \tag{1.24}$$

*The eigenvalues of the A matrix are $\lambda_{1,2} = 0 \pm j\sqrt{3}$. Figure 1.9 shows that the trajectories in this case are closed ellipses. This means that the dynamical system (1.24) is oscillatory. The amplitude of the oscillations is determined by the initial conditions.*  □

**Example 1.12**  *Consider the following system:*

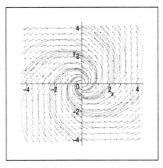

Figure 1.10: Trajectories for the system of Example 1.12

$$\left[ \begin{array}{c} \dot{x}_1 \\ \dot{x}_2 \end{array} \right] = \left[ \begin{array}{cc} 0.5 & -1 \\ 1 & 0.5 \end{array} \right] \left[ \begin{array}{c} x_1 \\ x_2 \end{array} \right].$$

*The eigenvalues of the A matrix are $\lambda_{1,2} = 0.5 \pm \jmath$; thus the origin is an unstable focus. Figure 1.10 shows the spiral behavior of the trajectories. The system in this case is also oscillatory, but the amplitude of the oscillations grow exponentially with time, because of the presence of the nonzero $\alpha$ term.*

□

The following table summarizes the different cases:

| Eigenvalues | Equilibrium point |
|---|---|
| $\lambda_1, \lambda_2$ real and negative | stable node |
| $\lambda_1, \lambda_2$ real and positive | unstable node |
| $\lambda_1, \lambda_2$ real, opposite signs | saddle |
| $\lambda_1, \lambda_2$ complex with negative real part | stable focus |
| $\lambda_1, \lambda_2$ complex with positive real part | unstable focus |
| $\lambda_1, \lambda_2$ imaginary | center |

As a final remark, we notice that the study of the trajectories of linear systems about the origin is important because, as we will see, in a neighborhood of an equilibrium point the behavior of a nonlinear system can often be determined by linearizing the nonlinear equations and studying the trajectories of the resulting linear system.

## 1.7    Phase-Plane Analysis of Nonlinear Systems

We mentioned earlier that nonlinear systems are more complex than their linear counterparts, and that their differences accentuate as the order of the state space realization
increases. The question then is: What features characterize second-order nonlinear equations not already seen the linear case? The answer is: *oscillations!* We will say that a system
oscillates when it has a nontrivial periodic solution, that is, a non-stationary trajectory for
which there exists $T > 0$ such that

$$x(t + T) = x(t) \quad \forall t \geq 0.$$

Oscillations are indeed a very important phenomenon in dynamical systems.

In the previous section we saw that if the eigenvalues of a second-order linear time-invaria
(LTI) system are imaginary, the equilibrium point is a center and the response is oscillatory. In practice however, LTI systems do not constitute oscillators of any practical use.
The reason is twofold: (1) as noticed in Example 1.11, the amplitude of the oscillations
is determined by the initial conditions, and (2) the very existence and maintenance of the
oscillations depend on the existence of purely imaginary eigenvalues of the $A$ matrix in the
state space realization of the dynamical equations. If the real part of the eigenvalues is
not identically zero, then the trajectories are not periodic. The oscillations will either be
damped out and eventually dissappear, or the solutions will grow unbounded. This means
that the oscillations in linear systems are not structurally stable. Small friction forces or
neglected viscous forces often introduce damping that, however small, adds a negative com-
ponent to the eigenvalues and consequently damps the oscillations. Nonlinear systems, on
the other hand, can have self-excited oscillations, known as *limit cycles*.

### 1.7.1    Limit Cycles

Consider the following system, commonly known as the *Van der Pol oscillator*:

$$\ddot{y} - \mu(1 - y^2)\dot{y} + y = 0 \qquad \mu > 0$$

defining state variables $x_1 = y$, and $x_2 = \dot{y}$ we obtain

$$\begin{aligned}
\dot{x}_1 &= x_2 & (1.25) \\
\dot{x}_2 &= x_1 + \mu(1 - x_1^2)x_2. & (1.26)
\end{aligned}$$

Notice that if $\mu = 0$ in equation (1.26), then the resulting system is

$$\begin{bmatrix} \dot{x}_1 \\ \dot{x}_2 \end{bmatrix} = \begin{bmatrix} 0 & 1 \\ -1 & 0 \end{bmatrix} \begin{bmatrix} x_1 \\ x_2 \end{bmatrix}$$

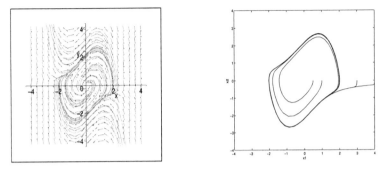

Figure 1.11: Stable limit cycle: (a) vector field diagram; (b) the closed orbit.

which is linear time-invariant. Moreover, the eigenvalues of the $A$ matrix are $\lambda_{1,2} = \pm j$, which implies that the equilibrium point $[0,0]$ is a center. The term $\mu(1-x_1^2)x_2$ in equation (1.26) provides additional dynamics that, as we will see, contribute to maintain the oscillations. Figure 1.11(a) shows the vector field diagram for the system (1.25)–(1.26) assuming $\mu = 1$. Notice the difference between the Van der Pol oscillator of this example and the center of Example 1.11. In Example 1.11 there is a continuum of closed orbits. A trajectory initiating at an initial condition $x_0$ at $t = 0$ is confined to the trajectory passing through $x_0$ for all future time. In the Van der Pol oscillator of this example there is only *one* isolated orbit. All trajectories converge to this trajectory as $t \to \infty$. An isolated orbit such as this is called a *limit cycle*. Figure 1.11(b) shows a clearer picture of the limit cycle.

We point out that the Van der Pol oscillator discussed here is not a theoretical example. These equations derive from simple electric circuits encountered in the first radios. Figure 1.12 shows a schematic of such a circuit, where $R$ in represents a nonlinear resistance. See Reference [84] for a detailed analysis of the circuit.

As mentioned, the Van der Pol oscillator of this example has the property that all trajectories converge toward the limit cycle. An orbit with this property is said to be a *stable limit cycle*. There are three types of limit cycles, depending on the behavior of the trajectories in the vicinity of the orbit: (1) stable, (2) unstable, and (3) semi stable. A limit cycle is said to be unstable if all trajectories in the vicinity of the orbit diverge from it as $t \to \infty$. It is said to be semi stable if the trajectories either inside or outside the orbit converge to it and diverge on the other side. An example of an unstable limit cycle can be obtained by modifying the previous example as follows:

$$\begin{aligned} \dot{x}_1 &= -x_2 \\ \dot{x}_2 &= x_1 - \mu(1 - x_1^2)x_2. \end{aligned}$$

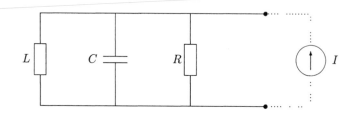

Figure 1.12: Nonlinear *RLC*-circuit.

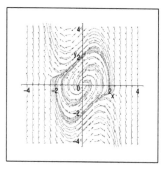

Figure 1.13: Unstable limit cycle.

Figure 1.13 shows the vector field diagram of this system with $\mu = 1$. As can be seen in the figure, all trajectories diverge from the orbit and the limit cycle is unstable.

## 1.8   Higher-Order Systems

When the order of the state space realization is greater than or equal to 3, nothing significantly different happens with linear time-invariant systems. The solution of the state equation with initial condition $x_0$ is still $x = e^{AT}x_0$, and the eigenvalues of the $A$ matrix still control the behavior of the trajectories. Nonlinear equations have much more room in which to maneuver. When the dimension of the state space realization increases from 2 to 3 a new phenomenon is encountered; namely, *chaos*.

### 1.8.1 Chaos

Consider the following system of nonlinear equations

$$
\begin{aligned}
\dot{x} &= \sigma(y - x) \\
\dot{y} &= rx - y - xz \\
\dot{z} &= xy - bz
\end{aligned}
$$

where $\sigma, r, b > 0$. This system was introduced by Ed Lorenz in 1963 as a model of convection rolls in the atmosphere. Since Lorenz' publication, similar equations have been found to appear in lasers and other systems. We now consider the following set of values: $\sigma = 10$, $b = \frac{8}{3}$, and $r = 28$, which are the original parameters considered by Lorenz. It is easy to prove that the system has three equilibrium points. A more detailed analysis reveals that, with these values of the parameters, none of these three equilibrium points are actually stable, but that nonetheless all trajectories are contained within a certain ellipsoidal region in $\mathbb{R}^3$.

Figure 1.14(a) shows a three dimensional view of a trajectory staring at a randomly selected initial condition, while Figure 1.14(b) shows a projection of the same trajectory onto the X–Z plane. It is apparent from both figures that the trajectory follows a *recurrent*, although not periodic, motion switching between two *surfaces*. It can be seen in Figure 1.14(a) that each of these 2 surfaces constitute a very thin set of points, almost defining a two-dimensional plane. This set is called an "strange attractor" and the two surphases, which together resemble a pair of *butterfly wings*, are much more complex than they appear in our figure. Each surface is in reality formed by an infinite number of complex surfaces forming what today is called a *fractal*.

It is difficult to define what constitutes a chaotic system; in fact, until the present time no universally accepted definition has been proposed. Nevertheless, the essential elements constituting chaotic behavior are the following:

A chaotic system is one where trajectories present aperiodic behavior and are critically sensitive with respect to initial conditions. Here aperiodic behavior implies that the trajectories never settle down to fixed points or to periodic orbits. Sensitive dependence with respect to initial conditions means that very small differences in initial conditions can lead to trajectories that deviate exponentially rapidly from each other.

Both of these features are indeed present in Lorenz' system, as is apparent in the Figures 1.14(a) and 1.14(b). It is of great theoretical importance that chaotic behavior cannot exist in autonomous systems of dimension less than 3. The justification of this statement comes from the well-known Poincaré-Bendixson Theorem which we state below without proof.

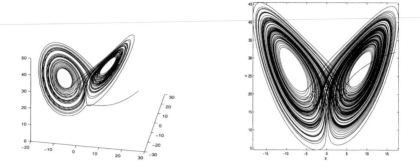

Figure 1.14: (a) Three-dimensional view of the trajectories of Lorenz' chaotic system; (b) two-dimensional projection of the trajectory of Lorenz' system.

**Theorem 1.1** *[76] Consider the two dimensional system*

$$\dot{x} = f(x)$$

*where $f : \mathbb{R}^2 \to \mathbb{R}^2$ is continuously differentiable in $D \subset \mathbb{R}^2$, and assume that*

*(1) $R \subset D$ is a closed and bounded set which contains no equilibrium points of $\dot{x} = f(x)$.*

*(2) There exists a trajectory $x(t)$ that is confined to $R$, that is, one that starts in $R$ and remains in $R$ for all future time.*

*Then either $R$ is a closed orbit, or converges toward a closed orbit as $t \to \infty$.*

According to this theorem, the Poincaré-Bendixson theorem predicts that, in two dimensions, a trajectory that is enclosed by a closed bounded region that contains no equilibrium points must eventually approach a limit cycle. In higher-order systems the new dimension adds an extra degree of freedom that allows trajectories to never settle down to an equilibrium point or closed orbit, as seen in the Lorenz system.

## 1.9   Examples of Nonlinear Systems

We conclude this chapter with a few examples of "real" dynamical systems and their nonlinear models. Our intention at this point is to simply show that nonlinear equation arise frequently in dynamical systems commonly encountered in real life. The examples in this section are in fact popular laboratory experiments used in many universities around the world.

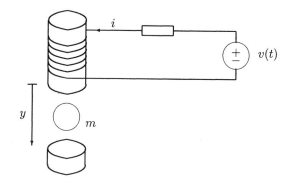

Figure 1.15: Magnetic suspension system.

## 1.9.1 Magnetic Suspension System

Magnetic suspension systems are a familiar setup that is receiving increasing attention in applications where it is essential to reduce friction force due to mechanical contact. Magnetic suspension systems are commonly encountered in high-speed trains and magnetic bearings, as well as in gyroscopes and accelerometers. The basic configuration is shown in Figure 1.15.

According to Newton's second law of forces, the equation of the motion of the ball is

$$m\ddot{y} = -f_k + mg + F \tag{1.27}$$

where $m$ is the mass of the ball, $f_k$ is the friction force, $g$ the acceleration due to gravity, and $F$ is the electromagnetic force due to the current $i$. To complete the model, we need to find a proper model for the magnetic force $F$. To this end we notice that the energy stored in the electromagnet is given by

$$E = \frac{1}{2}Li^2$$

where $L$ is the inductance of the electromagnet. This parameter is not constant since it depends on the position of the ball. We can approximate $L$ as follows:

$$L = L(y) = \frac{\lambda}{1 + \mu y}. \tag{1.28}$$

This model considers the fact that as the ball approaches the magnetic core of the coil, the flux in the magnetic circuit is affected, resulting in an increase of the value of the inductance. The energy in the magnetic circuit is thus $E = E(i, y) = \frac{1}{2}L(y)i^2$, and the force $F = F(i, y)$ is given by

$$F(i, y) = \frac{\partial E}{\partial y} = \frac{i^2}{2}\frac{\partial L(y)}{\partial y} = -\frac{1}{2}\frac{\lambda \mu i^2}{(1 + \mu y)^2}. \tag{1.29}$$

Assuming that the friction force has the form

$$f_k = k\dot{y} \tag{1.30}$$

where $k > 0$ is the viscous friction coefficient and substituting (1.30) and (1.29) into (1.27) we obtain the following equation of motion of the ball:

$$m\ddot{y} = -k\dot{y} + mg - \frac{1}{2}\frac{\lambda \mu i^2}{(1 + \mu y)^2}. \tag{1.31}$$

To complete the model, we recognize that the external circuit obeys the Kirchhoff's voltage law, and thus we can write

$$v = Ri + \frac{\mathrm{d}}{\mathrm{d}t}(Li) \tag{1.32}$$

where

$$
\begin{aligned}
\frac{\mathrm{d}}{\mathrm{d}t}(Li) &= \frac{\mathrm{d}}{\mathrm{d}t}\left(\frac{\lambda i}{1 + \mu y}\right) \\
&= \frac{\partial}{\partial y}\left(\frac{\lambda i}{1 + \mu y}\right)\frac{\mathrm{d}y}{\mathrm{d}t} + \frac{\partial}{\partial i}\left(\frac{\lambda i}{1 + \mu y}\right)\frac{\mathrm{d}i}{\mathrm{d}t} \\
&= -\frac{\lambda \mu i}{(1 + \mu y)^2}\frac{\mathrm{d}y}{\mathrm{d}t} + \left(\frac{\lambda}{1 + \mu y}\right)\frac{\mathrm{d}i}{\mathrm{d}t}.
\end{aligned}
\tag{1.33}
$$

Substituting (1.33) into (1.32), we obtain

$$v = Ri - \frac{\lambda \mu i}{(1 + \mu y)^2}\frac{\mathrm{d}y}{\mathrm{d}t} + \frac{\lambda}{1 + \mu y}\frac{\mathrm{d}i}{\mathrm{d}t}. \tag{1.34}$$

Defining state variables $x_1 = y, x_2 = \dot{y}, x_3 = i$, and substituting into (1.31) and (1.34), we obtain the following state space model:

$$
\begin{aligned}
\dot{x}_1 &= x_2 \\
\dot{x}_2 &= g - \frac{k}{m}x_2 - \frac{\lambda \mu x_3^2}{2m(1 + \mu x_1)^2} \\
\dot{x}_3 &= \frac{1 + \mu x_1}{\lambda}\left[-Rx_3 + \frac{\lambda \mu}{(1 + \mu x_1)^2}x_2 x_3 + v\right].
\end{aligned}
$$

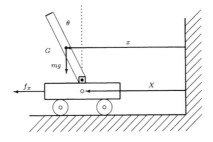

Figure 1.16: Pendulum-on-a-cart experiment.

## 1.9.2   Inverted Pendulum on a Cart

Consider the pendulum on a cart shown in Figure 1.16. We denote by $\theta$ the angle with respect to the vertical, $L = 2l, m$, and $J$ the length, mass, and moment of inertia about the center of gravity of the pendulum; $M$ represents the mass of the cart, and $G$ represents the center of gravity of the pendulum, whose horizontal and vertical coordinates are given by

$$x = X + \frac{L}{2}\sin\theta = X + l\sin\theta \tag{1.35}$$

$$y = \frac{l}{2}\cos\theta = l\cos\theta \tag{1.36}$$

The free-body diagrams of the cart and pendulum are shown in Figure 1.7, where $F_x$ and $F_y$ represent the reaction forces at the pivot point. Consider first the pendulum. Summing forces we obtain the following equations:

$$F_x = m\ddot{X} + ml\ddot{\theta}\cos\theta - ml\dot{\theta}^2\sin\theta \tag{1.37}$$

$$F_y - mg = -ml\ddot{\theta}\sin\theta - ml\dot{\theta}^2\cos\theta \tag{1.38}$$

$$F_y l\sin\theta - F_x l\cos\theta = J\ddot{\theta}. \tag{1.39}$$

Considering the horizontal forces acting on the cart, we have that

$$M\ddot{X} = f_x - F_x. \tag{1.40}$$

Substituting (1.40) into (1.37) taking account of (1.38) and (1.40), and defining state variables $x_1 = \theta, x_2 = \dot{\theta}$ we obtain

$$\dot{x}_1 = x_2$$

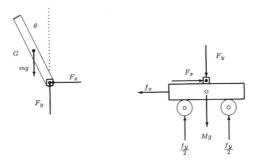

Figure 1.17: Free-body diagrams of the pendulum-on-a-cart system.

$$\dot{x}_2 = \frac{g \sin x_1 - amlx_2^2 \sin(2x_1) - 2a \cos(x_1) f_x}{4l/3 - 2aml \cos^2(x_1)}$$

where we have substituted

$$J = \frac{ml^2}{12}, \quad \text{and} \quad a \stackrel{def}{=} \frac{1}{2(m + M)}$$

### 1.9.3   The Ball-and-Beam System

The ball-and-beam system is another interesting and very familiar experiment commonly encountered in control systems laboratories in many universities. Figure 1.18 shows an schematic of this system. The beam can rotate by applying a torque at the center of rotation, and the ball can move freely along the beam. Assuming that the ball is always in contact with the beam and that rolling occurs without slipping, the Lagrange equations of motion are (see [28] for further details)

$$0 = (\frac{J_b}{R^2} + m)\ddot{r} + mg \sin \theta - mr\dot{\theta}^2$$
$$\tau = (mr^2 + J + J_b)\ddot{\theta} + 2mr\dot{r}\dot{\theta} + mgr \cos \theta$$

where $J$ represents the moment of inertia of the beam and $R$, $m$ and $J_b$ are the radius, mass and moment of inertia of the ball, respectively. The acceleration of gravity is represented by

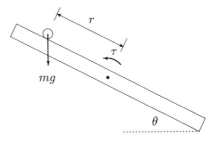

Figure 1.18: Ball-and-beam experiment.

$g$, and $r$ and $\theta$ are shown in Figure 1.18. Defining now state variables $x_1 = r, x_2 = \dot{r}, x_3 = \theta$, and $x_4 = \dot{\theta}$, we obtain the following state space realization:

$$\dot{x}_1 = x_2$$
$$\dot{x}_2 = \frac{-mg \sin x_3 + m x_1 x_4^2}{m + \frac{J_b}{R^2}}$$
$$\dot{x}_3 = x_4$$
$$\dot{x}_4 = \frac{\tau - mg x_1 \cos x_3 - 2m x_1 x_2 x_4}{m x_1^2 + J + J_b}.$$

## 1.10  Exercises

(1.1) For the following dynamical systems, plot $\dot{x} = f(x)$ versus $x$. From each graph, find the equilibrium points and analyze their stability:

(a) $\dot{x} = x^2 - 2$.

(b) $\dot{x} = x^3 - 2x^2 - x + 2$.

(c) $\dot{x} = \tan x \quad -\frac{\pi}{2} < x < \frac{\pi}{2}$.

(1.2) Given the following linear systems, you are asked to

(i) Find the eigenvalues of the $A$ matrix and classify the stability of the origin.

(ii) Draw the phase portrait and verify your conclusions in part (i).

(a)

$$\begin{bmatrix} \dot{x}_1 \\ \dot{x}_2 \end{bmatrix} = \begin{bmatrix} 2 & 1 \\ 0 & 4 \end{bmatrix} \begin{bmatrix} x_1 \\ x_2 \end{bmatrix}$$

(b)

$$\begin{bmatrix} \dot{x}_1 \\ \dot{x}_2 \end{bmatrix} = \begin{bmatrix} -2 & -1 \\ 0 & -4 \end{bmatrix} \begin{bmatrix} x_1 \\ x_2 \end{bmatrix}$$

(c)

$$\begin{bmatrix} \dot{x}_1 \\ \dot{x}_2 \end{bmatrix} = \begin{bmatrix} -1 & 3 \\ 0 & 4 \end{bmatrix} \begin{bmatrix} x_1 \\ x_2 \end{bmatrix}$$

(d)

$$\begin{bmatrix} \dot{x}_1 \\ \dot{x}_2 \end{bmatrix} = \begin{bmatrix} 1 & 3 \\ 0 & 4 \end{bmatrix} \begin{bmatrix} x_1 \\ x_2 \end{bmatrix}$$

(e)

$$\begin{bmatrix} \dot{x}_1 \\ \dot{x}_2 \end{bmatrix} = \begin{bmatrix} 2 & -1 \\ 2 & 0 \end{bmatrix} \begin{bmatrix} x_1 \\ x_2 \end{bmatrix}$$

(f)

$$\begin{bmatrix} \dot{x}_1 \\ \dot{x}_2 \end{bmatrix} = \begin{bmatrix} 0 & -1 \\ 2 & -2 \end{bmatrix} \begin{bmatrix} x_1 \\ x_2 \end{bmatrix}$$

(g)

$$\begin{bmatrix} \dot{x}_1 \\ \dot{x}_2 \end{bmatrix} = \begin{bmatrix} 1 & 1 \\ 2 & -1 \end{bmatrix} \begin{bmatrix} x_1 \\ x_2 \end{bmatrix}$$

(1.3) Repeat problem (1.2) for the following three-dimensional systems:

(a)

$$\begin{bmatrix} \dot{x}_1 \\ \dot{x}_2 \\ \dot{x}_3 \end{bmatrix} = \begin{bmatrix} 2 & -6 & 5 \\ 0 & -4 & 5 \\ 0 & 0 & 6 \end{bmatrix} \begin{bmatrix} x_1 \\ x_2 \\ x_3 \end{bmatrix}$$

(b)

$$\begin{bmatrix} \dot{x}_1 \\ \dot{x}_2 \\ \dot{x}_3 \end{bmatrix} = \begin{bmatrix} -2 & -2 & -1 \\ 0 & -4 & -1 \\ 0 & 0 & -6 \end{bmatrix} \begin{bmatrix} x_1 \\ x_2 \\ x_3 \end{bmatrix}$$

(c)

$$\begin{bmatrix} \dot{x}_1 \\ \dot{x}_2 \\ \dot{x}_3 \end{bmatrix} = \begin{bmatrix} -2 & 6 & 1 \\ 0 & 4 & 1 \\ 0 & 0 & 6 \end{bmatrix} \begin{bmatrix} x_1 \\ x_2 \\ x_3 \end{bmatrix}$$

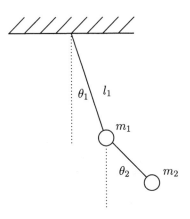

Figure 1.19: Double-pendulum of Exercise (1.5).

(1.4) For each of the following systems you are asked to (i) find the equilibrium points, (ii) find the phase portrait, and (iii) classify each equilibrium point as stable or unstable, based on the analysis of the trajectories:

(a)
$$\begin{cases} \dot{x}_1 &= x_1 - x_1^3 + x_2 \\ \dot{x}_2 &= -x_2 \end{cases}$$

(b)
$$\begin{cases} \dot{x}_1 &= -x_2 + 2x_1(x_1^2 + x_2^2) \\ \dot{x}_2 &= x_1 + 2x_2(x_1^2 + x_2^2) \end{cases}$$

(c)
$$\begin{cases} \dot{x}_1 &= \cos x_2 \\ \dot{x}_2 &= \sin x_1 \end{cases}$$

(1.5) Find a state space realization for the double-pendulum system shown in Figure 1.19. Define state variables $x_1 = \theta_1$, $x_2 = \dot{\theta}_1$, $x_3 = \theta_2$, $x_4 = \dot{\theta}_2$.

# Notes and References

We will often refer to linear time-invariant systems throughout the rest of the book. There are many good references on state space theory of LTI systems. See, for example, references [15], [2], [64], or [40]. Good sources on phase-plane analysis of second-order systems are References [32] and [59]. See also Arnold [3] for an excellent in-depth treatment of phase-plane analysis of nonlinear systems. First-order systems are harder to find in the literature. Section 1.4 follows Strogatz [76]. The literature on chaotic systems is very extensive. See Strogatz [76] for an inspiring, remarkably readable introduction to the subject. Section 1.8 is based mainly on this reference. Excellent sources on chaotic dynamical systems are References [26], [22] and [58]. The magnetic suspension system of Section 1.9.1 is a slightly modified version of a laboratory experiment used at the University of Alberta, prepared by Drs. A. Lynch and Q. Zhao. Our modification follows the model used in reference [81]. The pendulum-on-a-cart example follows References [9] and [13]. Our simple model of the ball-and-beam experiment was taken from reference [43]. This model is not very accurate since it neglects the moment of inertia of the ball. See Reference [28] for a more complete version of this model.

# Chapter 2

# Mathematical Preliminaries

This chapter collects some background material needed throughout the book. As the material is standard and is available in many textbooks, few proofs are offered. The emphasis has been placed in explaining the concepts, and pointing out their importance in later applications. More detailed expositions can be found in the references listed at the end of the chapter. This is, however, not essential for the understanding of the rest of the book.

## 2.1   Sets

We assume that the reader has some acquaintance with the notion of set. A *set* is a collection of objects, sometimes called *elements* or *points*. If $A$ is a set and $x$ is an element of $A$, we write $x \in A$. If $A$ and $B$ are sets and if every element of $A$ is also an element of $B$, we say that $B$ includes $A$, and that $A$ is a subset of $B$, and we write $A \subset B$ or $B \supset A$. As it has no elements, the empty set, denoted by $\emptyset$, is thus contained in every set, and we can write $\emptyset \subset A$. The union and intersection of $A$ and $B$ are defined by

$$A \cup B = \{x : x \in A \text{ or } x \in B\} \tag{2.1}$$
$$A \cap B = \{x : x \in A \text{ and } x \in B\}. \tag{2.2}$$

Assume now that $A$ and $B$ are non empty sets. Then the Cartesian product $A \times B$ of $A$ and $B$ is the set of all ordered pairs of the form $(a, b)$ with $a \in A$ and $b \in B$:

$$A \times B = \{(a, b) : a \in A, \text{ and } b \in B\}. \tag{2.3}$$

## 2.2   Metric Spaces

In real and complex analysis many results depend solely on the idea of distance between numbers $x$ and $y$. Metric spaces form a natural generalization of this concept. Throughout the rest of the book, $\mathbb{R}$ and $\mathbb{C}$ denote the field of real and complex numbers, respectively. $\mathbb{Z}$ represents the set of integers. $\mathbb{R}^+$ and $\mathbb{Z}^+$ represent the subsets of non negative elements of $\mathbb{R}$ and $\mathbb{Z}$, respectively. Finally $\mathbb{R}^{m \times n}$ denotes the set of real matrices with $m$ rows and $n$ columns.

**Definition 2.1** *A metric space is a pair* $(X, d)$ *of a non empty set* $X$ *and a metric or distance function* $d : X \times X \to \mathbb{R}$ *such that, for all* $x, y, z \in X$ *the following conditions hold:*

*(i)* $d(x, y) = 0$ *if and only if* $x = y$.

*(ii)* $d(x, y) = d(y, x)$.

*(iii)* $d(x, z) \leq d(x, y) + d(y, z)$.

Defining property (iii) is called the *triangle inequality*. Notice that, letting $x = z$ in (iii) and taking account of (i) and (ii) we have, $d(x, x) = 0 \leq d(x, y) + d(y, x) = 2d(x, y)$ from which it follows that $d(x, y) \geq 0 \; \forall x, y \in X$.

## 2.3   Vector Spaces

So far we have been dealing with metric spaces where the emphasis was placed on the notion of distance. The next step consists of providing the space with a proper algebraic structure. If we define addition of elements of the space and also multiplication of elements of the space by real or complex numbers, we arrive at the notion of vector space. Alternative names for vector spaces are *linear spaces* and *linear vector spaces*.

In the following definition $\mathbb{F}$ denotes a field of scalars that can be either the real or complex number system, $\mathbb{R}$, and $\mathbb{C}$.

**Definition 2.2** *A vector space over* $\mathbb{F}$ *is a non empty set* $X$ *with a function* "$+$": $X \times X \to X$, *and a function* "$.$": $\mathbb{F} \times X \to X$ *such that, for all* $\lambda, \mu \in \mathbb{F}$ *and* $x, y, z \in X$ *we have*

*(1)* $x + y = y + x$ *(addition is commutative).*

*(2)* $x + (y + z) = (x + y) + z$ *(addition is associative).*

*(3)* $\exists\, 0 \in X : x + 0 = x$ *("0" is the neutral element in the operation of addition).*

*(4)* $\exists\, -x \in X : x + (-x) = 0$ *(Every $x \in X$ has a negative $-x \in X$ such that their sum is the neutral element defined in (3)).*

*(5)* $\exists\, 1 \in \mathbb{F} : 1.x = x$ *("1" is the neutral element in the operation of scalar multiplication).*

*(6)* $\lambda(x + y) = \lambda x + \lambda y$ *(first distributive property).*

*(7)* $(\lambda + \mu)x = \lambda x + \mu x$ *(second distributive property).*

*(8)* $\lambda\,(\mu x) = (\lambda\mu)\,x$ *(scalar multiplication is associative).*

*A vector space is called real or complex according to whether the field $\mathbb{F}$ is the real or complex number system.*

We will restrict our attention to real vector spaces, so from now on we assume that $\mathbb{F} = \mathbb{R}$. According to this definition, a linear space is a structure formed by a set $X$ furnished with two operations: vector addition and scalar multiplication. The essential feature of the definition is that the set $X$ is closed under these two operations. This means that when two vectors $x, y \in X$ are added, the resulting vector $z = x + y$ is also an element of $X$. Similarly, when a vector $x \in X$ is multiplied by a scalar $\alpha \in \mathbb{R}$, the resulting *scaled* vector $\alpha x$ is also in $X$.

A simple and very useful example of a linear space is the $n$-dimensional "Euclidean" space $\mathbb{R}^n$ consistent of $n$-tuples of vectors of the following form:

$$x = \begin{bmatrix} x_1 \\ x_2 \\ \cdots \\ x_n \end{bmatrix}.$$

More precisely, if $X = \mathbb{R}^n$ and addition and scalar multiplication are defined as the usual coordinatewise operation

$$x + y = \begin{bmatrix} x_1 \\ x_2 \\ \cdots \\ x_n \end{bmatrix} + \begin{bmatrix} y_1 \\ y_2 \\ \cdots \\ y_n \end{bmatrix} \overset{def}{=} \begin{bmatrix} x_1 + y_1 \\ x_2 + y_2 \\ \cdots \\ x_n + y_n \end{bmatrix} , \quad \lambda x = \lambda \begin{bmatrix} x_1 \\ x_2 \\ \cdots \\ x_n \end{bmatrix} \overset{def}{=} \begin{bmatrix} \lambda x_1 \\ \lambda x_2 \\ \cdots \\ \lambda x_n \end{bmatrix}$$

then it is straightforward to show that $\mathbb{R}^n$ satisfies properties (1)- (8) in Definition 2.2. In the sequel, we will denote by $x^T$ the transpose of the vector $x$, that is, if

$$x = \begin{bmatrix} x_1 \\ x_2 \\ \cdots \\ x_n \end{bmatrix}$$

then $x^T$ is the "row vector" $x^T = [x_1, x_2, \cdots, x_n]$. The inner product of 2 vector in $x, y \in \mathbb{R}^n$ is $x^T y = \sum_{i=1}^n x_i y_i$.

Throughout the rest of the book we also encounter *function spaces*, namely spaces where the vectors in $X$ are functions of time. Our next example is perhaps the simplest space of this kind.

**Example 2.1** *Let $X$ be the space of continuous real functions $x = x(t)$ over the closed interval $0 \le t \le 1$.*

*It is easy to see that this $X$ is a (real) linear space. Notice that it is closed with respect to addition since the sum of two continuous functions is once again continuous.*  □

### 2.3.1   Linear Independence and Basis

We now look at the concept of vector space in more detail. The following definition introduces the fundamental notion of *linear independence*.

**Definition 2.3** *A finite set $\{x_i\}$ of vectors is said to be linearly dependent if there exists a corresponding set $\{\alpha_i\}$ of scalars, not all zero, such that*

$$\sum_i \alpha_i x_i = 0.$$

On the other hand, if $\sum_i \alpha_i x_i = 0$ implies that $\alpha_i = 0$ for each $i$, the set $\{x_i\}$ is said to be linearly *independent*.

**Example 2.2** *Every set containing a linearly dependent subset is itself linearly dependent.*
□

**Example 2.3** *Consider the space $\mathbb{R}^n$ and let*

$$e_i = \begin{bmatrix} 0 \\ \vdots \\ 1 \\ 0 \\ \vdots \end{bmatrix}$$

where the 1 element is in the ith row and there are zeros in the other $n - 1$ rows. Then $\{e_1, e_2, \cdots, e_n\}$ is called the set of unit vectors in $\mathbb{R}^n$. This set is linearly independent, since

$$\lambda_1 e_1 + \cdots + \lambda_n e_n = \begin{bmatrix} \lambda_1 \\ \vdots \\ \lambda_n \end{bmatrix}.$$

Thus,

$$\lambda_1 e_1 + \cdots + \lambda_n e_n = 0$$

is equivalent to

$$\lambda_1 = \lambda_2 = \cdots = \lambda_n = 0$$

$\square$

**Definition 2.4** A basis in a vector space $X$, is a set of linearly independent vectors $\mathcal{B}$ such that every vector in $X$ is a linear combination of elements in $\mathcal{B}$.

**Example 2.4** Consider the space $\mathbb{R}^n$. The set of unit vectors $e_i, i = 1, \cdots, n$ forms a basis for this space since they are linearly independent, and moreover, any vector $x \in \mathbb{R}^n$ can be obtain as linear combination of the $e_i$ values, since

$$\lambda_1 e_1 + \cdots + \lambda_n e_n = \begin{bmatrix} \lambda_1 \\ \vdots \\ \lambda_n \end{bmatrix}.$$

$\square$

It is an important property of any finite dimensional vector space with basis $\{b_1, b_2, \cdots, b_n\}$ that the linear combination of the basis vectors that produces a vector $x$ is unique. To prove that this is the case, assume that we have two different linear combinations producing the same $x$. Thus, we must have that

$$x = \sum_{i=1}^{n} \xi_i b_i = \sum_{i=1}^{n} \eta_i b_i.$$

But then, by subtraction we have that

$$\sum_{i=1}^{n} (\xi_i - \eta_i) b_i = 0$$

and since the $b_i$'s, being a basis, are linearly independent, we have that $\xi_i - \eta_i = 0$ for $i = 1, \cdots, n$, which means that the $\xi_i$'s are the same as the $\eta_i$'s.

In general, a finite-dimensional vector space can have an infinite number of basis.

**Definition 2.5** *The dimension of a vector space $X$ is the number of elements in any of its basis.*

For completeness, we now state the following theorem, which is a corollary of previous results. The reader is encouraged to complete the details of the proof.

**Theorem 2.1** *Every set of $n + 1$ vectors in an $n$-dimensional vector space $\mathcal{X}$ is linearly dependent. A set of $n$ vectors in $\mathcal{X}$ is a basis if and only if it is linearly independent.*

## 2.3.2  Subspaces

**Definition 2.6** *A non-empty subset $M$ of a vector space $X$ is a subspace if for any pair of scalars $\lambda$ and $\mu$, $\lambda x + \mu y \in M$ whenever $x$ and $y \in M$.*

According to this definition, a subspace $M$ in a vector space $X$ is itself a vector space.

**Example 2.5** *In any vector space $X$, $X$ is a subspace of itself.*                    □

**Example 2.6** *In the three-dimensional space $\mathbb{R}^3$, any two-dimensional plane passing through the origin is a subspace. Similarly, any line passing through the origin is a one-dimensional subspace of $\mathbb{R}^3$. The necessity of passing through the origin comes from the fact that along with any vector $x$, a subspace also contains $0 = 1 \cdot x + (-1) \cdot x$.*                    □

**Definition 2.7** *Given a set of vectors $S$, in a vector space $X$, the intersection of all subspaces containing $S$ is called the subspace generated or spanned by $S$, or simply the span of $S$.*

The next theorem gives a useful characterization of the span of a set of vectors.

**Theorem 2.2** *Let $S$ be a set of vectors in a vector space $X$. The subspace $M$ spanned by $S$ is the set of all linear combinations of the members of $S$.*

**Proof:** First we need to show that the set of linear combinations of elements of $\mathcal{S}$ is a subspace of $X$. This is straightforward since linear combinations of linear combinations of elements of $\mathcal{S}$ are again linear combinations of the elements of $\mathcal{S}$. Denote this subspace by $N$. It is immediate that $N$ contains every element of $\mathcal{S}$ and thus $N \subset M$. For the converse notice that $M$ is also a subspace which contains $\mathcal{S}$, and therefore contains all linear combinations of the elements of $\mathcal{S}$. Thus $M \subset N$, and the theorem is proved. $\square$

### 2.3.3 Normed Vector Spaces

As defined so far, vector spaces introduce a very useful algebraic structure by incorporating the operations of vector addition and scalar multiplication. The limitation with the concept of vector spaces is that the notion of distance associated with metric spaces has been lost. To recover this notion, we now introduce the concept of *normed vector space*.

**Definition 2.8** *A normed vector space (or simply a normed space) is a pair $(X, \| \cdot \|)$ consisting of a vector space $X$ and a norm $\| \cdot \| \colon X \to \mathbb{R}$ such that*

*(i) $\| x \| = 0$ if and only if $x = 0$.*

*(ii) $\| \lambda x \| = | \lambda | \| x \| \quad \forall \lambda \in \mathbb{R}, \forall x \in X$.*

*(iii) $\| x + y \| \leq \| x \| + \| y \| \quad \forall x, y \in X$.*

Notice that, letting $y = -x \neq 0$ in (iii), and taking account of (i) and (ii), we have

$$
\begin{aligned}
\| x + (-x) \| &\leq \| x \| + \| -x \| \\
\| 0 \| = 0 &\leq \| x \| + | -1 | \| x \| = 2 \| x \| \\
\Rightarrow \| x \| &\geq 0
\end{aligned}
$$

thus, the norm of a vector $X$ is nonnegative. Also, by defining property (iii), the triangle inequality,

$$
\| x - y \| = \| x - z + z - y \| \leq \| x - z \| + \| z - y \| \quad \forall x, y \in X \tag{2.4}
$$

holds. Equation (2.4) shows that *every normed linear space* may be regarded as a metric space with distance defined by $d(x, y) = \| x - y \|$.

The following example introduces the most commonly used norms in the Euclidean space $\mathbb{R}^n$.

**Example 2.7** *Consider again the vector space $\mathbb{R}^n$. For each $p, 1 \leq p \leq \infty$, the function $\|\cdot\|_p$, known as the p-norm in $\mathbb{R}^n$, makes this space a normed vector space, where*

$$\|x\|_p \overset{def}{=} (|x_1|^p + \cdots + |x_n|^p)^{1/p}. \tag{2.5}$$

*In particular*

$$\|x\|_1 \overset{def}{=} |x_1| + \cdots + |x_n|. \tag{2.6}$$

$$\|x\|_2 \overset{def}{=} \sqrt{|x_1|^2 + \cdots + |x_n|^2}. \tag{2.7}$$

*The 2-norm is the so-called Euclidean norm. Also, the $\infty$-norm is defined as follows:*

$$\|x\|_\infty \overset{def}{=} \max_i |x_i|. \tag{2.8}$$

□

By far the most commonly used of the $p$-norms in $\mathbb{R}^n$ is the 2-norm. Many of the theorems encountered throughout the book, as well as some of the properties of functions and sequences (such as continuity and convergence) depend only on the three defining properties of a norm, and not on the specific norm adopted. In these cases, to simplify notation, it is customary to drop the subscript $p$ to indicate that the norm can be any $p$-norm. The distinction is somewhat superfluous in that all p-norms in $\mathbb{R}^n$ are *equivalent* in the sense that given any two norms $\|\cdot\|_a$ and $\|\cdot\|_b$ on $\mathbb{R}^n$, there exist constants $k_1$ and $k_2$ such that (see exercise (2.6))

$$k_1\|x\|_a \leq \|x\|_b \leq k_2\|x\|_a, \qquad \forall x \in \mathbb{R}^n$$

Two frequently used inequalities involving $p$-norms in $\mathbb{R}^n$ are the following:

- **Holder's inequality**: Let $p \in \mathbb{R}$, $p > 1$ and let $q \in \mathbb{R}$ be such that

$$\frac{1}{p} + \frac{1}{q} = 1$$

  then

$$\|x^T y\|_1 \leq \|x\|_p \|y\|_q, \qquad \forall x, y \in \mathbb{R}^n. \tag{2.9}$$

- **Minkowski's inequality** Let $p \in \mathbb{R}$, $p \geq 1$. Then

$$\|x + y\|_p \leq \|x\|_p + \|y\|_p, \qquad \forall x, y \in \mathbb{R}^n. \tag{2.10}$$

## 2.4 Matrices

We assume that the reader has some acquaintance with the elementary theory of matrices and matrix operations. We now introduce some notation and terminology as well as some useful properties.

**Transpose:** If $A$ is an $m \times n$ matrix, its *transpose*, denoted $A^T$, is the $n \times m$ matrix obtained by interchanging rows and columns with $A$. The following properties are straightforward to prove:

- $(A^T)^T = A$.

- $(AB)^T = B^T A^T$ (transpose of the product of two matrices).

- $(A + B)^T = A^T + B^T$ (transpose of the sum of two matrices).

**Symmetric matrix:** A is symmetric if $A = A^T$.

**Skew symmetric matrix:** A is skew symmetric if $A = -A^T$.

**Orthogonal matrix:** A matrix $Q$ is orthogonal if $Q^T Q = QQ^T = I$. or equivalently, if

$$Q^T = Q^{-1}.$$

**Inverse matrix:** A matrix $A^{-1} \in \mathbb{R}^{n \times n}$ is said to be the inverse of the square matrix $A \in \mathbb{R}^{n \times n}$ if

$$AA^{-1} = A^{-1}A = I$$

It can be verified that

- $(A^{-1})^{-1} = A$.

- $(AB)^{-1} = B^{-1}A^{-1}$, provided that $A$ and $B$ are square of the same size, and invertible.

**Rank of a matrix:** The rank of a matrix $A$, denoted $\text{rank}(A)$, is the maximum number of linearly independent columns in $A$.

Every matrix $A \in \mathbb{R}^{n \times n}$ can be considered as a linear function $A : \mathbb{R}^n \to \mathbb{R}^n$, that is the mapping

$$y = Ax$$

maps the vector $x \in \mathbb{R}^n$ into the vector $y \in \mathbb{R}^n$.

**Definition 2.9** *The null space of a linear function $A : X \to Y$ is the set $\mathcal{N}(A)$, defined by*

$$\mathcal{N}(A) = \{x \in X : Ax = 0\}$$

It is straightforward to show that $\mathcal{N}(A)$ is a vector space. The dimension of this vector space, denoted $\dim\mathcal{N}(A)$, is important. We now state the following theorem without proof.

**Theorem 2.3** *Let $A \in \mathcal{F}^{m \times n}$. Then $A$ has the following property:*

$$rank(A) + dim\mathcal{N}(A) = n.$$

## 2.4.1   Eigenvalues, Eigenvectors, and Diagonal Forms

**Definition 2.10** *Consider a matrix $A \in \mathcal{F}^{n \times n}$. A scalar $\lambda \in \mathcal{F}$ is said to be an eigenvalue and a nonzero vector $x$ is an eigenvector of $A$ associated with this eigenvalue if*

$$Ax = \lambda x$$

*or*

$$(A - \lambda I)x = 0$$

thus $x$ is an eigenvector associated with $\lambda$ if and only if $x$ is in the null space of $(A - \lambda I)$.

Eigenvalues and eigenvectors are fundamental in matrix theory and have numerous applications. We first analyze their use in the most elementary form of diagonalization.

**Theorem 2.4** *If $A \in \mathbb{R}^{n \times n}$ with eigenvalues $\lambda_1, \lambda_2, \cdots, \lambda_n$ has $n$ linearly independent eigenvectors $v_1, v_2, \cdots, v_n$, then it can be expressed in the form*

$$A = SDS^{-1}$$

*where $D = diag\{\lambda_1, \lambda_2, \cdots, \lambda_n\}$, and*

$$S = \begin{bmatrix} \vdots & & \vdots \\ v_1 & \cdots & v_n \\ \vdots & & \vdots \end{bmatrix}.$$

**Proof:** By definition, the columns of the matrix $S$ are the eigenvectors of $A$. Thus, we have

$$AS = A \begin{bmatrix} \vdots & & \vdots \\ v_1 & \cdots & v_n \\ \vdots & & \vdots \end{bmatrix} = \begin{bmatrix} \vdots & & \vdots \\ \lambda_1 v_1 & \cdots & \lambda_n v_n \\ \vdots & & \vdots \end{bmatrix}$$

and we can re write the last matrix in the following form:

$$\begin{bmatrix} \vdots & & \vdots \\ \lambda_1 v_1 & \cdots & \lambda_n v_n \\ \vdots & & \vdots \end{bmatrix} = \begin{bmatrix} \vdots & & \vdots \\ v_1 & \cdots & v_n \\ \vdots & & \vdots \end{bmatrix} \begin{bmatrix} \lambda_1 & & \\ & \cdots & \\ & & \lambda_n \end{bmatrix}.$$

Therefore, we have that $AS = SD$. The columns of the matrix $S$ are, by assumption, linearly independent. It follows that $S$ is invertible and we can write

$$A = SDS^{-1}$$

or also

$$D = S^{-1}AS.$$

This completes the proof of the theorem. □

**Special Case: Symmetric Matrices**

Symmetric matrices have several important properties. Here we mention two of them, without proof:

(i) The eigenvalues of a symmetric matrix $A \in \mathbb{R}^{n \times n}$ are all real.

(ii) Every symmetric matrix $A$ is diagonalizable. Moreover, if $A$ is symmetric, then the diagonalizing matrix $S$ can be chosen to be an orthogonal matrix $P$; that is, if $A = A^T$, then there exist a matrix $P$ satisfying $P^T = P^{-1}$ such that

$$P^{-1}AP = P^T AP = D.$$

### 2.4.2 Quadratic Forms

Given a matrix $A \in \mathbb{R}^{n \times n}$, a function $q : \mathbb{R}^n \to \mathbb{R}$ of the form

$$q(x) = x^T Ax, \qquad x \in \mathbb{R}^n$$

is called a *quadratic form*. The matrix $A$ in this definition can be any real matrix. There is, however, no loss of generality in restricting this matrix to be *symmetric*. To see this, notice any matrix $A \in \mathbb{R}^{n \times n}$ can be rewritten as the sum of a symmetric and a skew symmetric matrix, as shown below:

$$A = \frac{1}{2}(A + A^T) + \frac{1}{2}(A - A^T)$$

Clearly,

$$B = \frac{1}{2}(A + A^T) = B^T, \text{ and}$$

$$C = \frac{1}{2}(A - A^T) = -C^T$$

thus $B$ is symmetric, whereas $C$ is skew symmetric. For the skew symmetric part, we have that

$$x^T C x = (x^T C x)^T = x^T C^T x = -x^T C x.$$

Hence, the real number $x^T C x$ must be identically zero. This means that the quadratic form associated with a skew symmetric matrix is identically zero.

**Definition 2.11** *Let $A \in \mathbb{R}^{n \times n}$ be a symmetric matrix and let $x \in \mathbb{R}^n$. Then $A$ is said to be:*

(i) *Positive definite if $x^T A x > 0 \forall x \neq 0$.*

(ii) *Positive semidefinite if $x^T A x \geq 0 \forall x \neq 0$.*

(iii) *Negative definite if $x^T A x < 0 \forall x \neq 0$.*

(iv) *Negative semidefinite if $x^T A x \leq 0 \forall x \neq 0$.*

(v) *Indefinite if $x^T A x$ can take both positive and negative values.*

It is immediate that the positive/negative character of a symmetric matrix is determined completely by its eigenvalues. Indeed, given a $A = A^T$, there exist $P$ such that $P^{-1} A P = P^T A P = D$. Thus defining $x = P y$, we have that

$$
\begin{aligned}
x^T A x &= y^T P^T A P y \\
&= y^T P^{-1} A P y \\
&= y^T D y \\
&= \lambda_1 y_1^2 + \lambda_2 y_2^2 + \cdots + \lambda_n y_n^2
\end{aligned}
$$

where $\lambda_i$, $i = 1, 2, n$ are the eigenvalues of $A$. From this construction, it follows that

(i) $A$ is positive definite if and only if all of its (real) eigenvalues are positive.

(ii) $A$ is positive semi definite if and only if $\lambda_i \geq 0$, $\forall i = 1, 2, \cdots n$.

(iii) $A$ is negative definite if and only if all of its eigenvalues are negative.

(iv) $A$ is negative semi definite if and only if $\lambda_i \leq 0$, $\forall i = 1, 2, \cdots n$.

(v) Indefinite if and only if it has positive and negative eigenvalues.

The following theorem will be useful in later sections.

**Theorem 2.5** *(Rayleigh Inequality) Consider a nonsingular symmetric matrix $Q \in \mathbb{R}^{n \times n}$, and let $\lambda_{min}$ and $\lambda_{max}$ be respectively the minimum and maximum eigenvalues of $Q$. Under these conditions, for any $x \in \mathbb{R}^n$,*

$$\lambda_{min}(Q)\|x\|^2 \leq x^T Q x \leq \lambda_{max}(Q)\|x\|^2. \tag{2.11}$$

**Proof:** The matrix $Q$, which is symmetric, is diagonalizable and it must have a full set of linearly independent eigenvectors. Moreover, we can always assume that the eigenvectors $u_1, u_2, \cdots, u_n$ associated with the eigenvalues $\lambda_1, \lambda_2, \cdots, \lambda_n$ are orthonormal. Given that the set of eigenvectors $\{u_1, u_2, \cdots, u_n\}$ form a basis in $\mathbb{R}^n$, every vector $x$ can be written as a linear combination of the elements of this set. Consider an arbitrary vector $x$. We can assume that $\|x\| = 1$ (if this is not the case, divide by $\|x\|$). We can write

$$x = x_1 u_1 + x_2 u_2 + \cdots + x_n u_n$$

for some scalars $x_1, x_2, \cdots, x_n$. Thus

$$
\begin{aligned}
x^T Q x &= x^T Q[x_1 u_1 + \cdots + x_n u_n] \\
&= x^T[\lambda_1 x_1 u_1 + \cdots + \lambda_n x_n u_n] \\
&= \lambda_1 x^T x_1 u_1 + \cdots + \lambda_n x^T x_n u_n \\
&= \lambda_1 |x_1|^2 + \cdots + \lambda_n |x_n|^2
\end{aligned}
$$

which implies that

$$\lambda_{min}(Q) \leq x^T Q x \leq \lambda_{max}(Q) \tag{2.12}$$

since

$$\sum_{i=1}^{n} |x_1|^2 = 1.$$

Finally, it is worth nothing that (2.12) is a special case of (2.11) when the norm of $x$ is 1.

$\square$

## 2.5 Basic Topology

A few elements of basic topology will be needed throughout the book. Let $X$ be a metric space. We say that

(a) A *neighborhood* of a point $p \in X$ is a set $N_r(p) \subset X$ consisting of all points $q \in X$ such that $d(p, q) < r$.

(b) Let $A \subset X$ and consider a point $p \in X$. Then $p$ is said to be a *limit point* of $A$ if every neighborhood of $p$ contains a point $q \neq p$ such that $q \in A$.

   It is important to notice that $p$ itself needs not be in the set $A$.

(c) A point $p$ is an interior point of a set $A \subset X$ if there exist a neighborhood $N$ of $p$ such that $N \subset E$.

(d) A set $A \subset X$ is said to be *open* if every point of $A$ is an interior point.

(e) The complement of $A \subset X$ is the set $A^c = \{p \in X : p \notin A\}$.

(f) A set $A \subset X$ is said to be *closed* if it contains all of its limit points. Equivalently, $A$ is closed if and only if and only if $A^c$ is open.

(g) A set $A \subset X$ is bounded if there exist a real number $\delta$ and a point $q \in A$ such that $d(p, q) < \delta, \forall p \in A$.

### 2.5.1 Basic Topology in $\mathbb{R}^n$

All the previous concepts can be specialized to the Euclidean space $\mathbb{R}^n$. We emphasize those concepts that we will use more frequently. Now consider a set $A \subset \mathbb{R}^n$.

- **Neighborhood**: A neighborhood of a point $p \in A \subset \mathbb{R}^n$ is the set $B_r(p)$ defined as follows:
$$B_r(p) = \{x \in \mathbb{R}^n : \|x - p\| < r\}.$$
  Neighborhoods of this form will be used very frequently and will sometimes be referred to as an *open ball* with center $p$ and radius $r$.

- **Open set**: A set $A \subset \mathbb{R}^n$ is said to be open if for every $p \in A$ one can find a neighborhood $B_r(p) \subset A$.

- **Bounded set**: A set $A \subset \mathbb{R}^n$ is said to be bounded if there exists a real number $M > 0$ such that
$$\|x\| < M \qquad \forall x \in A.$$

- **Compact set**: A set $A \subset \mathbb{R}^n$ is said to be compact if it is closed and bounded.

- **Convex set**: A set $A \subset \mathbb{R}^n$ is said to be convex if, whenever $x, y \in A$, then

$$\theta x_1 + (1 - \theta)x_2, \quad 0 \leq \theta \leq 1$$

also belongs to $A$.

## 2.6  Sequences

**Definition 2.12** *A sequence of vectors $x_0, x_1, x_2, \cdots$ in a metric space $(X, d)$, denoted $\{x_n\}$ is said to converge if there is a point $x_0 \in X$ with the property that for every real number $\xi > 0$ there is an integer $N$ such that $n \geq N$ implies that $d(x_n, x_0) < \xi$. We then write $x_0 = \lim x_n$, or $x_n \to x_0$ and call $x$ the limit of the sequence $\{x_n\}$.*

It is important to notice that in Definition 2.12 convergence must be taking place in the metric space $(X, d)$. In other words, if a sequence has a limit, $\lim x_n = x^*$ such that $x^* \notin X$, then $\{x_n\}$ is not convergent (see Example 2.10).

**Example 2.8** *Let $X_1 = \mathbb{R}$, $d(x, y) = |x - y|$, and consider the sequence*

$$\{x_n\} = \{1, 1.4, 1.41, 1.414, \cdots\}$$

*(each term of the sequence is found by adding the corresponding digit in $\sqrt{2}$). We have that*

$$|\sqrt{2} - x(n)| < \epsilon \quad \text{for } n \geq N$$

*and since $\sqrt{2} \in \mathbb{R}$ we conclude that $x_n$ is convergent in $(X_1, d)$.* ☐

**Example 2.9** *Let $X_2 = \mathbb{Q}$, the set of rational numbers $(x \in \mathbb{Q} \Rightarrow x = \frac{a}{b}$, with $a, b \in \mathbb{Z}, b \neq 0)$. Let $d(x, y) = |x - y|$, and consider the sequence of the previous example. Once again we have that $x_n$ is trying to converge to $\sqrt{2}$. However, in this case $\sqrt{2} \notin \mathbb{Q}$, and thus we conclude that $x_n$ is not convergent in $(X_2, d)$.* ☐

**Definition 2.13** *A sequence $\{x_n\}$ in a metric space $(X, d)$ is said to be a Cauchy sequence if for every real $\xi > 0$ there is an integer $N$ such that $d(x_n, x_m) < \xi$, provided that $n, m \geq N$.*

It is easy to show that every convergent sequence is a Cauchy sequence. The converse is, however, not true; that is a Cauchy sequence in not necessarily convergent, as shown in the following example.

**Example 2.10** Let $X = (0, 1)$ ( i.e., $X = \{x \in \mathbb{R} : 0 < x < 1\}$), and let $d(x, y) =\mid x - y \mid$. Consider the sequence $\{x_n\} = 1/n$. We have

$$d(x_n, x_m) =\mid \frac{1}{n} - \frac{1}{m} \mid \leq \frac{1}{n} + \frac{1}{m} < \frac{2}{N}$$

where $N = \min(n, m)$. It follows that $\{x_n\}$ is a Cauchy sequence since $d(x_n, x_m) < \epsilon$, provided that $n, m > 2/\epsilon$. It is not, however, convergent in $X$ since $\lim_{n \to \infty} 1/n = 0$, and $0 \notin X$.                                                                                      □

An important class of metric spaces are the so-called *complete* metric spaces.

**Definition 2.14** *A metric space* $(X, d)$ *is called complete if and only if every Cauchy sequence converges (to a point of* $X$*). In other words,* $(X, d)$ *is a complete metric space if for every sequence* $\{x_n\}$ *satisfying* $d(x_n, x_m) < \epsilon$ *for* $n, m > N$ *there exists* $x \in X$ *such that* $d(x_n, x) \to 0$ *as* $n \to \infty$*.*

If a space is incomplete, then it has "holes". In other words, a sequence might be "trying" to converge to a point that does not belong to the space and thus not converging. In incomplete spaces, in general, one needs to "guess" the limit of a sequence to prove convergence. If a space is known to be complete, on the other hand, then to check the convergence of a sequence to some point of the space, it is sufficient to check whether the sequence is Cauchy. The simplest example of a complete metric space is the real-number system with the metric $d =\mid x - y \mid$. We will encounter several other important examples in the sequel.

## 2.7   Functions

**Definition 2.15** *Let A and B be abstract sets. A function from A to B is a set f of ordered pairs in the Cartesian product* $A \times B$ *with the property that if* $(a, b)$ *and* $(a, c)$ *are elements of f, then* $b = c$*.*

In other words, a function is a subset of the Cartesian product between $A$ and $B$ where each argument can have one and only one image. Alternative names for functions used in this book are map, mapping, operator, and transformation. The set of elements of $A$ that can occur as first members of elements in $f$ is called the *domain* of $f$. The set of elements of $B$ that can occur as first members of elements of $f$ is called the *range* of $f$. A function $f$ is called *injective* if $f(x_1) = f(x_2)$ implies that $x_1 = x_2$ for every $x_1, x_2 \in A$. A function $f$ is called *surjective* if the range of $f$ is the whole of $B$. It is called *bijective* if it is both injective and surjective.

If $f$ is a function with domain $D$ and $D_1$ is a subset of $D$, then it is often useful to define a new function $f_1$ with domain $D_1$ as follows:

$$f_1 = \{(a, b) \in f : a \in D_1\}.$$

The function $f_1$ is called a restriction of $f$ to the set $D_1$.

**Definition 2.16** *Let $D_1, D_2 \subset \mathbb{R}^n$, and consider functions $f_1$ and $f_2$ of the form $f_1 : D_1 \to \mathbb{R}^n$ and $f_2 : D_1 \to \mathbb{R}^n$. If $f_1(D_1) \subset D_2$, then the composition of $f_2$ and $f_1$ is the function $f_2 \circ f_1$, defined by*

$$(f_2 \circ f_1)(x) = f_2[f_1(x)] \qquad x \in D_1$$

**Definition 2.17** *Let $(X, d_1), (Y, d_2)$ be metric spaces and consider a function $f : X \to Y$. We say that $f$ is continuous at $x_0$ if for every real $\epsilon > 0$ there exists a real $\delta = \delta(\epsilon, x_0)$ such that $d(x, x_0) < \delta$ implies that $d_2(f(x), f(x_0)) < \epsilon$.*

Equivalently, $f : X \to Y$ is continuous if for every sequence $\{x_n\}$ that converges to $x$, the corresponding sequence $\{f(x_n)\}$ converges to $y = f(x)$. In the special case of functions of the form $f : \mathbb{R}^n \to \mathbb{R}^m$, that is, functions mapping Euclidean spaces, then we say that $f$ is continuous at $x \in \mathbb{R}^n$ if given $\epsilon > 0$ there exists $\delta > 0$ such that

$$\|x - y\| < \delta \Rightarrow \|f(x) - f(y)\| < \epsilon.$$

This definition is clearly local, that is, it corresponds to pointwise convergence. If $f$ is continuous at every point of $X$, then $f$ is said to be continuous on $X$.

**Definition 2.18** *Let $(X, d_1)$ and $(Y, d_2)$ be metric spaces. Then $f : X \to Y$ is called uniformly continuous on $X$ if for every $\epsilon > 0$ there exists $\delta = \delta(\epsilon) > 0$, such that $d_1(x, y) < \delta$ implies that $d_2(f(x), f(y)) < \epsilon$, where $x, y \in X$.*

**Remarks:** The difference between ordinary continuity and uniform continuity is that in the former $\delta(\epsilon, x_0)$ depends on both $\epsilon$ and the particular $x_0 \in X$, while in the latter $\delta(\epsilon)$ is only a function of $\epsilon$. Clearly, uniform continuity is stronger than continuity, and if a function is uniformly continuous on a set $A$, then it is continuous on $A$. The converse is in general not true; that is, not every continuous function is uniformly continuous on the same set. Consider for example the function $f(x) = \frac{1}{x}$, $x > 0$, is continuous over $(0, \infty)$ but not uniformly continuous. The exception to this occurs when working with compact sets. Indeed, consider a function $f$ mapping a *compact set $X$* into a *metric space $Y$*. Then $f$ is uniformly continuous if and only if it is continuous.

## 2.7.1   Bounded Linear Operators and Matrix Norms

Now consider a function $L$ mapping vector spaces $X$ into $Y$.

**Definition 2.19** *A function $L : X \to Y$ is said to be a linear operator (or a linear map, or a linear transformation) if and only if given any $x_1, x_2 \in X$ and any $\lambda, \mu \in \mathbb{R}$*

$$L(\lambda x_1 + \mu x_2) = \lambda L(x_1) + \mu L(x_2). \tag{2.13}$$

*The function $L$ is said to be a bounded linear operator if there exist a constant $M$ such that*

$$\|L(x)\| \leq M \|x\| \qquad \forall x \in X. \tag{2.14}$$

The constant $M$ defined in (2.14) is called the *operator norm*. A special case of interest is that when the vector spaces $X$ and $Y$ are $\mathbb{R}^n$ and $\mathbb{R}^m$, respectively. In this case all linear functions $A : \mathbb{R}^n \to \mathbb{R}^m$ are of the form

$$y = Ax, \qquad x \in \mathbb{R}^n, \ y \in \mathbb{R}^m$$

where $A$ is a $m \times n$ of real elements. The operator norm applied to this case originates a *matrix norm*. Indeed, given $A \in \mathbb{R}^{m \times n}$, this matrix defines a linear mapping $A : \mathbb{R}^n \to \mathbb{R}^m$ of the form $y = Ax$. For this mapping, we define

$$\|A\|_p \stackrel{def}{=} \sup_{x \neq 0} \frac{\|Ax\|_p}{\|x\|_p} = \max_{\|x\|_p = 1} \|Ax\|_p. \tag{2.15}$$

Where all the norms on the right-hand side of (2.15) are vector norms. This norm is sometimes called the *induced* norm because it is "induced" by the $p$ vector norm. Important special cases are $p = 1, 2$, and $\infty$. It is not difficult to show that

$$\|A\|_1 = \max_{\|x\|_1 = 1} \|Ax\|_1 = \max_{j} \sum_{i=1}^{m} |a_{ij}| \tag{2.16}$$

$$\|A\|_2 = \max_{\|x\|_2 = 1} \|Ax\|_2 = \sqrt{\lambda_{max}(A^T A)} \tag{2.17}$$

$$\|A\|_\infty = \max_{\|x\|_\infty = 1} \|Ax\|_\infty = \max_{i} \sum_{j=1}^{n} |a_{ij}| \tag{2.18}$$

where $\lambda_{max}(A^T A)$ represents the maximum eigenvalue of $A^T A$.

## 2.8 Differentiability

**Definition 2.20** *A function $f : \mathbb{R} \to \mathbb{R}$ is said to be differentiable at $x$ if $f$ is defined in an open interval $(a, b) \subset \mathbb{R}$ and*

$$f'(x) = \lim_{h \to 0} \frac{f(x+h) - f(x)}{h}$$

*exists.*

The limit $f'(x)$ is called the *derivative* of $f$ at $x$. The function $f$ is said to be differentiable if it is differentiable at each $x$ in its domain. If the derivative exists, then

$$f(x+h) - f(x) = f'(x)h + r(h)$$

where the "remainder" $r(h)$ is small in the sense that

$$\lim_{h \to 0} \frac{r(h)}{h} = 0.$$

Now consider the case of a function $f : \mathbb{R}^n \to \mathbb{R}^m$.

**Definition 2.21** *A function $f : \mathbb{R}^n \to \mathbb{R}^m$ is said to be differentiable at a point $x$ if $f$ is defined in an open set $D \subset \mathbb{R}^n$ containing $x$ and the limit*

$$\lim_{h \to 0} \frac{\|f(x+h) - f(x) - f'(x)h\|}{\|h\|} = 0$$

*exists.*

Notice, of course, in Definition 2.21 that $h \in \mathbb{R}^n$. If $\|h\|$ is small enough, then $x + h \in D$, since $D$ is open and $f(x+h) \in \mathbb{R}^m$.

The derivative $f'(x)$ defined in Definition 2.21 is called the *differential* or the *total derivative* of $f$ at $x$, to distinguish it from the partial derivatives that we discuss next. In the following discussion we denote by $f_i, 1 \leq i \leq m$ the components of the function $f$, and by $\{e_1, e_2, \cdots, e_n\}$ the standard basis in $\mathbb{R}^n$:

$$f(x) = \begin{bmatrix} f_1(x) \\ f_2(x) \\ \cdots \\ f_m(x) \end{bmatrix}, e_1 = \begin{bmatrix} 1 \\ 0 \\ \cdots \\ 0 \end{bmatrix}, e_2 = \begin{bmatrix} 0 \\ 1 \\ \cdots \\ 0 \end{bmatrix}, \quad e_n = \begin{bmatrix} 0 \\ 0 \\ \cdots \\ 1 \end{bmatrix}.$$

**Definition 2.22** *Consider a function* $f : \mathbb{R}^n \to \mathbb{R}^m$ *and let* $D$ *be an open set in* $\mathbb{R}^n$. *For* $x \in D \subset \mathbb{R}^n$ *and* $1 \leq i \leq m$, $1 \leq j \leq n$ *we define*

$$\frac{\partial f_i}{\partial x_j} = D_j f_i \stackrel{def}{=} \lim_{\Delta \to 0} \frac{f_i(x + \Delta e_j) - f_i(x)}{\Delta} = 0, \quad \Delta \in \mathbb{R}$$

*provided that the limit exists.*

The functions $D_j f_i$ (or $\frac{\partial f_i}{\partial x_j}$) are called the *partial derivatives of* $f$.

Differentiability of a function, as defined in Definition 2.21 is not implied by the existence of the partial derivatives of the function. For example, the function $f : \mathbb{R}^2 \to \mathbb{R}$

$$f(x_1, x_2) = \frac{x_1 x_2^2}{x_1^2 + x_2^4}$$

is not continuous at $(0, 0)$ and so it is not differentiable. However, both $\frac{\partial f}{\partial x_1}$ and $\frac{\partial f}{\partial x_2}$ exist at every point in $\mathbb{R}^2$. Even for continuous functions the existence of all partial derivatives does not imply differentiability in the sense of Definition 2.21. On the other hand, if $f$ is known to be differentiable at a point $x$, then the partial derivatives exist at $x$, and they determine $f'(x)$. Indeed, if $f : \mathbb{R}^n \to \mathbb{R}^m$ is differentiable, then $f'(x)$ is given by the *Jacobian matrix* or *Jacobian transformation* $[f'(x)]$:

$$[f'(x)] = \begin{bmatrix} \frac{\partial f_1}{\partial x_1} & \cdots & \frac{\partial f_1}{\partial x_n} \\ \vdots & & \vdots \\ \frac{\partial f_m}{\partial x_1} & \cdots & \frac{\partial f_m}{\partial x_n} \end{bmatrix}.$$

If a function $f : \mathbb{R}^n \to \mathbb{R}^m$ is differentiable on an open set $D \subset \mathbb{R}^n$, then it is continuous on $D$. The derivative of the function, $f'$, on the other hand, may or may not be continuous. The following definition introduces the concept of *continuously differentiable function*.

**Definition 2.23** *A differentiable mapping* $f$ *of an open set* $D \subset \mathbb{R}^n$ *into* $\mathbb{R}^m$ *is said to be continuously differentiable in* $D$ *if* $f'$ *is such that*

$$\|f'(y) - f'(x)\| < \epsilon$$

*provided that* $x, y \in D$ *and* $\|x - y\| < \delta$.

The following theorem, stated without proof, implies that the continuously differentiable property can be evaluated directly by studying the partial derivatives of the function.

**Theorem 2.6** *Suppose that $f$ maps an open set $D \subset \mathbb{R}^n$ into $\mathbb{R}^m$. Then $f$ is continuously differentiable in $D$ if and only if the partial derivatives $\frac{\partial f_i}{\partial x_j}$, $1 \leq i \leq m$, $1 \leq j \leq n$ exist and are continuous on $D$.*

Because it is relatively easier to work with the partial derivatives of a function, Definition 2.23 is often restated by saying that

> *A function $f : \mathbb{R}^n \to \mathbb{R}^m$ is said to be continuously differentiable at a point $x_0$ if the partial derivatives $\frac{\partial f_i}{\partial x_j}$, $1 \leq i \leq m$, $1 \leq j \leq n$ exist and are continuous at $x_0$. A function $f : \mathbb{R}^n \to \mathbb{R}^m$ is said to be continuously differentiable on a set $D \subset \mathbb{R}^n$, if it is continuously differentiable at every point of $D$.*

**Summary:** Given a function $f : \mathbb{R}^n \to \mathbb{R}^m$ with continuous partial derivatives, abusing the notation slightly, we shall use $f'(x)$ or $Df(x)$ to represent both the Jacobian matrix and the total derivative of $f$ at $x$. If $f : \mathbb{R}^n \to \mathbb{R}$, then the Jacobian matrix is the row vector

$$f'(x) = \left[ \frac{\partial f}{\partial x_1}, \frac{\partial f}{\partial x_2}, \cdots, \frac{\partial f}{\partial x_n} \right].$$

This vector is called the *gradient* of $f$ because it identifies the direction of steepest ascent of $f$. We will frequently denote this vector by either

$$\frac{\partial f}{\partial x} \quad \text{or} \quad \nabla f(x)$$

i.e.

$$\nabla f(x) = \frac{\partial f}{\partial x} = \left[ \frac{\partial f}{\partial x_1}, \frac{\partial f}{\partial x_2}, \cdots, \frac{\partial f}{\partial x_n} \right].$$

It is easy to show that the set of functions $f : \mathbb{R}^n \to \mathbb{R}^m$ with continuous partial derivatives, together with the operations of addition and scalar multiplication, form a vector space. This vector space is denoted by $C^1$. In general, if a function $f : \mathbb{R}^n \to \mathbb{R}^m$ has continuous partial derivatives up to order $k$, the function is said to be in $C^k$, and we write $f \in C^k$. If a function $f$ has continuous partial derivatives of any order, then it is said to be *smooth*, and we write $f \in C^\infty$. Abusing this terminology and notation slightly, $f$ is said to be *sufficiently smooth* where $f$ has continuous partial derivatives of any *required* order.

## 2.8.1 Some Useful Theorems

We collect a number of well known results that are often useful.

**Property 2.1** *(Chain Rule) Let $f_1 : D_1 \subset \mathbb{R}^n \to \mathbb{R}^n$, and $f_2 : D_2 \subset \mathbb{R}^n \to \mathbb{R}^n$. If $f_2$ is differentiable at $a \in D_2$ and $f_1$ is differentiable at $f_2(a)$, then $f_1 \circ f_2$ is differentiable at $a$, and*

$$D(f_1 \circ f_2)(a) = DF_1(f_2(a))Df_2(a).$$

**Theorem 2.7** *(Mean-Value Theorem) Let $f : [a, b] \to \mathbb{R}$ be continuous on the closed interval $[a, b]$ and differentiable in the open interval $(a, b)$. Then there exists a point $c$ in $(a, b)$ such that*

$$f(b) - f(a) = f'(c)(b - a).$$

A useful extension of this result to functions $f : \mathbb{R}^n \to \mathbb{R}^m$ is given below. In the following theorem $\Omega$ represents an open subset of $\mathbb{R}^n$.

**Theorem 2.8** *Consider the function $f : \Omega \subset \mathbb{R}^n \to \mathbb{R}^m$ and suppose that the open set $\omega$ contains the points $a$, and $b$ and the line segment $S$ joining these points, and assume that $f$ is differentiable at every point of $S$. The there exists a point $c$ on $S$ such that*

$$\|f(b) - f(a)\| = \|f'(c)(b - a)\|.$$

**Theorem 2.9** *(Inverse Function Theorem) Let $f : \mathbb{R}^n \to \mathbb{R}^n$ be continuously differentiable in an open set $D$ containing the point $x_0 \in \mathbb{R}^n$, and let $f'(x_0) \neq 0$. Then there exist an open set $U_0$ containing $x_0$ and an open set $W_0$ containing $f(x_0)$ such that $f : U_0 \to W_0$ has a continuous inverse $f^{-1} : W_0 \to U_0$ that is differentiable and for all $y = f(x) \in W_0$ satisfies*

$$Df^{-1}(y) = [Df(x)]^{-1} = [Df^{-1}(y)]^{-1}.$$

## 2.9 Lipschitz Continuity

We defined continuous functions earlier. We now introduce a stronger form of continuity, known as *Lipschitz continuity*. As will be seen in Section 2.10, this property will play a major role in the study of the solution of differential equations.

**Definition 2.24** *A function $f(x) : \mathbb{R}^n \to \mathbb{R}^m$ is said to be locally Lipschitz on $D$ if every point of $D$ has a neighborhood $D_0 \subset D$ over which the restriction of $f$ with domain $D_1$ satisfies*

$$\|f(x_1) - f(x_2)\| \leq L\|x_1 - x_2\|. \tag{2.19}$$

*It is said to be Lipschitz on an open set $D \subset \mathbb{R}^n$ if it satisfies (2.19) for all $x_1, x_2 \in D$ with the same Lipschitz constant. Finally, $f$ is said to be globally Lipschitz if it satisfies (2.19) with $D = \mathbb{R}^n$.*

Notice that if $f : \mathbb{R}^n \to \mathbb{R}^m$ is Lipschitz on $D \subset \mathbb{R}^n$, then, given $\epsilon > 0$, we can define $\delta = \epsilon/L$, and we have that

$$\|x_1 - x_2\| < \delta \;\Rightarrow\; \|f(x_1) - f(x_2)\| < L\|x_1 - x_2\| < L\frac{\epsilon}{L} = \epsilon, \qquad x_1, x_2 \in D$$

which implies that $f$ is uniformly continuous. However, the converse is not true; that is, not every uniformly continuous function is Lipschitz. It is in fact very easy to find counterexamples that show that this in not the case. The next theorem gives an important sufficient condition for Lipschitz continuity.

**Theorem 2.10** *If a function* $f : \mathbb{R}^n \times \mathbb{R}^n$ *is continuously differentiable on an open set* $D \subset \mathbb{R}^n$, *then it is locally Lipschitz on* $D$.

**Proof:** Consider $x_0 \in D$ and let $r > 0$ be small enough to ensure that $B_r(x_0) \subset D$, where

$$B_r(x_0) = \{x \in D : |x - x_0| \le r\}$$

and consider two arbitrary points $x_1, x_2 \in B_r$. Noticing that the closed ball $B_r$ is a convex set, we conclude that the line segment $\gamma(\theta) = \theta x_1 + (1 - \theta)x_2$, $0 \le \theta \le 1$, is contained in $B_r(x_0)$. Now define the function $\phi : [0, 1] \to \mathbb{R}^n$ as follows:

$$\phi(\theta) = f \circ \gamma = f(\gamma(\theta)).$$

By the mean-value theorem 2.8, there exist $\theta_1 \in (0, 1)$ such that

$$\|\phi(1) - \phi(0)\| = \|\phi'(\theta_1)\|.$$

Thus, calculating $\phi'(\theta_1)$ using the chain rule, we have that

$$\|\phi(1) - \phi(0)\| = \left\|\frac{\partial f}{\partial x}(x_1 - x_2)\right\|$$

and, substituting $\phi(1) = f(x_1)$, $\phi(0) = f(x_2)$

$$\|f(x_1) - f(x_2)\| \le \left\|\frac{\partial f}{\partial x}\right\| \|x_1 - x_2\| = L\|x_1 - x_2\|. \qquad (2.20)$$

$\square$

Theorem 2.10 provides a mechanism to calculate the Lipschitz constant (equation (2.20). Indeed, $L$ can be estimated as follows:

$$\left\|\frac{\partial f}{\partial x}\right\| \le L, \qquad \forall x \in B_r(x_0).$$

If the function $f$ is also a function of $t$, that is, $f = f(x, t)$, then the previous definition can be extended as follows

**Definition 2.25** *A function $f(x,t) : \mathbb{R}^n \times \mathbb{R} \to \mathbb{R}^n$ is said to be locally Lipschitz in $x$ on an open set $D \times [t_0, T] \subset \mathbb{R}^n \times \mathbb{R}$ if every point of $D$ has a neighborhood $D_1 \subset D$ over which the restriction of $f$ with domain $D_1 \times [t_0, T]$ satisfies (2.20). It is said to be locally Lipschitz on $D \subset \mathbb{R}^n \times [t_0, \infty)$ if it is locally Lipschitz in $x$ on every $D_1 \times [t_0, T] \subset D \times [t_0, \infty)$. It is said to be Lipschitz in $x$ on $D \times [t_0, T]$ if it satisfies (2.20) for all $x_1, x_2 \in D$ and all $t \in [t_0, T]$.*

**Theorem 2.11** *Let $f : \mathbb{R}^n \times \mathbb{R} \to \mathbb{R}^m$ be continuously differentiable on $D \times [t_0, T]$ and assume that the derivative of $f$ satisfies*

$$\left\| \frac{\partial f}{\partial x}(x,t) \right\| \leq L \tag{2.21}$$

*on $D \times [t_0, T]$. The $f$ is Lipschitz continuous on $D$ with constant $L$:*

$$\|f(x,t) - f(y,t)\| \leq L\|x - y\|, \quad \forall x, y \in D, \ \forall t \in [t_0, T].$$

## 2.10   Contraction Mapping

In this section we discuss the contraction mapping principle, which we use later to analyze the existence and uniqueness of solutions of a class of nonlinear differential equations.

**Definition 2.26** *Let $(X, d)$ be a metric space, and let $S \subset X$. A mapping $f : S \to S$ is said to be a contraction on $S$ if there exists a number $\xi < 1$ such that*

$$d(f(x), f(y)) \leq \xi \, d(x, y) \tag{2.22}$$

*for any two points $x, y \in S$.*

It is an straightforward exercise that every contraction is continuous (in fact, uniformly continuous) on $X$.

**Theorem 2.12** *(Contraction Mapping Principle) Let $S$ be a closed subset of the complete metric space $(X, d)$. Every contraction mapping $f : S \to S$ has one and only one $x \in S$ such that $f(x) = x$.*

**Proof:** A point $x_0 \in X$ satisfying $f(x_0) = x_0$ is called a *fixed point*. Thus, the contraction mapping principle is sometimes called the *fixed-point theorem*. Let $x_0$ be an arbitrary point in $S$, and denote $x_1 = f(x_0)$, $x_2 = f(x_1), \cdots, x_{n+1} = f(x_n)$. This construction defines a sequence $\{x_n\} = \{f(x_{n-1})\}$. Since $f$ maps $S$ into itself, $x_k \in S \ \forall k$. We have

$$d(x_{n+1}, x_n) = d(f(x_n), f(x_{n-1}) \leq \xi d(x_n, x_{n-1}) \tag{2.23}$$

and then, by induction

$$d(x_{n+1}, x_n) \leq \xi^n d(x_1, x_0). \tag{2.24}$$

Now suppose $m > n > N$. Then, by successive applications of the triangle inequality we obtain

$$
\begin{aligned}
d(x_n, x_m) &\leq \sum_{i=n+1}^{m} d(x_i, x_{i-1}) \\
&\leq (\xi^n + \xi^{n+1} + \cdots + \xi^{m-1}) \, d(x_1, x_0) \\
\Rightarrow d(x_n, x_m) &\leq \xi^n (1 + \xi^1 + \cdots + \xi^{m-n-1}) \, d(x_1, x_0). \tag{2.25}
\end{aligned}
$$

The series $(1 + x + x^2 + \cdots)$ converges to $1/(1-x)$, for all $x < 1$. Therefore, noticing that all the summands in equation (2.25) are positive, we have:

$$1 + \xi + \cdots + \xi^{m-n-1} \leq \frac{1}{(1-\xi)}$$

$$\Rightarrow d(x_n, x_m) \leq \frac{\xi^n}{(1-\xi)} d(x_1, x_0)$$

defining $\frac{\xi^n}{(1-\xi)} d(x_1, x_0) = \epsilon$, we have that

$$d(x_n, x_m) \leq \epsilon \quad n, m > N.$$

In other words, $\{x_n\}$ is a Cauchy sequence. Since the metric space $(X, d)$ is complete, $\{x_n\}$ has a limit, $\lim_{n \to \infty}(x_n) = x$, for some $x \in X$. Now, since $f$ is a contraction, $f$ is continuous. It follows that

$$f(x) = \lim_{n \to \infty} f(x_n) = \lim_{n \to \infty} (x_{n+1}) = x.$$

Moreover, we have seen that $x_n \in S \subset X, \forall n$, and since $S$ is closed it follows that $x \in S$. Thus, the existence of a fixed point is proved. To prove uniqueness suppose $x$ and $y$ are two different fixed points. We have

$$f(x) = x, \quad f(y) = y \tag{2.26}$$

$$
\begin{aligned}
d(f(x), f(y)) &= d(x, y) & \text{by (2.26)} \tag{2.27} \\
\Rightarrow d(x, y) &\leq \xi \, d(x, y) & \text{because } f \text{ is a contraction} \tag{2.28}
\end{aligned}
$$

where $\xi < 1$. Since (2.28) can be satisfied if and only if $d(x, y) = 0$, we have $x = y$. This completes the proof. $\square$

## 2.11 Solution of Differential Equations

When dealing with ordinary linear differential equations, it is usually possible to derive a closed-form expression for the solution of a differential equation. For example, given the state space realization of a linear time-invariant system

$$\dot{x} = Ax + Bu, \quad x(0) = x_0$$

we can find the following closed form solution for a given $u$:

$$x(t) = e^{At}x_0 + \int_0^t e^{A(t-\tau)} Bu(\tau) \, d\tau.$$

In general, this is not the case for nonlinear differential equations. Indeed, when dealing with nonlinear differential equations two issues are of importance: (1) existence, and (2) uniqueness of the solution.

In this section we derive sufficient conditions for the existence and uniqueness of the solution of a differential equation of the form

$$\dot{x} = f(t, x). \quad x(t_0) = x_0 \tag{2.29}$$

**Theorem 2.13** *(Local Existence and Uniqueness) Consider the nonlinear differential equation*

$$\dot{x} = f(x, t), \quad x(t_0) = x_0 \tag{2.30}$$

*and assume that $f(x, t)$ is piecewise continuous in $t$ and satisfies*

$$\|f(x_1, t) - f(x_2, t)\| \le L \|x_1 - x_2\|$$

$\forall x_1, x_2 \in B = \{x \in \mathbb{R}^n : \|x - x_0\| \le r\}, \forall t \in [t_0, t_1]$. *Thus, there exist some $\delta > 0$ such that (2.30) has a unique solution in $[t_0, t_0 + \delta]$.*

**Proof**: Notice in the first place that if $x(t)$ is a solution of (2.30), then $x(t)$ satisfies

$$x(t) = x_0 + \int_{t_0}^t f[x(\tau), \tau] \, d\tau \tag{2.31}$$

which is of the form

$$x(t) = (Fx)(t) \tag{2.32}$$

where $Fx$ is a continuous function of $t$. Equation (2.32) implies that a solution of the differential equation (2.30) is a fixed point of the mapping $F$ that maps $x$ into $Fx$. The existence of a fixed point of this map, in turn, can be verified by the contraction mapping theorem. To start with, we define the sets $X$ and the $S \subset X$ as follows:

$$X = C[t_0, t_0 + \delta]$$

thus $X$ is the set of all continuous functions defined on the interval $[t_0, t_0 + \delta]$. Given $x \in X$, we denote

$$\|x\|_c = \max_{t \in [t_0, t_0 + \delta]} \|x(t)\| \tag{2.33}$$

and finally

$$S = \{x \in X : \|x - x_0\|_c \le r\}.$$

Clearly, $S \subset X$. It can also be shown that $S$ is closed and that $X$ with the norm (2.33) is a complete metric space. To complete the proof, we will proceed in three steps. In step (1) we show that $F : S \to S$. In step (2) we show that $F$ is a contraction from $S$ into $S$. This, in turn implies that there exists one and only one fixed point $x = Fx \in S$, and thus a unique solution in $S$. Because we are interested in solutions of (2.30) in $X$ (not only in $S$). The final step, step (3), consists of showing that any possible solution of (2.30) in $X$ must be in $S$.

*Step (1)*: From (2.31), we obtain

$$
\begin{aligned}
(Fx)(t) - x_0 &= \int_{t_0}^{t} f(\tau, x(\tau)) \, d\tau \\
&= \int_{t_0}^{t} [f(x(\tau), \tau) - f(x_0, \tau) + f(x_0, \tau)] \, d\tau
\end{aligned}
$$

The function $f$ is bounded on $[t_0, t_1]$ (since it is piecewise continuous). It follows that we can find $\xi$ such that

$$\xi = \max_{t \in [t_0, t_1]} \|f(x_0, t)\|$$

$$
\begin{aligned}
\|Fx - x_0\| &\le \int_{t_0}^{t} [\, \|f(x(\tau), \tau) - f(x_0, \tau)\| + \|f(x_0, \tau)\| \,] \, d\tau \\
&\le \int_{t_0}^{t} [\, L(\|x(\tau) - x_0\|) + \xi \,] \, d\tau
\end{aligned}
$$

and since for each $x \in S, \|x - x_0\| \le r$ we have

$$
\begin{aligned}
\|Fx - x_0\| &\le \int_{t_0}^{t} [\, Lr + \xi \,] \, d\tau \\
&\le (t - t_0)[Lr + \xi].
\end{aligned}
$$

It follows that

$$\|Fx - x_0\|_c = \max_{t \in [T_0, t_0 + \delta]} \|Fx - x_0\| \leq \delta(Lr + \xi)$$

and then, choosing $\delta \leq r/[Lr + \xi]$ means that $F$ maps $S$ into $S$.

*Step (2)*:To show that $F$ is a contraction on $S$, we consider $x_1, x_2 \in S$ and proceed as follows:

$$
\begin{aligned}
\|(Fx_1)(t) - (Fx_2)(t)\| &= \left\| \int_{t_0}^{t} [\, f(x_1(\tau), \tau) - f(x_2(\tau), \tau) \,] \, d\tau \right\| \\
&\leq \int_{t_0}^{t} \| \, f(x_1(\tau), \tau) - f(x_2(\tau), \tau) \, \| \, d\tau \\
&\leq \int_{t_0}^{t} L \, \|x_1(\tau) - x_2(\tau)\| \, d\tau \\
&\leq L \, \|x_1 - x_2\|_c \int_{t_0}^{t} d\tau
\end{aligned}
$$

$$\Rightarrow \|Fx_1 - Fx_2\|_c \leq L\delta \|x_1 - x_2\|_c \leq \rho \|x_1 - x_2\|_c \quad \text{for } \delta \leq \frac{\rho}{L}.$$

Choosing $\rho < 1$ and $\delta \leq \frac{\rho}{L}$, we conclude that $F$ is a construction. This implies that there is a unique solution of the nonlinear equation (2.30) in $S$.

*Step (3)*: Given that $S \subset X$, to complete the proof, we must show that any solution of (2.30) in $X$ must lie in $S$. Starting at $x_0$ at $t_0$, a solution can leave $S$ if and only if at some $t = t_1, x(t)$ crosses the border $B$ for the first time. For this to be the case we must have

$$\|x(t_1) - x_0\| = r.$$

Thus, for all $t \leq t_1$ we have that

$$
\begin{aligned}
\|x(t) - x_0\| &\leq \int_{t_0}^{t} [\, \|f(x(\tau), \tau) - f(x_0, \tau)\| \, + \, \|f(x_0, \tau)\| \,] \, d\tau \\
&\leq \int_{t_0}^{t} [\, L(\|x(\tau) - x_0\|) + \xi \,] \, d\tau \\
&\leq \int_{t_0}^{t} [\, Lr + \xi \,] \, d\tau.
\end{aligned}
$$

It follows that

$$r = \|x(t_1) - x_0\| \leq (Lr + \xi)(t_1 - t_0).$$

Denoting $t_1 = t_0 + \mu$, we have that if $\mu$ is such that

$$\mu \geq \frac{r}{Lr + \xi}$$

then the solution $x(t)$ is confined to $B$. This completes the proof. $\qquad\square$

It is important to notice that Theorem 2.13 provides a *sufficient* but not *necessary* condition for the existence and uniqueness of the solution of the differential equation (2.30). According to the theorem, the solution is guaranteed to exist only *locally*, i.e. in the interval $[t_0, t_0+\delta]$. For completeness, below we state (but do not prove) a slightly modified version of this theorem that provides a condition for the *global* existence and uniqueness of the solution of the same differential equation. The price paid for this generalization is a stronger and much more conservative condition imposed on the differential equation.

**Theorem 2.14** (*Global Existence and Uniqueness*) *Consider again the nonlinear differential equation (2.30) and assume that $f(x,t)$ is piecewise continuous in $t$ and satisfies*

$$\|f(x_1,t) - f(x_2,t)\| \leq L\|x_1 - x_2\|$$
$$\|f(x_0,t)\| \leq \Delta$$

$\forall x_1, x_2 \in \mathbb{R}^n, \forall t \in [t_0, t_1]$. *Then, (2.30) has a unique solution in $[t_0, t_1]$.*

Theorem 2.14 is usually too conservative to be of any practical use. Indeed, notice that the conditions of the theorem imply the existence of a *global* Lipschitz constant, which is very restrictive. The local Lipschitz condition of theorem 2.13 on the other hand, is not very restrictive and is satisfied by any function $f(x,t)$ satisfying somewhat mild smoothness conditions, as outlined in Theorem 2.11

## 2.12 Exercises

(2.1) Consider the set of $2 \times 2$ real matrices. With addition and scalar multiplication defined in the usual way, is this set a vector space over the field of real numbers? Is so, find a basis.

(2.2) Let $x$, $y$ and $z$ be linearly independent vectors in the vector space $X$. Is it correct to infer that $x + y, y + z$, and $z + x$ are also linearly independent?

(2.3) Under what conditions on the scalars $\alpha$ and $\beta$ in $\mathbb{C}^2$ are the vectors $[1, \alpha]^T$, and $[1, \beta]^T$ linearly dependent?

(2.4) Prove Theorem 2.1.

(2.5) For each of the norms $\|x\|_1$, $\|x\|_2$, and $\|x\|_\infty$ in $\mathbb{R}^2$, sketch the "open unit ball" centered at $0 = [0,0]^T$, namely, the sets $\|x\|_1 < 1$, $\|x\|_2 < 1$, and $\|x\|_\infty < 1$.

(2.6) Show that the vector norms $\|\cdot\|_1$, $\|\cdot\|_2$, and $\|\cdot\|_\infty$ satisfy the following:

$$
\begin{array}{ccccc}
\|x\|_2 & \leq & \|x\|_1 & \leq & \sqrt{n}\,\|x\|_2 \\
\|x\|_\infty & \leq & \|x\|_1 & \leq & n\,\|x\|_2 \\
\|x\|_\infty & \leq & \|x\|_2 & \leq & \sqrt{n}\,\|x\|_2
\end{array}
$$

(2.7) Consider a matrix $A \in \mathbb{R}^{n \times n}$ and let $\lambda_1, \cdots, \lambda_n$ be its eigenvalues and $x_1, \cdots, x_n$ its (not necessarily linearly independent) eigenvectors.

   (i) Assuming that $A$ is nonsingular, what can be said about its eigenvalues?

   (ii) Under these assumptions, is it possible to express the eigenvalues and eigenvectors of $A^{-1}$ in terms of those of $A$? If the answer is negative, explain why. If the answer is positive, find the eigenvalues and eigenvectors of $A^{-1}$.

(2.8) Consider a matrix $A \in \mathbb{R}^{n \times n}$.

   (i) Show that the functions $\|A\|_1, \|A\|_2$, and $\|A\|_\infty$ defined in equations (2.16), (2.17), and (2.18), respectively, are *norms*, i.e. satisfy properties (i)–(iii) in Definition 2.8.

   (ii) Show that $\|I\|_p = 1$ for *any* induced norm $\|\cdot\|_p$.

   (iii) Show that the function

$$
\|A\| = \sqrt{\operatorname{trace}(A^T A)} = \sqrt{\sum_{i=j=1}^{n} |a_{ij}|^2}
$$

is a matrix norm[1] (i.e. it satisfies properties (i)–(iii) in Definition 2.8). Show that this norm, however, in not an operator norm since it is not induced by any vector norm.

(2.9) Let $(X, d)$ be a metric space. Show that

   (i) $X$ and the empty set $\emptyset$ are open.

   (ii) The intersection of a finite number of open sets is open.

   (iii) The union of any collection of open sets is open.

---

[1] This norm is called the *Frobenius norm*.

(2.10) Show that a set $A$ in a metric space $(X, d)$ is closed if and only if its complement is open.

(2.11) Let $(X, d)$ be a metric space. Show that

   (i) $X$ and the empty set $\emptyset$ are closed.

   (ii) The intersection of any number of closed sets is closed.

   (iii) The union of a finite collection of closed sets is closed.

(2.12) Let $(X, d_1)$ and $(Y, d_2)$ be metric spaces and consider a function $f : X \to Y$. Show that $f$ is continuous if and only if the inverse image of every open set in $Y$ is open in $X$.

(2.13) Determine the values of the following limits, whenever they exist, and determine whether each function is continuous at $(0, 0)$:

   (i) $\displaystyle \lim_{x \to 0, y \to 0} \frac{x^2 - y^2}{1 + x^2 + y^2}$
   (ii) $\displaystyle \lim_{x \to 0, y \to 0} \frac{x}{x^2 + y^2}$
   (iii) $\displaystyle \lim_{x \to 0, y \to 0} \frac{(1 + y^2) \sin x}{x}$

(2.14) Consider the function

$$
f(x, y) = \begin{cases} \frac{xy}{x^2 + y^2} & \text{for} \quad (x, y) \neq (0, 0) \\ 0 & \text{for} \quad x = y = 0 \end{cases}
$$

Show that $f(x, y)$ is not continuous (and thus not differentiable) at the origin. Proceed as follows:

   (i) Show that
$$
\lim_{x \to 0, y = x^2 \to 0} f(x, y) = 0
$$
   that is, if $x \to 0$ and $y \to 0$ along the parabola $y = x^2$, then $\lim_{x, y \to 0} = 0$.
   (ii) Show that
$$
\lim_{x \to 0, y = x \to 0} f(x, y) = \frac{1}{2}
$$

thus the result.

(2.15) Given the function $f(x, y)$ of exercise (2.13), show that the partial derivatives $\frac{\partial f}{\partial x}$ and $\frac{\partial f}{\partial y}$ both exist at $(0, 0)$. This shows that existence of the partial derivatives does not imply continuity of the function.

(2.16) Determine whether the function

$$f(x,y) = \begin{cases} \frac{x^2 y^2}{x^2+y^2} & \text{for} \quad (x,y) \neq (0,0) \\ 0 & \text{for} \quad x = y = 0 \end{cases}$$

is continuous at $(0,0)$. (*Suggestion*: Notice that $\frac{x^2 y^2}{x^2+y^2} \leq x^2$.)

(2.17) Given the following functions, find the partial derivatives $\frac{\partial f}{\partial x}$ and $\frac{\partial f}{\partial y}$:

(i) $f(x,y) = e^{xy} \cos x \sin y$ \qquad (ii) $f(x,y) = \dfrac{x^2 y^2}{x^2 + y^2}$

(2.18) Use the chain rule to obtain the indicated partial derivatives.

(i) $z = x^3 + y^3$, $x = 2r + s$, $y = 3r - 2s$, $\frac{\partial z}{\partial r}$, and $\frac{\partial z}{\partial s}$.

(ii) $z = \frac{x}{x^2+y^2}$, $x = r \cos s$, $y = r \sin s$, $\frac{\partial z}{\partial r}$, and $\frac{\partial z}{\partial s}$.

(2.19) Show that the function $f = 1/x$ is not uniformly continuous on $E = (0,1)$.

(2.20) Show that $f = 1/x$ does not satisfy a Lipschitz condition on $E = (0,1)$.

(2.21) Given the following functions $f : \mathbb{R} \to \mathbb{R}$, determine in each case whether $f$ is (a) continuous at $x = 0$, (b) continuously differentiable at $x = 0$, (c) locally Lipschitz at $x = 0$.

(i) $f(x) = e^{x^2}$.
(ii) $f(x) = \cos x$.
(iii) $f(x) = \text{sat} x$.
(iv) $f(x) = \sin(\frac{1}{x})$.

(2.22) For each of the following functions $f : \mathbb{R}^2 \to \mathbb{R}^2$, determine whether $f$ is (a) continuous at $x = 0$, (b) continuously differentiable at $x = 0$, (c) locally Lipschitz at $x = 0$, (d) Lipschitz on some $D \subset \mathbb{R}^2$.

(i)
$$\begin{cases} \dot{x}_1 = x_2 - x_1(x_1^2 + x_2^2) \\ \dot{x}_2 = -x_1 - x_2(x_1^2 + x_2^2) \end{cases}$$

(ii)
$$\begin{cases} \dot{x}_1 = x_2 + x_1(\beta^2 - x_1^2 - x_2^2) \\ \dot{x}_2 = -x_1 + x_2(\beta^2 - x_1^2 - x2^2) \end{cases}$$

(iii)

$$\begin{cases} \dot{x}_1 = x_2 \\ \dot{x}_2 = -\frac{k}{m}x_1 - \frac{\beta}{m}x_2 \end{cases}$$

(iv)

$$\begin{cases} \dot{x}_1 = x_2 \\ \dot{x}_2 = -\frac{g}{l}\sin x_1 - \frac{k}{m}x_2 \end{cases}$$

## Notes and References

There are many good references for the material in this chapter. For a complete, remarkably well-written account of vector spaces, see Halmos [33]. For general background in mathematical analysis, we refer to Bartle [7], Maddox [51], and Rudin [62]. Section 2.9, including Theorem 2.10, is based on Hirsch [32]. The material on existence and uniqueness of the solution of differential equations can be found on most textbooks on ordinary differential equations. We have followed References [32], [88] and [41]. See also References [59] and [55].

# Chapter 3

# Lyapunov Stability I: Autonomous Systems

In this chapter we look at the important notion of stability in the sense of Lyapunov. Indeed, there are many definitions of stability of systems. In all cases the idea is: given a set of dynamical equations that represent a physical system, try to determine whether such a system is *well behaved* in some conceivable sense. Exactly *what* constitutes a meaningful notion of good behavior is certainly a very debatable topic. The problem lies in how to convert the intuitive notion a good behavior into a precise mathematical definition that can be applied to a given dynamical system. In this chapter, we explore the notion of stability in the sense of Lyapunov, which applies to *equilibrium points*. Other notions of stability will be explored in Chapters 6 and 7. Throughout this chapter we restrict our attention to autonomous systems. The more general case of nonautonomous systems is treated in the next chapter.

## 3.1 Definitions

Consider the autonomous system[1]

$$\dot{x} = f(x) \qquad\qquad f : D \to \mathbb{R}^n \qquad\qquad (3.1)$$

where $D$ is an open and connected subset of $\mathbb{R}^n$ and $f$ is a locally Lipschitz map from $D$ into $\mathbb{R}^n$. In the sequel we will assume that $x = x_e$ is an equilibrium point of (3.1). In other words, $x_e$ is such that

$$f(x_e) = 0.$$

---

[1]Notice that (3.1) represents an *unforced* system

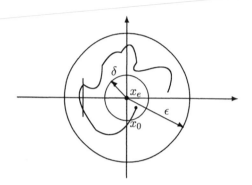

Figure 3.1: Stable equilibrium point.

We now introduce the following definition.

**Definition 3.1** *The equilibrium point $x = x_e$ of the system (3.1) is said to be stable if for each $\epsilon > 0$, $\exists \delta = \delta(\epsilon) > 0$*

$$\|x(0) - x_e\| < \delta \quad \Rightarrow \quad \|x(t) - x_e\| < \epsilon \quad \forall t \geq t_0$$

*otherwise, the equilibrium point is said to be unstable .*

This definition captures the following concept–we want the solution of (3.1) to be *near* the equilibrium point $x_e$ for all $t \geq t_0$. To this end, we start by measuring proximity in terms of the norm $\| \cdot \|$, and we say that we want the solutions of (3.1) to remain inside the open region delimited by $\|x(t) - x_e\| < \epsilon$. If this objective is accomplished by starting from an initial state $x(0)$ that is close to the equilibrium $x_e$, that is, $\|x(0) - x_e\| < \delta$, then the equilibrium point is said to be stable (see Figure 3.1).

　　This definition represents the weakest form of stability introduced in this chapter. Recall that what we are trying to capture is the concept of *good behavior* in a dynamical system. The main limitation of this concept is that solutions are not required to converge to the equilibrium $x_e$. Very often, staying close to $x_e$ is simply not enough.

**Definition 3.2** *The equilibrium point $x = x_e$ of the system (3.1) is said to be convergent if there exists $\delta_1 > 0$ :*

$$\|x(0) - x_e\| < \delta_1 \quad \Rightarrow \quad \lim_{t \to \infty} x(t) = x_e.$$

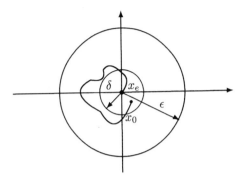

Figure 3.2: Asymptotically stable equilibrium point.

Equivalently, $x_e$ is convergent if for any given $\epsilon_1 > 0, \exists T$ such that

$$\|x(0) - x_e\| < \delta_1 \quad \Rightarrow \quad \|x(t) - x_e\| < \epsilon_1 \quad \forall t \geq t_0 + T.$$

A convergent equilibrium point $x_e$ is one where every solution starting sufficiently close to $x_e$ will eventually approach $x_e$ as $t \to \infty$. It is important to realize that stability and convergence, as defined in 3.1 and 3.2 are two different concepts and neither one of them implies the other. Indeed, it is not difficult to construct examples where an equilibrium point is convergent, yet does not satisfy the conditions of Definition 3.1 and is therefore not stable in the sense of Lyapunov.

**Definition 3.3** *The equilibrium point $x = x_e$ of the system (3.1) is said to be asymptotically stable if it is both stable and convergent.*

Asymptotic stability (Figure 3.2) is the desirable property in most applications. The principal weakness of this concept is that it says nothing about how fast the trajectories approximate the equilibrium point. There is a stronger form of asymptotic stability, referred to as *exponential stability*, which makes precise this idea.

**Definition 3.4** *The equilibrium point $x = x_e$ of the system (3.1) is said to be (locally) exponentially stable if there exist two real constants $\alpha, \lambda > 0$ such that*

$$\|x(t) - x_e\| \leq \alpha \|x(0) - x_e\| e^{-\lambda t} \quad \forall t > 0 \tag{3.2}$$

*whenever $\|x(0) - x_e\| < \delta$. It is said to be globally exponentially stable if (3.2) holds for any $x \in \mathbb{R}^n$.*

Clearly, exponential stability is the strongest form of stability seen so far. It is also immediate that exponential stability implies asymptotic stability. The converse is, however, not true.

The several notions of stability introduced so far refer to *stability of equilibrium points*. In general, the same dynamical system can have more than one isolated equilibrium point. Very often, in the definitions and especially in the proofs of the stability theorems, it is assumed that the equilibrium point under study is the origin $x_e = 0$. There is no loss of generality in doing so. Indeed, if this is *not* the case, we can perform a change of variables and define a new system with an equilibrium point at $x = 0$. To see this, consider the equilibrium point $x_e$ of the system (3.1) and define,

$$
\begin{aligned}
y &= x - x_e \\
\Rightarrow \dot{y} &= \dot{x} = f(x) \\
\Rightarrow f(x) &= f(y + x_e) \stackrel{def}{=} g(y).
\end{aligned}
$$

Thus, the equilibrium point $y_e$ of the new systems $\dot{y} = g(y)$ is $y_e = 0$, since

$$
g(0) = f(0 + x_e) = f(x_e) = 0.
$$

According to this, studying the stability of the equilibrium point $x_e$ for the system $\dot{x} = f(x)$ is equivalent to studying the stability of the origin for the system $\dot{y} = g(y)$.

*Given this property, in the sequel we will state the several stability theorems assuming that $x_e = 0$.*

**Example 3.1** *Consider the mass–spring system shown in Figure 3.3. We have*

$$
m\ddot{y} + \beta\dot{y} + ky = mg
$$

*defining states $x_1 = y, x_2 = \dot{y}$, we obtain the following state space realization,*

$$
\begin{cases}
\dot{x}_1 = x_2 \\
\dot{x}_2 = -\frac{k}{m}x_1 - \frac{\beta}{m}x_2 + g
\end{cases}
$$

*which has a unique equilibrium point $x_e = (\frac{mg}{k}, 0)$. Now define the transformation $z = x - x_e$. According to this*

$$
\begin{aligned}
z_1 &= x_1 - \frac{mg}{k}, \Rightarrow & \dot{z}_1 &= \dot{x}_1 \\
z_2 &= x_2, & \Rightarrow & \dot{z}_2 &= \dot{x}_2.
\end{aligned}
$$

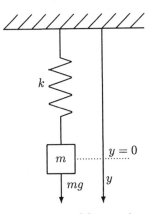

Figure 3.3: Mass–spring system.

*Thus,*

$$\begin{cases} \dot{z}_1 = z_2 \\ \dot{z}_2 = -\frac{k}{m}\left(z_1 + \frac{mg}{k}\right) - \frac{\beta}{m}z_2 + g \end{cases}$$

*or*

$$\begin{cases} \dot{z}_1 = z_2 \\ \dot{z}_2 = -\frac{k}{m}z_1 - \frac{\beta}{m}z_2 \end{cases}$$

*Thus, $\dot{z} = g(x)$, and $g(x)$ has a single equilibrium point at the origin.* □

## 3.2  Positive Definite Functions

Now that the concept of stability has been defined, the next step is to study how to analyze the stability properties of an equilibrium point. This is the center of the Lyapunov stability theory. The core of this theory is the analysis and construction of a class of functions to be defined and its derivatives along the trajectories of the system under study. We start by introducing the notion of *positive definite functions*. In the following definition, $D$ represents an open and connected subset of $\mathbb{R}^n$.

**Definition 3.5** *A function $V : D \to \mathbb{R}$ is said to be* positive semi definite *in $D$ if it satisfies the following conditions:*

*(i) $0 \in D$ and $V(0) = 0$.*

(ii) $V(x) \geq 0, \quad \forall x$ in $D - \{0\}$.

$V : D \to \mathbb{R}$ is said to be positive definite in $D$ if condition (ii) is replaced by (ii')

(ii') $V(x) > 0$ in $D - \{0\}$.

Finally, $V : D \to \mathbb{R}$ is said to be negative definite (semi definite)in $D$ if $-V$ is positive definite (semi definite).

We will often abuse the notation slightly and write $V > 0$, $V \geq 0$, and $V < 0$ in $D$ to indicate that $V$ is positive definite, semi definite, and negative definite in $D$, respectively.

**Example 3.2** *The simplest and perhaps more important class of positive definite function is defined as follow:s,*

$$V(x) : \mathbb{R}^n \to \mathbb{R} = x^T Q x , \quad Q \in \mathbb{R}^{n \times n}, \ Q = Q^T.$$

*In this case, $V(\cdot)$ defines a quadratic form. Since by assumption, $Q$ is symmetric (i.e., $Q = Q^T$), we know that its eigenvalues $\lambda_i, i = 1, \cdots n$, are all real. Thus we have that*

$$V(\cdot) \text{ positive definite} \iff x^T Q x > 0, \forall x \neq 0 \iff \lambda_i > 0, \forall i = 1, \cdots, n$$
$$V(\cdot) \text{ positive semi definite} \iff x^T Q x \geq 0, \forall x \neq 0 \iff \lambda_i \geq 0, \forall i = 1, \cdots, n$$
$$V(\cdot) \text{ negative definite} \iff x^T Q x < 0, \forall x \neq 0 \iff \lambda_i < 0, \forall i = 1, \cdots, n$$
$$V(\cdot) \text{ negative semi definite} \iff x^T Q x \leq 0, \forall x \neq 0 \iff \lambda_i \leq 0, \forall i = 1, \cdots, n$$

*Thus, for example:*

$$V_1(x) : \quad \mathbb{R}^2 \to \mathbb{R} = ax_1^2 + bx_2^2 = [x_1, x_2] \begin{bmatrix} a & 0 \\ 0 & b \end{bmatrix} \begin{bmatrix} x_1 \\ x_2 \end{bmatrix} > 0, \quad \forall a, b > 0$$

$$V_2(x) : \quad \mathbb{R}^2 \to \mathbb{R} \ = \ ax_1^2 = [x_1, x_2] \begin{bmatrix} a & 0 \\ 0 & 0 \end{bmatrix} \begin{bmatrix} x_1 \\ x_2 \end{bmatrix} \geq 0, \quad \forall a \geq 0.$$

*$V_2(\cdot)$ is not positive definite since for any $x_2 \neq 0$, any $x$ of the form $x^* = [0, x_2]^T \neq 0$; however, $V_2(x^*) = 0$.* $\square$

Positive definite functions (PDFs) constitute the basic building block of the Lyapunov theory. PDFs can be seen as an abstraction of the total "energy" stored in a system, as we will see. All of the Lyapunov stability theorems focus on the study of the time derivative of a positive definite function along the trajectories of 3.1. In other words, given an autonomous system of the form 3.1, we will first construct a positive definite function $V(x)$ and study $\dot{V}(x)$ given by

$$\dot{V}(x) = \frac{dV}{dt} = \frac{\partial V}{\partial x}\frac{dx}{dt} = \nabla V \cdot f(x)$$

$$= \left[\frac{\partial V}{\partial x_1}, \frac{\partial V}{\partial x_2}, \cdots, \frac{\partial V}{\partial x_n}\right]\begin{bmatrix} f_1(x) \\ \vdots \\ f_n(x) \end{bmatrix}.$$

The following definition introduces a useful and very common way of representing this derivative.

**Definition 3.6** Let $V : D \to \mathbb{R}$ and $f : D \to \mathbb{R}^n$. The Lie derivative of $V$ along $f$, denoted by $L_f V$, is defined by

$$L_f V(x) = \frac{\partial V}{\partial x} f(x).$$

Thus, according to this definition, we have that

$$\dot{V}(x) = \frac{\partial V}{\partial x} f(x) = \nabla V \cdot f(x) = L_f V(x). \tag{3.3}$$

**Example 3.3** Let

$$\dot{x} = \begin{bmatrix} ax_1 \\ bx_2 + \cos x_1 \end{bmatrix}$$

and define $V = x_1^2 + x_2^2$. Thus, we have

$$\dot{V}(x) = L_f V(x) = [2x_1, 2x_2]\begin{bmatrix} ax_1 \\ bx_2 + \cos x_1 \end{bmatrix}$$

$$= 2ax_1^2 + 2bx_2^2 + 2x_2 \cos x_1.$$

□

It is clear from this example that the $\dot{V}(x)$ depends on the system's equation $f(x)$ and thus it will be different for different systems.

## 3.3 Stability Theorems

**Theorem 3.1** (Lyapunov Stability Theorem) Let $x = 0$ be an equilibrium point of $\dot{x} = f(x)$, $f : D \to \mathbb{R}^n$, and let $V : D \to \mathbb{R}$ be a continuously differentiable function such that *(i)* $V(0) = 0$,

*(ii)* $V(x) > 0$    *in* $D - \{0\}$,

*(iii)* $\dot{V}(x) \leq 0$    *in* $D - \{0\}$,

*thus* $x = 0$ *is stable.*

In other words, the theorem implies that a *sufficient condition* for the stability of the equilibrium point $x = 0$ is that there exists a continuously differentiable–positive definite function $V(x)$ such that $\dot{V}(x)$ is negative semi definite in a neighborhood of $x = 0$.

As mentioned earlier, positive definite functions can be seen as generalized energy functions. The condition $V(x) = c$ for constant $c$ defines what is called a *Lyapunov surface*. A Lyapunov surface defines a region of the state space that contains all Lyapunov surfaces of lesser value, that is, given a Lyapunov function $V(\cdot)$ and defining

$$\begin{aligned}
\Omega_1 &= \{x \in B_r : V(x) \leq c_1\} \\
\Omega_2 &= \{x \in B_r : V(x) \leq c_2\}
\end{aligned}$$

where $B_r = \{x \in \mathbb{R}^n : \|x\| \leq r\|\}$, and $c_1 > c_2$ are chosen such that $\Omega_i \subset B_r, i = 1, 2$, then we have that $\Omega_2 \subset \Omega_1$. The condition $\dot{V} \leq 0$ implies that when a trajectory crosses a Lyapunov surface $V(x) = c$, it can never come out again. Thus a trajectory satisfying this condition is actually confined to the closed region $\Omega = \{x : V(x) \leq c\}$. This implies that the equilibrium point is stable, and makes Theorem 3.1 intuitively very simple.

Now suppose that $\dot{V}(x)$ is assumed to be negative definite. In this case, a trajectory can only move from a Lyapunov surface $V(x) = c$ into an inner Lyapunov surface with smaller $c$. This clearly represents a stronger stability condition.

**Theorem 3.2** *(Asymptotic Stability Theorem) Under the conditions of Theorem 3.1, if $V(\cdot)$ is such that*

*(i)* $V(0) = 0$,

*(ii)* $V(x) > 0$    *in* $D - \{0\}$,

*(iii)* $\dot{V}(x) < 0$    *in* $D - \{0\}$,

*thus* $x = 0$ *is asymptotically stable.*

In other words, the theorem says that asymptotic stability is achieved if the conditions of Theorem 3.1 are strengthened by requiring $\dot{V}(x)$ to be negative definite, rather than semi definite.

The discussion above is important since it elaborates on the ideas and motivation behind all the Lyapunov stability theorems. We now provide a proof of Theorems 3.1 and 3.2. These proofs will clarify certain technicalities used later on to distinguish between local and global stabilities, and also in the discussion of region of attraction.

**Proof of theorem 3.1**: Choose $r > 0$ such that the closed ball

$$B_r = \{x \in \mathbb{R}^n : \|x\| \leq r\}$$

is contained in $D$. So $f$ is well defined in the compact set $B_r$. Let

$$\alpha = \min_{\|x\|=r} V(x) \qquad \text{(thus } \alpha > 0, \text{ by the fact that } V(x) > 0 \in D).$$

Now choose $\beta \in (0, \alpha)$ and denote

$$\Omega_\beta = \{x \in B_r : V(x) \leq \beta\}.$$

Thus, by construction, $\Omega_\beta \subset B_r$. Now suppose that $x(0) \in \Omega_\beta$. By assumption (iii) of the theorem we have that

$$\dot{V}(x) \leq 0 \quad \Rightarrow \quad V(x) \leq V(x(0)) \leq \beta \quad \forall t \geq 0.$$

It then follows that any trajectory starting in $\Omega_\beta$ at $t = 0$ stays inside $\Omega_\beta$ for all $t \geq 0$. Moreover, by the continuity of $V(x)$ it follows that $\exists \delta > 0$ such that

$$\|x\| < \delta \Rightarrow V(x) < \beta \qquad (B_\delta \subset \Omega_\beta \subset B_r).$$

It then follows that

$$\|x(0)\| < \delta \Rightarrow x(t) \in \Omega_\beta \subset B_r \qquad \forall t > 0$$

and then

$$\|x(0)\| < \delta \quad \Rightarrow \quad \|x(t)\| < r \leq \epsilon \quad \forall t \geq 0$$

which means that the equilibrium $x = 0$ is stable. $\qquad\square$

**Proof of theorem 3.2**: Under the assumptions of the theorem, $V(x)$ actually decreases along the trajectories of $f(x)$. Using the same argument used in the proof of Theorem 3.1, for every real number $a > 0$ we can find $b > 0$ such that $\Omega_b \subset B_a$, and from Theorem 3.1 whenever the initial condition is inside $\Omega_b$, the solution will remain inside $\Omega_b$. Therefore, to prove asymptotic stability, all we need to show is that $\Omega_b$ reduces to 0 in the limit. In

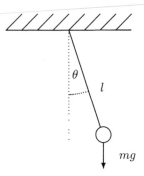

Figure 3.4: Pendulum without friction.

other words, $\Omega_b$ shrinks to a single point as $t \to \infty$. However, this is straightforward, since
by assumption

$$\dot{V}(x) < 0 \quad \text{in } D.$$

Thus, $V(x)$ tends steadily to zero along the solutions of $f(x)$. This completes the proof. □

**Remarks**: The first step when studying the stability properties of an equilibrium point
consists of choosing a positive definite function $V(\cdot)$. Finding a positive definite function is
fairly easy; what is rather tricky is to select a $V(\cdot)$ whose derivative along the trajectories
near the equilibrium point is either negative definite, or semi definite. The reason, of course,
is that $V(\cdot)$ is independent of the dynamics of the differential equation under study, while
$\dot{V}$ depends on this dynamics in an essential manner. For this reason, when a function $V(\cdot)$
is proposed as possible *candidate* to prove any form of stability, such a $V(\cdot)$ is said to be a
*Lyapunov function candidate*. If in addition $\dot{V}(\cdot)$ happens to be negative definite, then $V$
is said to be a *Lyapunov function* for that particular equilibrium point.


## 3.4   Examples

**Example 3.4**  *(Pendulum Without Friction)*

*Using Newton's second law of motion we have,*

$$
\begin{aligned}
ma &= -mg\sin\theta \\
a = l\alpha &= l\ddot{\theta}
\end{aligned}
$$

*where l is the length of the pendulum, and α is the angular acceleration. Thus*

$$ml\ddot{\theta} + mg\sin\theta = 0$$
$$or \quad \ddot{\theta} + \frac{g}{l}\sin\theta = 0$$

*choosing state variables*

$$\begin{cases} x_1 = \theta \\ x_2 = \dot{\theta} \end{cases}$$

*we have*

$$\begin{cases} \dot{x}_1 = x_2 \\ \dot{x}_2 = -\frac{g}{l}\sin x_1 \end{cases}$$

*which is of the desired form $\dot{x} = f(x)$. The origin is an equilibrium point (since $f(0) = 0$). To study the stability of the equilibrium at the origin, we need to propose a Lyapunov function candidate $V(x)$ and show that satisfies the properties of one of the stability theorems seen so far. In general, choosing this function is rather difficult; however, in this case we proceed inspired by our understanding of the physical system. Namely, we compute the total energy of the pendulum (which is a positive function), and use this quantity as our Lyapunov function candidate. We have*

$$E = K + P \qquad \text{(kinetic plus potential energy)}$$
$$= \frac{1}{2}m(\omega l)^2 + mgh$$

*where*

$$\omega = \dot{\theta} = x_2$$
$$h = l(1 - \cos\theta) = l(1 - \cos x_1).$$

*Thus*

$$E = \frac{1}{2}ml^2 x_2^2 + mgl(1 - \cos x_1).$$

*We now define $V(x) = E$ and investigate whether $V(\cdot)$ and its derivative $\dot{V}(\cdot)$ satisfy the conditions of Theorem 3.1 and/or 3.2. Clearly, $V(0) = 0$; thus, defining property (i) is satisfied in both theorems. With respect to (ii), we see that because of the periodicity of $\cos(x_1)$, we have that $V(x) = 0$ whenever $x = (x_1, x_2)^T = (2k\pi, 0)^T$, $k = 1, 2, \cdots$. Thus, $V(\cdot)$ is not positive definite. This situation, however, can be easily remedied by restricting the domain of $x_1$ to the interval $(-2\pi, 2\pi)$; i.e., we take $V : D \to \mathbb{R}$, with $D = ((-2\pi, 2\pi), \mathbb{R})^T$.*

*With this restriction, $V : D \to \mathbb{R}$ is indeed positive definite. There remains to evaluate the derivative of $V(\cdot)$ along the trajectories of $f(t)$. We have*

$$
\begin{aligned}
\dot{V}(x) &= \nabla V \cdot f(x) \\
&= \left[ \frac{\partial V}{\partial x_1}, \frac{\partial V}{\partial x_2} \right] [f_1(x), f_2(x)]^T \\
&= [mgl \sin x_1, ml^2 x_2][x_2, -\frac{g}{l} \sin x_1]^T \\
&= mglx_2 \sin x_1 - mglx_2 \sin x_1 = 0.
\end{aligned}
$$

*Thus $\dot{V}(x) = 0$ and the origin is stable by Theorem 3.1.* □

The result of Example 3.4 is consistent with our physical observations. Indeed, a simple pendulum without friction is a conservative system. This means that the sum of the kinetic and potential energy remains constant. The pendulum will continue to balance without changing the amplitude of the oscillations and thus constitutes a stable system. In our next example we add friction to the dynamics of the pendulum. The added friction leads to a loss of energy that results in a decrease in the amplitude of the oscillations. In the limit, all the initial energy supplied to the pendulum will be dissipated by the friction force and the pendulum will remain at rest. Thus, this version of the pendulum constitutes an asymptotically stable equilibrium of the origin.

**Example 3.5** *(Pendulum with Friction) We now modify the previous example by adding the friction force $kl\dot{\theta}$*

$$
ma = -mg \sin \theta - kl\dot{\theta}
$$

*defining the same state variables as in example 3.4 we have*

$$
\begin{cases}
\dot{x}_1 &= x_2 \\
\dot{x}_2 &= -\frac{g}{l} \sin x_1 - \frac{k}{m} x_2.
\end{cases}
$$

*Again $x = 0$ is an equilibrium point. The energy is the same as in Example 3.4. Thus*

$$
V(x) = \frac{1}{2} ml^2 x_2^2 + mgl(1 - \cos x_1) \;>0 \quad in \; D - \{0\}
$$

$$
\begin{aligned}
\dot{V}(x) &= \nabla V \cdot f(x) \\
&= \left[ \frac{\partial V}{\partial x_1}, \frac{\partial V}{\partial x_2} \right] [f_1(x), f_2(x)]^T \\
&= [mgl \sin x_1, ml^2 x_2][x_2, -\frac{g}{l} \sin x_1 - \frac{k}{m} x_2]^T \\
&= -kl^2 x_2^2.
\end{aligned}
$$

*Thus $\dot{V}(x)$ is negative semi-definite. It is not negative definite since $\dot{V}(x) = 0$ for $x_2 = 0$, regardless of the value of $x_1$ (thus $\dot{V}(x) = 0$ along the $x_1$ axis). According to this analysis, we conclude that the origin is stable by Theorem 3.1, but cannot conclude asymptotic stability as suggested by our intuitive analysis, since we were not able to establish the conditions of Theorem 3.2. Namely, $\dot{V}(x)$ is not negative definite in a neighborhood of $x = 0$. The result is indeed disappointing since we know that a pendulum with friction has an asymptotically stable equilibrium point at the origin.* $\qquad\square$

This example emphasizes the fact that *all* of the theorems seen so far provide *sufficient* but by no means *necessary* conditions for stability.

**Example 3.6** *Consider the following system:*

$$\dot{x}_1 = x_1(x_1^2 + x_2^2 - \beta^2) + x_2$$
$$\dot{x}_2 = -x_1 + x_2(x_1^2 + x_2^2 - \beta^2).$$

*To study the equilibrium point at the origin, we define $V(x) = 1/2(x_1^2 + x_2^2)$. We have*

$$
\begin{aligned}
\dot{V}(x) &= \nabla V \cdot f(x) \\
&= [x_1, x_2][x_1(x_1^2 + x_2^2 - \beta^2) + x_2, -x_1 x_2(x_1^2 + x_2^2 - \beta^2)]^T \\
&= x_1^2(x_1^2 + x_2^2 - \beta^2) + x_2^2(x_1^2 + x_2^2 - \beta^2) \\
&= (x_1^2 + x_2^2)(x_1^2 + x_2^2 - \beta^2).
\end{aligned}
$$

*Thus, $V(x) > 0$ and $\dot{V}(x) < 0$, provided that $(x_1^2 + x_2^2) < \beta^2$, and it follows that the origin is an asymptotically stable equilibrium point.* $\qquad\square$

## 3.5 Asymptotic Stability in the Large

A quick look at the definitions of stability seen so far will reveal that all of these concepts are *local* in character. Consider, for example, the definition of stability. The equilibrium $x_e$ is said to be stable if

$$\|x(t) - x_e\| < \epsilon, \quad \text{provided that} \quad \|x(0) - x_e\| < \delta$$

or in words, this says that starting "near" $x_e$, the solution will remain "near" $x_e$. More important is the case of asymptotic stability. In this case the solution not only stays within $\epsilon$ but also converges to $x_e$ in the limit. When the equilibrium *is* asymptotically stable, it is often important to know under what conditions an initial state will converge to the equilibrium point. In the best possible case, *any* initial state will converge to the equilibrium point. An equilibrium point that has this property is said to be *globally asymptotically stable*, or *asymptotically stable in the large*.

**Definition 3.7** *The equilibrium state $x_e$ is said to be asymptotically stable in the large, or globally asymptotically stable, if it is stable and every motion converges to the equilibrium as $t \to \infty$.*

At this point it is tempting to infer that if the conditions of Theorem 3.2 hold in the whole space $\mathbb{R}^n$, then the asymptotic stability of the equilibrium is global. While this condition is clearly necessary, it is however not sufficient. The reason is that the proof of Theorem 3.1 (and so also that of theorem 3.2) relies on the fact that the positive definiteness of the function $V(x)$ coupled with the negative definiteness of $\dot{V}(x)$ ensure that $V(x) < V(x_0)$. This property, however, holds in a compact region of the space defined in Theorem 3.1 by $\Omega_\beta = \{x \in B_r : V(x) \leq \beta\}$. More precisely, in Theorem 3.1 we started by choosing a ball $B_r = \{x \in \mathbb{R}^n : \|x\| \leq r\}$ and then showed that $\Omega_\beta \subset B_r$. Both sets $\Omega_\beta$ and $B_r$ are closed set (and so compact, since they are also bounded). If now $B_r$ is allowed to be the entire space $\mathbb{R}^n$ the situation changes since the condition $V(x) \leq \beta$ does not, in general, define a closed and bounded region. This in turn implies that $\Omega_\beta$ is not a closed region and so it is possible for state trajectories to drift away from the equilibrium point, even if $V(x) \leq \beta$. The following example shows precisely this.

**Example 3.7** *Consider the following positive definite function:*

$$V(x) = \frac{x_1^2}{1 + x_1^2} + x_2^2.$$

*The region $V(x) \leq \beta$ is closed for values of $\beta < 1$. However, if $\beta > 1$, the surface is open. Figure 3.5 shows that an initial state can diverge from the equilibrium state at the origin while moving towards lower energy curves.*                                            □

The solution to this problem is to include an extra condition that ensures that $V(x) = \beta$ is a closed curve. This can be achieved by considering only functions $V(\cdot)$ that grow unbounded when $x \to \infty$. This functions are called *radially unbounded.*

**Definition 3.8** *Let $V : D \to \mathbb{R}$ be a continuously differentiable function. Then $V(x)$ is said to be radially unbounded if*

$$V(x) \to \infty \quad as \quad \|x\| \to \infty.$$

**Theorem 3.3** *(Global Asymptotic Stability) Under the conditions of Theorem 3.1, if $V(\cdot)$ is such that*

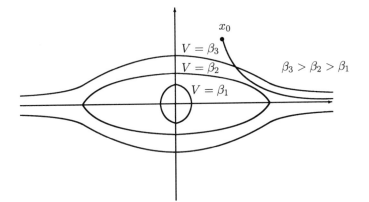

Figure 3.5: The curves $V(x) = \beta$.

(i) $V(0) = 0$.

(ii) $V(x) > 0 \qquad \forall x \neq 0$.

(iii) $V(x)$ is radially unbounded.

(iii) $\dot{V}(x) < 0 \qquad \forall x \neq 0$.

then $x = 0$ is globally asymptotically stable.

**Proof**: The proof is similar to that of Theorem 3.2. We only need to show that given an arbitrary $\beta > 0$, the condition

$$\Omega_\beta = \{x \in \mathbb{R}^n : V(x) \leq \beta\}$$

defines a set that is contained in the ball $B_r = \{x \in \mathbb{R}^n : \|x\| \leq r\}$, for some $r > 0$. To see this notice that the radial unboundedness of $V(\cdot)$ implies that for any $\beta > 0, \exists r > 0$ such that $V(x) > \beta$ whenever $\|x\| > r$, for some $r > 0$. Thus, $\Omega_\beta \subset B_r$, which implies that $\Omega_\beta$ is bounded. $\qquad \square$

**Example 3.8** *Consider the following system*

$$\dot{x}_1 = x_2 - x_1(x_1^2 + x_2^2)$$

$$\dot{x}_2 = -x_1 - x_2(x_1^2 + x_2^2).$$

*To study the equilibrium point at the origin, we define* $V(x) = x_1^2 + x_2^2$. *We have*

$$
\begin{aligned}
\dot{V}(x) &= \frac{\partial V}{\partial x} f(x) \\
&= 2[x_1, x_2][x_2 - x_1(x_1^2 + x_2^2), -x_1 - x_2(x_1^2 + x_2^2)]^T \\
&= -2(x_1^2 + x_2^2)^2.
\end{aligned}
$$

*Thus,* $V(x) > 0$ *and* $\dot{V}(x) < 0$ *for all* $x \in \mathbb{R}^2$. *Moreover, since* $V(\cdot)$ *is radially unbounded, it follows that the origin is globally asymptotically stable.* □

## 3.6   Positive Definite Functions Revisited

We have seen that positive definite functions play an important role in the Lyapunov theory. We now introduce a new class of functions, known as *class* $\mathcal{K}$, and show that positive definite functions can be characterized in terms of this class of functions. This new characterization is useful in many occasions.

**Definition 3.9**  *A continuous function* $\alpha : [0, a) \to \mathbb{R}^+$ *is said to be in the class* $\mathcal{K}$  *if*

(i) $\alpha(0) = 0$.

(ii) *It is strictly increasing.*

$\alpha$ *is said to be in the class* $\mathcal{K}_\infty$ *if in addition* $\alpha : \mathbb{R}^+ \to \mathbb{R}^+$ *and* $\alpha(r) \to \infty$ *as* $r \to \infty$.

In the sequel, $B_r$ represents the ball

$$B_r = \{x \in \mathbb{R}^n : \|x\| \le r\}.$$

**Lemma 3.1**  $V : D \to \mathbb{R}$ *is positive definite if and only if there exists class* $\mathcal{K}$ *functions* $\alpha_1$ *and* $\alpha_2$ *such that*

$$\alpha_1(\|x\|) \le V(x) \le \alpha_2(\|x\|) \quad \forall x \in B_r \subset D.$$

*Moreover, if* $D = \mathbb{R}^n$ *and* $V(\cdot)$ *is radially unbounded then* $\alpha_1$ *and* $\alpha_2$ *can be chosen in the class* $\mathcal{K}_\infty$.

**Proof:** See the Appendix.

**Example 3.9** *Let* $V(x) = x^T P x$, *where* $P$ *is a symmetric matrix. This function is positive definite if and only if the eigenvalues of the symmetric matrix* $P$ *are strictly positive. Denote* $\lambda_{min}(P)$ *and* $\lambda_{max}(P)$ *the minimum and maximum eigenvalues of* $P$, *respectively. It then follows that*

$$\lambda_{min}(P)\|x\|^2 \leq x^T P x \leq \lambda_{max}(P)\|x\|^2$$
$$\lambda_{min}(P)\|x\|^2 \leq V(x) \leq \lambda_{max}(P)\|x\|^2.$$

*Thus,* $\alpha_1, \alpha_2 : [0, \infty) \rightarrow \mathbb{R}^+$, *and are defined by*

$$\alpha_1(x) = \lambda_{min}(P)\|x\|^2$$
$$\alpha_2(x) = \lambda_{max}(P)\|x\|^2.$$

□

For completeness, we now show that it is possible to re state the stability definition in terms of class $\mathcal{K}$ of functions.

**Lemma 3.2** *The equilibrium* $x_e$ *of the system (3.1) is stable if and only if there exists a class* $\mathcal{K}$ *function* $\alpha(\cdot)$ *and a constant* $\epsilon$ *such that*

$$\|x(0) - x_e\| < \delta \quad \Rightarrow \quad \|x(t) - x_e\| \leq \alpha(\|x(0) - x_e\|) \quad \forall t \geq 0. \tag{3.4}$$

**Proof:** See the Appendix.

A stronger class of functions is needed in the definition of asymptotic stability.

**Definition 3.10** *A continuous function* $\beta : [0, a) \times \mathbb{R}^+ \rightarrow \mathbb{R}^+$ *is said to be in the class* $\mathcal{KL}$ *if*

(i) *For fixed* $s$, $\beta(r, s)$ *is in the class* $\mathcal{K}$ *with respect to* $r$.

(ii) *For fixed* $r$, $\beta(r, s)$ *is decreasing with respect to* $s$.

(iii) $\beta(r, s) \rightarrow 0$ *as* $s \rightarrow \infty$.

**Lemma 3.3** *The equilibrium* $x_e$ *of the system (3.1) is asymptotically stable if and only if there exists a class* $\mathcal{KL}$ *function* $\beta(\cdot, \cdot)$ *and a constant* $\epsilon$ *such that*

$$\|x(0) - x_e\| < \delta \quad \Rightarrow \quad \|x(t) - x_e\| \leq \beta(\|x(0) - x_e\|, t) \quad \forall t \geq 0. \tag{3.5}$$

**Proof:** See the Appendix.

## 3.6.1   Exponential Stability

As mentioned earlier, *exponential stability* is the strongest form of stability seen so far. The advantage of this notion is that it makes precise the rate at which trajectories converge to the equilibrium point. Our next theorem gives a sufficient condition for exponential stability.

**Theorem 3.4** *Suppose that all the conditions of Theorem 3.2 are satisfied, and in addition assume that there exist positive constants $K_1$, $K_2$, $K_3$ and $p$ such that*

$$\begin{aligned} K_1\|x\|^p &\leq & V(x) &\leq & K_2\|x\|^p \\ \dot{V}(x) &\leq & -K_3\|x\|^p. \end{aligned}$$

*Then the origin is exponentially stable. Moreover, if the conditions hold globally, the $x = 0$ is globally exponentially stable.*

**Proof:** According to the assumptions of Theorem 3.4, the function $V(x)$ satisfies Lemma 3.1 with $\alpha_1(\cdot)$ and $\alpha_2(\cdot)$, satisfying somewhat strong conditions. Indeed, by assumption

$$\begin{aligned} K_1\|x\|^p &\leq & V(x) &\leq & K_2\|x\|^p \\ \dot{V}(x) &\leq & -K_3\|x\|^p \\ &\leq & -\frac{K_3}{K_2}V(x) \end{aligned}$$

i.e.

$$\begin{aligned} \dot{V}(x) &\leq & -\frac{K_3}{K_2}V(x) \\ \Rightarrow V(x) &\leq & V(x_0)e^{-(K_3/K_2)t} \\ \Rightarrow \|x\| &\leq & [\frac{V(x)}{K_1}]^{1/p} &\leq & [\frac{V(x_0)e^{-(K_3/K_2)t}}{K_1}]^{1/p} \end{aligned}$$

or

$$\|x(t)\| \leq \|x_0\| \, [\frac{K_2}{K_1}]^{1/p} \, e^{-(K_3/2K_2)t}.$$

$\square$

## 3.7   Construction of Lyapunov Functions

The main shortcoming of the Lyapunov theory is the difficulty associated with the construction of suitable Lyapunov functions. In this section we study one approach to this problem, known as the "variable gradient." This method is applicable to autonomous systems and often but not always leads to a desired Lyapunov function for a given system.

**The Variable Gradient**: The essence of this method is to assume that the *gradient* of the (unknown) Lyapunov function $V(\cdot)$ is known up to some adjustable parameters, and then finding $V(\cdot)$ itself by integrating the assumed gradient. In other words, we start out by assuming that

$$\nabla \cdot V(x) = g(x), \qquad (\Rightarrow \dot{V}(x) = \nabla V(x) \cdot f(x) = g(x) \cdot f(x))$$

and propose a possible function $g(x)$ that contains some adjustable parameters. An example of such a function, for a dynamical system with 2 states $x_1$ and $x_2$, could be

$$g(x) = [g_1, g_2] \quad = \quad [h_1^1 x_1 + h_1^2 x_2 \,,\, h_2^1 x_1 + h_2^2 x_2]. \tag{3.6}$$

The power of this method relies on the following fact. Given that

$$\nabla V(x) = g(x) \qquad \text{it follows that}$$

$$g(x) dx \;=\; \nabla V(x) dx = dV(x)$$

thus, we have that

$$V(x_b) - V(x_a) = \int_{x_a}^{x_b} \nabla V(x)\, dx = \int_{x_a}^{x_b} g(x)\, dx$$

that is, the difference $V(x_b) - V(x_a)$ depends on the initial and final states $x_a$ and $x_b$ and *not* on the particular path followed when going from $x_a$ to $x_b$. This property is often used to obtain $V$ by integrating $\nabla V(x)$ along the coordinate axis:

$$
\begin{aligned}
V(x) \;&=\; \int_0^x g(x)\, dx \;=\; \int_0^{x_1} g_1(s_1, 0, \cdots, 0)\, ds_1 \\
&+ \int_0^{x_2} g_2(x_1, s_2, 0, \cdots, 0)\, ds_2 + \cdots + \int_0^{x_n} g_n(x_1, x_2, \cdots, s_n)\, ds_n. \tag{3.7}
\end{aligned}
$$

The free parameters in the function $g(x)$ are constrained to satisfy certain symmetry conditions, satisfied by all gradients of a scalar function. The following theorem details these conditions.

**Theorem 3.5** *A function $g(x)$ is the gradient of a scalar function $V(x)$ if and only if the matrix*

$$
g(x) =
\begin{bmatrix}
\dfrac{\partial g_1}{\partial x_1} & \dfrac{\partial g_2}{\partial x_1} & \cdots & \dfrac{\partial g_n}{\partial x_1} \\[1mm]
\dfrac{\partial g_1}{\partial x_2} & \dfrac{\partial g_2}{\partial x_2} & \cdots & \dfrac{\partial g_n}{\partial x_2} \\[1mm]
\vdots & & & \vdots \\[1mm]
\dfrac{\partial g_1}{\partial x_n} & \dfrac{\partial g_2}{\partial x_n} & \cdots & \dfrac{\partial g_n}{\partial x_n}
\end{bmatrix}
$$

*is symmetric.*

**Proof:** See the Appendix.

We now put these ideas to work using the following example.

**Example 3.10** *Consider the following system:*

$$\begin{cases} \dot{x}_1 &= -ax_1 \\ \dot{x}_2 &= bx_2 + x_1x_2^2. \end{cases}$$

*Clearly, the origin is an equilibrium point. To study the stability of this equilibrium point, we proceed to find a Lyapunov function as follows.*

*Step 1: Assume that* $\nabla V(x) = g(x)$ *has the form*

$$g(x) = [h_1^1 x_1 + h_1^2 x_2 \, , \; h_2^1 x_1 + h_2^2 x_2]. \tag{3.8}$$

*Step 2: Impose the symmetry conditions,*

$$\frac{\partial^2 V}{\partial x_i \partial x_j} = \frac{\partial^2 V}{\partial x_j \partial x_i} \quad or, \; equivalently \quad \frac{\partial g_i}{\partial x_j} = \frac{\partial g_j}{\partial x_i}.$$

*In our case we have*

$$\begin{aligned} \frac{\partial g_1}{\partial x_2} &= x_1 \frac{\partial h_1^1}{\partial x_2} + h_1^2 + x_2 \frac{\partial h_1^2}{\partial x_2} \\ \frac{\partial g_2}{\partial x_1} &= h_2^1 + x_1 \frac{\partial h_2^1}{\partial x_1} + x_2 \frac{\partial h_2^2}{\partial x_1}. \end{aligned}$$

*To simplify the solution, we attempt to solve the problem assuming that the* $g_i^j$ *'s are constant. If this is the case, then*

$$\frac{\partial h_1^1}{\partial x_2} = \frac{\partial h_1^2}{\partial x_2} = \frac{\partial h_2^1}{\partial x_1} = \frac{\partial h_2^2}{\partial x_1} = 0$$

*and we have that:*

$$\frac{\partial g_1}{\partial x_2} = \frac{\partial g_2}{\partial x_1} \quad \Longleftrightarrow \quad h_1^2 = h_2^1 = k$$

$$\Rightarrow \quad g(x) = [h_1^1 x_1 + kx_2 \, , \; kx_1 + h_2^2 x_2].$$

*In particular, choosing* $k = 0$, *we have*

$$g(x) = [g_1, g_2] \;=\; [h_1^1 x_1, h_2^2 x_2].$$

*Step 3: Find $\dot{V}$:*

$$
\begin{aligned}
\dot{V}(x) &= \nabla V \cdot f(x) \\
&= g(x) \cdot f(x) \\
&= [h_1^1 x_1, h_2^2 x_2] f(x) \\
&= -ah_1^1 x_1^2 + h_2^2 (b + x_1 x_2) x_2^2.
\end{aligned}
$$

*Step 4: Find V from $\nabla V$ by integration. Integrating along the axes, we have that*

$$
\begin{aligned}
V(x) &= \int_0^{x_1} g_1(s_1, 0) \, ds_1 \;+\; \int_0^{x_2} g_2(x_1, s_2) \, ds_2 \\
&= \int_0^{x_1} h_1^1 s_1 \, ds_1 \;+\; \int_0^{x_2} h_2^2 s_2 \, ds_2 \\
&= \frac{1}{2} h_1^1 x_1^2 \;+\; \frac{1}{2} h_2^2 x_2^2.
\end{aligned}
$$

*Step 5: Verify that $V > 0$ and $\dot{V} < 0$. we have that*

$$
\begin{aligned}
V(x) &= \frac{1}{2} h_1^1 x_1^2 \;+\; \frac{1}{2} h_2^2 x_2^2 & (3.9) \\
\dot{V}(x) &= -ah_1^1 x_1^2 + h_2^2 (b + x_1 x_2) x_2^2 & (3.10)
\end{aligned}
$$

*From (3.9), $V(x) > 0$ if and only if $h_1^1, h_2^2 > 0$. Assume then that $h_1^1 = h_2^2 = 1$. In this case*

$$
\dot{V}(x) = -ax_1^2 + (b + x_1 x_2) x_2^2
$$

*assume now that $a > 0$, and $b < 0$. In this case*

$$
\dot{V}(x) = -ax_1^2 - (|b| - x_1 x_2) x_2^2
$$

*and we conclude that, under these conditions, the origin is (locally) asymptotically stable.*

□

## 3.8   The Invariance Principle

Asymptotic stability is always more desirable that stability. However, it is often the case that a Lyapunov function candidate fails to identify an asymptotically stable equilibrium point by having $\dot{V}(x)$ *negative semi definite*. An example of this is that of the pendulum with friction (Example 3.5). This shortcoming was due to the fact that when studying the properties of the function $\dot{V}$ we assumed that the variables $x_1$ and $x_2$ are *independent*. Those variables, however, are related by the pendulum equations and so they are not independent of one another. An extension of Lyapunov's theorem due to LaSalle studies this problem in great detail. The central idea is a generalization of the concept of equilibrium point called *invariant set*.

**Definition 3.11**  *A set $M$ is said to be an invariant set with respect to the dynamical system $\dot{x} = f(x)$ if:*

$$x(0) \in M \quad \Rightarrow \quad x(t) \in M \quad \forall t \in \mathbb{R}^+.$$

In other words, $M$ is the set of points such that if a solution of $\dot{x} = f(x)$ belongs to $M$ at *some instant*, initialized at $t = 0$, then it belongs to $M$ for all future time.

**Remarks:** In the dynamical system literature, one often views a differential equation as being defined for all $t$ rather than just all the nonnegative $t$, and a set satisfying the definition above is called *positively invariant*.

The following are some examples of invariant sets of the dynamical system $\dot{x} = f(x)$.

**Example 3.11**  *Any equilibrium point is an invariant set, since if at $t = 0$ we have $x(0) = x_e$, then $x(t) = x_e$  $\forall t \geq 0$.*  □

**Example 3.12**  *For autonomous systems, any trajectory is an invariant set.*  □

**Example 3.13**  *A limit cycle is an invariant set (this is a special case of Example 3.12).*

□

**Example 3.14**  *If $V(x)$ is continuously differentiable (not necessarily positive definite) and satisfies $\dot{V}(x) < 0$ along the solutions of $\dot{x} = f(x)$, then the set $\Omega_l$ defined by*

$$\Omega_l = \{x \in \mathbb{R}^n : V(x) \leq l\}$$

*is an invariant set. Notice that the condition $\dot{V} \leq 0$ implies that if a trajectory crosses a Lyapunov surface $V(x) = c$ it can never come out again.*  □

**Example 3.15**  *The whole space $\mathbb{R}^n$ is an invariant set.*  □

**Definition 3.12** *Let $x(t)$ be a trajectory of the dynamical system $\dot{x} = f(x)$. The set $N$ is called the limit set (or positive limit set) of $x(t)$ if for any $p \in N$ there exist a sequence of times $\{t_n\} \in [0, \infty)$ such that*

$$x(t_n) \to p \quad as \quad t_n \to \infty$$

*or, equivalently*

$$\lim_{n \to \infty} \|x(t_n) - p\| = 0.$$

Roughly speaking, the limit set $N$ of $x(t)$ is whatever $x(t)$ tends to in the limit.

**Example 3.16** *An asymptotically stable equilibrium point is the limit set of any solution starting sufficiently near the equilibrium point.*

**Example 3.17** *A stable limit cycle is the positive limit set of any solution starting sufficiently near it.* □

**Lemma 3.4** *If the solution $x(t, x_0, t_0)$ of the system (3.1) is bounded for $t > t_0$, then its (positive) limit set $N$ is (i) bounded, (ii) closed, and (iii) nonempty. Moreover, the solution approaches $N$ as $t \to \infty$.*

**Proof:** See the Appendix.

The following lemma can be seen as a corollary of Lemma 3.4.

**Lemma 3.5** *The positive limit set $N$ of a solution $x(t, x_0, t_0)$ of the autonomous system (3.1) is invariant with respect to (3.1).*

**Proof:** See the Appendix.

Invariant sets play a fundamental role in an extension of Lyapunov's work produced by LaSalle. The problem is the following: recall the example of the pendulum with friction. Following energy considerations we constructed a Lyapunov function that turned out to be useful to prove that $x = 0$ is a stable equilibrium point. However, our analysis, based on this Lyapunov function, failed to recognize that $x = 0$ is actually asymptotically stable, something that we know thanks to our understanding of this rather simple system. LaSalle's invariance principle removes this problem and it actually allows us to prove that $x = 0$ is indeed asymptotically stable.

We start with the simplest and most useful result in LaSalle's theory. Theorem 3.6 can be considered as a corollary of LaSalle's theorem as will be shown later. The difference between theorems 3.6 and theorem 3.2 is that in Theorem 3.6 $\dot{V}(\cdot)$ is allowed to be only positive semi-definite, something that will remove part of the conservativism associated with certain Lyapunov functions.

**Theorem 3.6** *The equilibrium point $x = 0$ of the autonomous system (3.1) is asymptotically stable if there exists a function $V(x)$ satisfying*

(i) *$V(x)$ positive definite $\forall x \in D$, where we assume that $0 \in D$.*

(ii) *$\dot{V}(x)$ is negative semi definite in a bounded region $R \subset D$.*

(iii) *$\dot{V}(x)$ does not vanish identically along any trajectory in $R$, other than the null solution $x = 0$.*

**Example 3.18** *Consider again the pendulum with friction of Example 3.5:*

$$\dot{x}_1 = x_2 \tag{3.11}$$
$$\dot{x}_2 = -\frac{g}{l}\sin x_1 - \frac{k}{m}x_2. \tag{3.12}$$

*Again*

$$V(x) > 0 \quad \forall x \in (-\pi, \pi) \times \mathbb{R},$$
$$\dot{V}(x) = -kl^2 x_2^2 \tag{3.13}$$

*which is negative semi definite since $\dot{V}(x) = 0$ for all $x = [x_1, 0]^T$. Thus, with $\dot{V}$ short of being negative definite, the Lyapunov theory fails to predict the asymptotic stability of the origin expected from the physical understanding of the problem. We now look to see whether application of Theorem 3.6 leads to a better result. Conditions (i) and (ii) of Theorem 3.6 are satisfied in the region*

$$R = \begin{bmatrix} x_1 \\ x_2 \end{bmatrix}$$

*with $-\pi < x_1 < \pi$, and $-a < x_2 < a$, for any $a \in \mathbb{R}^+$. We now check condition (iii) of the same theorem, that is, we check whether $\dot{V}$ can vanish identically along the trajectories trapped in $R$, other than the null solution. The key of this step is the analysis of the condition $\dot{V} = 0$ using the system equations (3.11)–(3.12). Indeed, assume that $\dot{V}(x)$ is identically zero over a nonzero time interval. By (3.13) we have*

$$\dot{V}(x) = 0 \quad \Rightarrow \quad 0 = -kl^2 x_2^2 \quad \Longleftrightarrow x_2 = 0$$

$$thus \quad x_2 = 0 \quad \forall t \;\Rightarrow\; \dot{x}_2 = 0$$

and by (3.12), we obtain

$$0 = \frac{g}{l}\sin x_1 - \frac{k}{m}x_2 \quad and \; thus \; x_2 = 0 \Rightarrow \sin x_1 = 0$$

restricting $x_1$ to the interval $x_1 \in (-\pi, \pi)$ the last condition can be satisfied if and only if $x_1 = 0$. It follows that $\dot{V}(x)$ does not vanish identically along any solution other than $x = 0$, and the origin is (locally) asymptotically stable by Theorem 3.6. $\qquad\square$

**Proof of Theorem 3.6**: By the Lyapunov stability theorem (Theorem 3.1), we know that for each $\epsilon > 0$ there exist $\delta > 0$

$$\|x_0\| < \delta \quad \Rightarrow \quad \|x(t)\| < \epsilon$$

that is, any solution starting inside the closed ball $B_\delta$ will remain within the closed ball $B_\epsilon$. Hence any solution $x(t, x_0, t_0)$ of (3.1) that starts in $B_\delta$ is bounded and tends to its limit set $N$ that is contained in $B_\epsilon$ (by Lemma 3.4). Also $V(x)$ is continuous on the compact set $B_\epsilon$ and thus is bounded from below in $B_\epsilon$. It is also non increasing by assumption and thus tends to a non negative limit $L$ as $t \to \infty$. Notice also that $V(x)$ is continuous and thus, $V(x) = L \; \forall x$ in the limit set $N$. Also by lemma 3.5, $N$ is an invariant set with respect to (3.1), which means that any solution that starts in $N$ will remain there for all future time. But along that solution, $\dot{V}(x) = 0$ since $V(x)$ is constant $(= L)$ in $N$. Thus, by assumption, $N$ is the origin of the state space and we conclude that any solution starting in $R \subset B_\delta$ converges to $x = 0$ as $t \to \infty$. $\qquad\square$

**Theorem 3.7** *The null solution $x = 0$ of the autonomous system (3.1) is asymptotically stable in the large if the assumptions of theorem 3.6 hold in the entire state space (i.e., $R = \mathbb{R}^n$), and $V(\cdot)$ is radially unbounded.*

**Proof**: The proof follows the same argument used in the proof of Theorem 3.3 and is omitted. $\qquad\square$

**Example 3.19** *Consider the following system:*

$$\dot{x}_1 = x_2$$
$$\dot{x}_2 = -x_2 - \alpha x_1 - (x_1 + x_2)^2 x_2.$$

*To study the equilibrium point at the origin we define $V(x) = \alpha x_1^2 + x_2^2$. We have*

$$\dot{V}(x) = \frac{\partial V}{\partial x} f(x)$$
$$= 2[\alpha x_1, x_2][x_2, -x_2 - \alpha x_1 - (x_1 + x_2)^2 x_2]^T$$
$$= -2x_2^2[1 + (x_1 + x_2)^2].$$

*Thus, $V(x) > 0$ and $\dot{V}(x) \leq 0$ since $\dot{V}(x) = 0$ for $x = (x_1, 0)$. Proceeding as in the previous example, we assume that $\dot{V} = 0$ and conclude that*

$$\dot{V} = 0 \quad \Longleftrightarrow \quad x_2 = 0 , \quad x_2 = 0 \ \forall t \ \Rightarrow \ \dot{x}_2 = 0$$

$$\dot{x}_2 = 0 \ \Rightarrow \quad -x_2 - \alpha x_1 - (x_1 + x_2)^2 x_2 = 0$$

*and considering the fact that $x_2 = 0$, the last equation implies that $x_1 = 0$. It follows that $\dot{V}(x)$ does not vanish identically along any solution other than $x = [0,0]^T$. Moreover, since $V(\cdot)$ is radially unbounded, we conclude that the origin is globally asymptotically stable.* $\square$

**Theorem 3.8** *(LaSalle's theorem) Let $V : D \to \mathbb{R}$ be a continuously differentiable function and assume that*

(i) *$M \subset D$ is a compact set, invariant with respect to the solutions of (3.1).*

(ii) *$\dot{V} \leq 0$ in $M$.*

(iii) *$E : \{x : x \in M, \text{and } \dot{V} = 0\}$; that is, $E$ is the set of all points of $M$ such that $\dot{V} = 0$.*

(iv) *$N$: is the largest invariant set in $E$.*

*Then every solution starting in $M$ approaches $N$ as $t \to \infty$.*

**Proof**: Consider a solution $x(t)$ of (3.1) starting in $M$. Since $\dot{V}(x) \leq 0 \in M$, $V(x)$ is a decreasing function of $t$. Also, since $V(\cdot)$ is a continuous function, it is bounded from below in the compact set $M$. It follows that $V(x(t))$ has a limit as $t \to \infty$. Let $\omega$ be the limit set of this trajectory. It follows that $\omega \subset M$ since $M$ is (an invariant) closed set. For any $p \in \omega \exists$ a sequence $t_n$ with $t_n \to \infty$ and $x(t_n) \to p$. By continuity of $V(x)$, we have that

$$V(p) = \lim_{n \to \infty} V(x(t_n)) = a \quad \text{(a constant)}.$$

Hence, $V(x) = a$ on $\omega$. Also, by Lemma 3.5 $\omega$ is an invariant set, and moreover $\dot{V}(x) = 0$ on $\omega$ (since $V(x)$ is constant on $\omega$). It follows that

$$\omega \subset N \subset E \subset M.$$

Since $x(t)$ is bounded, Lemma 3.4 implies that $x(t)$ approaches $\omega$ (its positive limit set) as $t \to \infty$. Hence $x(t)$ approaches $N$ as $t \to \infty$. $\square$

**Remarks**: LaSalle's theorem goes beyond the Lyapunov stability theorems in two important aspects. In the first place, $V(\cdot)$ is required to be continuously differentiable (and so bounded), but it is not required to be positive definite. Perhaps more important, LaSalle's result applies not only to equilibrium points as in all the Lyapunov theorems, but also to more general dynamic behaviors such as limit cycles. Example 3.20, at the end of this section, emphasizes this point.

Before looking at some examples, we notice that some useful corollaries can be found if $V(\cdot)$ is assumed to be positive definite.

**Corollary 3.1** *Let* $V : D \rightarrow \mathbb{R}$ *be a continuously differentiable positive definite function in a domain* $D$ *containing the origin* $x = 0$, *and assume that* $\dot{V}(x) \leq 0 \in D$. *Let* $S = \{x \in D : \dot{V}(x) = 0\}$ *and suppose that no solution can stay identically in* $S$ *other than the trivial one. Then the origin is asymptotically stable.*

**Proof**: Straightforward (see Exercise 3.11).

**Corollary 3.2** *If* $D = \mathbb{R}^n$ *in Corollary 3.1, and* $V(\cdot)$ *is radially unbounded, then the origin is globally asymptotically stable.*

**Proof**: See Exercise 3.12.

**Example 3.20** *[68] Consider the system defined by*

$$\begin{aligned}
\dot{x}_1 &= x_2 + x_1(\beta^2 - x_1^2 - x_2^2) \\
\dot{x}_2 &= -x_1 + x_2(\beta^2 - x_1^2 - x_2^2)
\end{aligned}$$

*It is immediate that the origin* $x = (0, 0)$ *is an equilibrium point. Also, the set of points defined by the circle* $x_1^2 + x_2^2 = \beta^2$ *constitute an invariant set. To see this, we compute the time derivative of points on the circle, along the solution of* $\dot{x} = f(x)$:

$$\begin{aligned}
\frac{d}{dt}\left[x_1^2 + x_2^2 - \beta^2\right] &= (2x_1, 2x_2)f(x) \\
&= 2(x_1^2 + x_2^2)(b^2 - x_1^2 - x_2^2)
\end{aligned}$$

*for all points on the set. It follows that any trajectory initiating on the circle stays on the circle for all future time, and thus the set of points of the form* $\{x \in \mathbb{R}^2 : x_1^2 + x_2^2 = \beta^2\}$ *constitute an invariant set. The trajectories on this invariant set are described by the solutions of*

$$\dot{x} = f(x)|_{x : x_1^2 + x_2^2 = \beta^2} \quad \Rightarrow \quad \begin{cases} \dot{x}_1 &= x_2 \\ \dot{x}_2 &= -x_1. \end{cases}$$

*Thus, the circle is actually a limit cycle along which the trajectories move in the clockwise direction.*

*We now investigate the stability of this limit cycle using LaSalle's theorem. To this end, consider the following function:*

$$V(x) = \frac{1}{4}(x_1^2 + x_2^2 - \beta^2)^2$$

*Clearly, $V(\cdot)$ is positive semi definite in $\mathbb{R}^2$. Also*

$$\dot{V}(x) = \left[\frac{\partial V}{\partial x_1}, \frac{\partial V}{\partial x_1}\right] f(x)$$
$$= -(x_1^2 + x_2^2)(x_1^2 + x_2^2 - \beta^2)^2 \leq 0.$$

*Moreover, $\dot{V}(x) = 0$ if and only if one of the following conditions is satisfied:*

*(a) $x_1^2 + x_2^2 = 0$*
*(b) $x_1^2 + x_2^2 - \beta^2 = 0$.*

*In other words, $\dot{V} = 0$ either at the origin or on the circle of radius $\beta$. We now apply LaSalle's theorem.*

*Step #1: Given any real number $c > \beta$, define the set $M$ as follows:*

$$M = \{x \in \mathbb{R}^2 : V(x) \leq c\}.$$

*By construction $M$ is closed and bounded (i.e., compact). Also $\dot{V}(x) \leq 0 \forall x \in M$, and thus any trajectory staring from an arbitrary point $x_0 \in M$ will remain inside $M$ and $M$ is therefore an invariant set.*

*Step #2: Find $E = \{x \in M : \dot{V}(x) = 0\}$. Clearly, we have*

$$E = (0,0) \cup \{x \in \mathbb{R}^2 : x_1^2 + x_2^2 = \beta^2\}$$

*that is, $E$ is the union of the origin and the limit cycle.*

*Step #3: Find $N$, the largest invariant set in $E$.*

*This is trivial, since $E$ is the union of the origin (an invariant set, since $(0,0)$ is an equilibrium point), and the invariant set $x_1^2 + x_2^2 = \beta^2$. Thus, we conclude that $N = E$, and by LaSalle's theorem, every motion staring in $M$ converges to either the origin or the limit cycle.*

We can refine our argument by noticing that the function $V$ was designed to measure the distance from a point to the limit cycle:

$$V(x) = \frac{1}{2}(x_1^2 + x_2^2 - \beta^2)^2$$

$$\Rightarrow \quad \begin{cases} V(x) = 0 & \text{whenever } x_2^2 + x_2^2 = \beta^2 \\ V(x) = \frac{1}{2}\beta^4 & \text{when } x = (0,0). \end{cases}$$

Thus, choosing $c : 0 < c < 1/2\beta^4$ we have that $M = \{x \in \mathbb{R}^2 : V(x) \le c\}$ includes the limit cycle but not the origin. Thus application of LaSalle's theorem with any $c$ satisfying $c : \epsilon < c < 1/2\beta^4$ with $\epsilon$ arbitrarily small shows that any motion starting in $M$ converges to the limit cycle, and the limit cycle is said to be convergent, or attractive. The same argument also shows that the origin is unstable since any motion starting arbitrarily near $(0,0)$ converges to the limit cycle, thus diverging from the origin. □

## 3.9 Region of Attraction

As discussed at the beginning of this chapter, asymptotic stability if often the most desirable form of stability and the focus of the stability analysis. Throughout this section we restrict our attention to this form of stability and discuss the possible application of Theorem 3.2 to estimate the region of asymptotic stability. In order to do so, we consider the following example,

**Example 3.21** *Consider the system defined by*

$$\begin{cases} \dot{x}_1 &= 3x_2 \\ \dot{x}_2 &= -5x_1 + x_1^3 - 2x_2. \end{cases}$$

This system has three equilibrium points: $x_e^1 = (0,0)$, $x_e^2 = (-\sqrt{5}, 0)$, and $x_e^3 = (\sqrt{5}, 0)$. We are interested in the stability of the equilibrium point at the origin. To study this equilibrium point, we propose the following Lyapunov function candidate:

$$V(x) = ax_1^2 - bx_1^4 + cx_1x_2 + dx_2^2$$

with $a, b, c, d$ constants to be determined. Differentiating we have that

$$\dot{V} = (3c - 4d)x_2^2 + (2d - 12b)x_1^3x_2 + (6a - 10d - 2c)x_1x_2 + cx_1^4 - 5cx_1^2$$

we now choose

$$\begin{cases} 2d - 12b &= 0 \\ 6a - 10d - 2c &= 0 \end{cases} \tag{3.14}$$

*which can be satisfied choosing* $a = 12, b = 1, c = 6,$ *and* $d = 6.$ *Using these values, we obtain*

$$
\begin{aligned}
V(x) &= 12x_1^2 - x_1^4 + 6x_1x_2 + 6x_2^2 \\
&= 3(x_1 + 2x_2)^2 + 9x_1^2 + 3x_2^2 - x_1^4 \quad &(3.15) \\
\dot{V}(x) &= -6x_2^2 - 30x_1^2 + 6x_1^4. \quad &(3.16)
\end{aligned}
$$

*So far, so good. We now apply Theorem 3.2. According to this theorem, if $V(x) > 0$ and $\dot{V} < 0$ in $D - \{0\}$, then the equilibrium point at the origin is "locally" asymptotically stable. The question here is: What is the meaning of the word "local"? To investigate this issue, we again check (3.15) and (3.16) and conclude that defining $D$ by*

$$
D = \{x \in \mathbb{R}^2 : -1.6 < x_1 < 1.6\} \quad (3.17)
$$

*we have that $V(x) > 0$ and $\dot{V} < 0, \forall x \in D - \{0\}$. It is therefore "tempting" to conclude that the origin is locally asymptotically stable and that any trajectory starting in $D$ will move from a Lyapunov surface $V(x_0) = c_1$ to an inner Lyapunov surface $V(x_1) = c_2$ with $c_1 > c_2$, thus suggesting that any trajectory initiating within $D$ will converge to the origin.*

*To check these conclusions, we plot the trajectories of the system as shown in Figure 3.6. A quick inspection of this figure shows, however, that our conclusions are incorrect. For example, the trajectory initiating at the point $x_1 = 0, x_2 = 4$ is quickly divergent from the origin even though the point $(0,4) \in D$. The problem is that in our example we tried to infer too much from Theorem 3.2. Strictly speaking, this theorem says that the origin is locally asymptotically stable, but the region of the plane for which trajectories converge to the origin cannot be determined from this theorem alone. In general, this region can be a very small neighborhood of the equilibrium point. The point neglected in our analysis is as follows: Even though trajectories starting in $D$ satisfy the conditions $V(x) > 0$ and $\dot{V} < 0$, thus moving to Lyapunov surfaces of lesser values, $D$ is not an invariant set and there are no guarantees that these trajectories will stay within $D$. Thus, once a trajectory crosses the border $|x_1| = \sqrt{5}$ there are no guarantees that $\dot{V}(x)$ will be negative.*          □

**In summary**: Estimating the so-called "region of attraction" of an asymptotically stable equilibrium point is a difficult problem. Theorem 3.2 simply guarantees existence of a possibly small neighborhood of the equilibrium point where such an attraction takes place. We now study how to estimate this region. We begin with the following definition.

**Definition 3.13** *Let $\psi(x,t)$ be the trajectories of the systems (3.1) with initial condition $x$ at $t = 0$. The region of attraction to the equilibrium point $x_e$, denoted $R_A$, is defined by*

$$
R_A = \{x \in D : \psi(x,t) \to x_e, \text{ as } t \to \infty\}.
$$

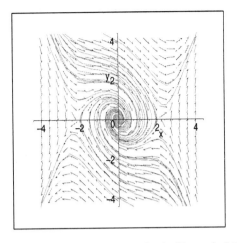

Figure 3.6: System trajectories in Example 3.21.

In general, the exact determination of this region can be a very difficult task. In this section we discuss one way to "estimate" this region. The following theorem, which is based entirely on the LaSalle's invariance principle (Theorem 3.8), outlines the details.

**Theorem 3.9** *Let $x_e$ be an equilibrium point for the system (3.1). Let $V : D \to \mathbb{R}$ be a continuous differentiable function and assume that*

*(i) $M \subset D$ is a compact set containing $x_e$, invariant with respect to the solutions of (3.1).*

*(ii) $\dot{V}$ is such that*

$$\dot{V} \; < \; 0 \;\; \forall x \neq x_e \in M.$$
$$\dot{V} \; = \; 0 \;\; if \, x = x_e.$$

*Under these conditions we have that*

$$M \subset R_A.$$

In other words, Theorem 3.9 states that if $M$ is an invariant set and $V$ is such that $\dot{V} < 0$ inside $M$, then $M$ itself provides an "estimate" of $R_A$.

**Proof:** Under the assumptions, we have that $E = \{x : x \in M, \text{ and } \dot{V} = 0\} = x_e$. It then follows that $N = $ largest invariant set in $E$ is also $x_e$, and the result follows from LaSalle's Theorem 3.8. $\qquad\square$

**Example 3.22** *Consider again the system of Example 3.21:*

$$\begin{cases} \dot{x}_1 &= 3x_2 \\ \dot{x}_2 &= -5x_1 + x_1^3 - 2x_2 \end{cases}$$

$$\begin{aligned} V(x) &= 12x_1^2 - x_1^4 + 6x_1 x_2 + 6x_2^2 \\ \dot{V}(x) &= -6x_2^2 - 30x_1^2 + 6x_1^4 \end{aligned}$$

*We know that $V > 0$ and $\dot{V} < 0$ for all $\{x \in \mathbb{R}^2 : -1.6 < x_1 < 1.6\}$. To estimate the region of attraction $R_A$ we now find the minimum of $V(x)$ at the very edge of this condition (i.e., $x = \pm 1.6$). We have*

$$\begin{aligned} V|_{x_1=1.6} &= 24.16 + 9.6x_2 + 6x_2^2 = z_1 \\ \frac{dz_1}{dx_2} &= 9.6 + 12x_2 = 0 \quad \Leftrightarrow \quad x_2 = -0.8 \end{aligned}$$

*Similarly*

$$\begin{aligned} V|_{x_1=-1.6} &= 24.16 - 9.6x_2 + 6x_2^2 = z_2 \\ \frac{dz_2}{dx_2} &= -9.6 + 12x_2 = 0 \quad \Leftrightarrow \quad x_2 = 0.8. \end{aligned}$$

*Thus, the function $V(\pm 1.6, x_2)$ has a minimum when $x_2 = \pm 0.8$. It is immediate that $V(1.6, -0.8) = V(-1.6, 0.8) = 20.32$. From here we can conclude that given any $\epsilon > 0$, the region defined by*

$$M = \{x \in \mathbb{R}^2 : V(x) \le 20.32 - \epsilon\}$$

*is an invariant set and satisfies the conditions of Theorem 3.9. This means that $M \subset R_A$.*
□

## 3.10   Analysis of Linear Time-Invariant Systems

Linear time-invariant systems constitute an important class, which has been extensively analyzed and is well understood. It is well known that given an LTI system of the form

$$\dot{x} = Ax, \quad A \in \mathbb{R}^{n \times n}, \qquad x(0) = x_0 \tag{3.18}$$

the origin is *stable* if and only if all eigenvalues $\lambda$ of $A$ satisfy $\Re e(\lambda_i) \le 0$, where $\Re e(\lambda_i)$ represents the real part of the eigenvalue $\lambda_i$, and every eigenvalue with $\Re e(\lambda_i) = 0$ has an associated Jordan block of order 1. The equilibrium point $x = 0$ is exponentially stable if and only if all the eigenvalues of the matrix $A$ satisfy $\Re e(\lambda_i) < 0$. Moreover, the solution of the differential equation (3.18) can be expressed in a rather simple closed form

$$x(t) = e^{At} x_0.$$

Given these facts, it seems unnecessary to investigate the stability of LTI systems via Lyapunov methods. This is, however, exactly what we will do in this section! There is more than one good reason for doing so. In the first place, the Lyapunov analysis permits studying linear and nonlinear systems under the same formalism, where LTI is a special case. Second, we will introduce a very useful class of Lyapunov functions that appears very frequently in the literature. Finally, we will study the stability of nonlinear systems via the linearization of the state equation and try to get some insight into the limitations associated with this process.

Consider the autonomous linear time-invariant system given by

$$\dot{x} = Ax, \qquad A \in \mathbb{R}^{n \times n} \tag{3.19}$$

and let $V(\cdot)$ be defined as follows
$$V(x) = x^T P x \tag{3.20}$$

where $P \in \mathbb{R}^{n \times n}$ is (i) symmetric and (ii) positive definite. With these assumptions, $V(\cdot)$ is positive definite. We also have that

$$\dot{V} = \dot{x}^T P x + x^T P \dot{x}$$

by (3.19), $\dot{x}^T = x^T A^T$. Thus

$$\begin{aligned} \dot{V} &= x^T A^T P x + x^T P A x \\ &= x^T (A^T P + P A) x \end{aligned}$$

or

$$\begin{aligned} \dot{V} &= -x^T Q x \tag{3.21} \\ P A + A^T P &= -Q. \tag{3.22} \end{aligned}$$

Here the matrix $Q$ is symmetric, since

$$Q^T = -(PA + A^T P)^T = -(A^T P + AP) = Q$$

If $Q$ is positive definite, then $\dot{V}(\cdot)$ is negative definite and the origin is (globally) asymptotically stable. Thus, analyzing the asymptotic stability of the origin for the LTI system (3.19) reduces to analyzing the positive definiteness of the pair of matrices $(P, Q)$. This is done in two steps:

(i) Choose an arbitrary symmetric, positive definite matrix $Q$.

(ii) Find $P$ that satisfies equation (3.22) and verify that it is positive definite.

Equation (3.22) appears very frequently in the literature and is called *Lyapunov equation.*

**Remarks**: There are two important points to notice here. In the first place, the approach just described may seem unnecessarily complicated. Indeed, it seems to be easier to first select a positive definite $P$ and use this matrix to find $Q$, thus eliminating the need for solving the Lyapunov equation. This approach may however lead to inconclusive results. Consider, for example, the system with the following $A$ matrix,

$$A = \begin{bmatrix} 0 & 4 \\ -8 & -12 \end{bmatrix}$$

taking $P = I$, we have that

$$-Q = PA + A^T P = \begin{bmatrix} 0 & -4 \\ -4 & -24 \end{bmatrix}$$

and the resulting $Q$ *is not* positive definite. Therefore no conclusion can be drawn from this $V(\cdot)$ regarding the stability of the origin of this system.

The second point to notice is that clearly the procedure described above for the stability analysis based on the pair $(P, Q)$ depends on the existence of a unique solution of the Lyapunov equation for a given matrix $A$. The following theorem guarantees the existence of such a solution.

**Theorem 3.10** *The eigenvalues $\lambda_i$ of a matrix $A \in \mathbb{R}^{n \times n}$ satisfy $\Re(\lambda_i) < 0$ if and only if for any given symmetric positive definite matrix $Q$ there exists a unique positive definite symmetric matrix $P$ satisfying the Lyapunov equation (3.22)*

**Proof**: Assume first that given $Q > 0$, $\exists P > 0$ satisfying (3.22). Thus $V = x^T P x > 0$ and $\dot{V} = -x^T Q x < 0$ and asymptotic stability follows from Theorem 3.2. For the converse assume that $\Re(\lambda_i) < 0$ and given $Q$, define $P$ as follows:

$$P = \int_0^\infty e^{A^T t} Q e^{At} \, dt$$

this $P$ is well defined, given the assumptions on the eigenvalues of $A$. The matrix $P$ is also symmetric, since $(e^{A^T t})^T = e^{At}$. We claim that $P$ so defined is positive definite. To see that this is the case, we reason by contradiction and assume that the opposite is true, that is, $\exists x \neq 0$ such that $x^T P x = 0$. But then

$$
\begin{aligned}
x^T P x = 0 \quad &\Rightarrow \quad \int_0^\infty x^T e^{A^T t} Q e^{At} x \; dt = 0 \\
&\Rightarrow \quad \int_0^\infty y^T Q y \; dt = 0 \quad \text{with } y = e^{At} x \\
&\Longleftrightarrow \quad y = e^{At} x = 0 \quad \forall t \geq 0 \\
&\Longleftrightarrow \quad x = 0
\end{aligned}
$$

since $e^{at}$ is nonsingular $\forall t$. This contradicts the assumption, and thus we have that $P$ is indeed positive definite. We now show that $P$ satisfies the Lyapunov equation

$$
\begin{aligned}
P A + A^T P &= \int_0^\infty e^{A^T t} Q e^{At} A \; dt + \int_0^\infty A^T e^{A^T t} Q e^{At} \; dt \\
&= \int_0^\infty \frac{d}{dt} (e^{A^T t} Q e^{At}) \; dt \\
&= e^{A^T t} Q e^{At} \Big|_0^\infty = -Q
\end{aligned}
$$

which shows that $P$ is indeed a solution of the Lyapunov equation. To complete the proof, there remains to show that this $P$ is unique. To see this, suppose that there is another solution $\tilde{P} \neq P$. Then

$$
\begin{aligned}
(P - \tilde{P}) A + A^T (P - \tilde{P}) &= 0 \\
\Rightarrow e^{A^T t} \left[ (P - \tilde{P}) A + A^T (P - \tilde{P}) \right] e^{At} &= 0 \\
\Rightarrow \frac{d}{dt} \left[ e^{A^T t} (P - \tilde{P}) e^{At} \right] &= 0
\end{aligned}
$$

which implies that $e^{A^T t}(P - \tilde{P}) e^{At}$ is constant $\forall t$. This can be the case if and only if $P - \tilde{P} = 0$, or equivalently, $P = \tilde{P}$. This completes the proof. $\qquad\square$

### 3.10.1 Linearization of Nonlinear Systems

Now consider the nonlinear system

$$
\dot{x} = f(x), \qquad\qquad f : D \to \mathbb{R}^n \qquad\qquad (3.23)
$$

assume that $x = x_e \in D$ is an equilibrium point and assume that $f$ is continuously differentiable in $D$. The Taylor series expansion about the equilibrium point $x_e$ has the form

$$
f(x) = f(x_e) + \frac{\partial f}{\partial x}(x_e)(x - x_e) + \text{higher-order terms}
$$

Neglecting the higher-order terms (HOTs) and recalling that, by assumption, $f(x_e) = 0$, we have that

$$f(x) = \frac{\partial f}{\partial x}(x_e)(x - x_e). \tag{3.24}$$

Now defining

$$\bar{x} = x - x_e, \quad A = \frac{\partial f}{\partial x}(x_e) \tag{3.25}$$

we have that $\dot{\bar{x}} = \dot{x}$, and moreover

$$\dot{\bar{x}} \approx A\bar{x}. \tag{3.26}$$

We now whether it is possible to investigate the local stability of the nonlinear system (3.23) about the equilibrium point $x_e$ by analyzing the properties of the linear time-invariant system (3.26). The following theorem, known as *Lyapunov's indirect method*, shows that if the linearized system (3.26) is exponentially stable, then it is indeed the case that for the original system (3.23) the equilibrium $x_e$ is locally exponentially stable. To simplify our notation, we assume that the equilibrium point is the origin.

**Theorem 3.11** *Let $x = 0$ be an equilibrium point for the system (3.23). Assume that $f$ is continuously differentiable in $D$, and let $A$ be defined as in (3.25). Then if the eigenvalues $\lambda_i$ of the matrix $A$ satisfy $\Re e(\lambda_i) < 0$, the origin is an exponentially stable equilibrium point for the system (3.23).*

The proof is omitted since it is a special case of Theorem 4.7 in the next chapter (see Section 4.5 for the proof of the time-varying equivalent of this result).

## 3.11   Instability

So far we have investigated the problem of *stability*. All the results seen so far are, however, *sufficient conditions* for stability. Thus the usefulness of these results is limited by our ability to find a function $V(\cdot)$ that satisfies the conditions of one of the stability theorems seen so far. If our attempt to find this function fails, then no conclusions can be drawn with respect to the stability properties of the particular equilibrium point under study. In these circumstances it is useful to study the opposite problem, namely; whether it is possible to show that the origin is actually *unstable*. The literature on instability is almost as extensive as that on stability. Perhaps the most famous and useful result is a theorem due to Chetaev given next.

**Theorem 3.12** *(Chetaev) Consider the autonomous dynamical systems (3.1) and assume that $x = 0$ is an equilibrium point. Let $V : D \to \mathbb{R}$ have the following properties:*

(i) $V(0) = 0$

(ii) $\exists x_0 \in \mathbb{R}^n$, arbitrarily close to $x = 0$, such that $V(x_0) > 0$

(iii) $\dot{V} > 0 \forall x \in U$, where the set $U$ is defined as follows:

$$U = \{x \in D : \|x\| \leq \epsilon, \text{and } V(x) > 0\}.$$

Under these conditions, $x = 0$ is unstable.

**Remarks:** Before proving the theorem, we briefly discuss conditions (ii) and (iii). According to assumption (ii), $V(\cdot)$ is such that $V(x_0) > 0$ for points inside the ball $B_\delta = \{x \in D : \|x\| \leq \delta\}$, where $\delta$ can be chosen arbitrarily small. No claim however was made about $V(\cdot)$ being positive definite in a neighborhood of $x = 0$. Assumptions (iii) says that $\dot{V}(\cdot)$ is positive definite in the set $U$. This set consists of all those points inside the ball $B_\epsilon$ (i.e., the set of points satisfying $\|x\| \leq \epsilon$), which, in addition, satisfy $V(x) > 0$.

**Proof:** The proof consists of showing that a trajectory initiating at a point $x_0$ arbitrarily close to the origin in the set $U$ will eventually cross the sphere defined by $\|x\| = \epsilon$. Given that $\epsilon$ is arbitrary, this implies that given $\epsilon > 0$, we cannot find $\delta > 0$ such that

$$\|x_0\| < \delta \quad \Rightarrow \quad \|x(t)\| < \epsilon$$

Notice first that condition (ii) guarantees that the set $U$ is not empty. $U$ is clearly bounded and moreover, its boundary consists of the points on the sphere $\|x\| = \epsilon$, and the surface defined by $V(x) = 0$. Now consider an interior point $x_0 \in U$. By assumption (ii) $V(x_0) > 0$, and taking account of assumption (iii) we can conclude that

$$\begin{aligned} V(x_0) &= \xi > 0 \\ \dot{V}(x_0) &> 0 \end{aligned}$$

but then, the trajectory $x(t)$ starting at $x_0$ is such that $V(x(t)) > \xi, \forall t > 0$. This conditions must hold *as long as* $x(t)$ *is inside the set* $U$. Define now the set of points $Q$ as follows:

$$Q = \{x \in U : \|x\| \leq \epsilon \text{ and } V(x) \geq \xi\}.$$

This set is compact (it is clearly bounded and is also closed since it contains its boundary points $\|x\| = \epsilon$ and $V(x) = \xi$). It then follows that $V(x)$, which is a continuous function in a compact set $Q$, has a minimum value and a maximum value in $Q$. This argument also shows that $V(x)$ is bounded in $Q$, and so in $U$. Define

$$\gamma = \min\{\dot{V}(x) : x \in Q\} > 0.$$

This minimum exists since $\dot{V}(x)$ is a continuous function in the compact set $Q$. Thus, for the trajectory starting at $x_0$, we can write

$$V(x(t)) \;=\; V(x_0) + \int_0^t \dot{V}(x(\tau))\,d\tau$$

$$\geq\; V(x_0) + \int_0^t \gamma\,d\tau = V(x_0) + \gamma t.$$

It then follows that $x(t)$ cannot stay forever inside the set $U$ since $V(x)$ is bounded in $U$. Thus a trajectory $x(t)$ initiating arbitrarily close to the origin must intersect the boundary of $U$. The boundaries of this set are $\|x\| = \epsilon$ and the surface $V(x) = 0$. However, the trajectory $x(t)$ is such that $V(x) > \xi$, and we thus conclude that $x(t)$ leaves the set $U$ through the sphere $\|x\| = \epsilon$. Thus, $x = 0$ is unstable since given $\epsilon > 0$, we cannot find $\delta > 0$ such that

$$\|x_0\| < \delta \quad \Rightarrow \quad \|x(t)\| < \epsilon.$$

This completes the proof.                                                  □

**Example 3.23**  *Consider again the system of Example 3.20*

$$\dot{x}_1 \;=\; x_2 + x_1(\beta^2 - x_1^2 - x_2^2)$$
$$\dot{x}_2 \;=\; -x_1 + x_2(\beta^2 - x_1^2 - x_2^2)$$

*We showed in Section 3.8 that the origin of this system is an unstable equilibrium point. We now verify this result using Chetaev's result. Let $V(x) = 1/2(x_1^2 + x_2^2)$. Thus we have that $V(0) = 0$, and moreover $V(x) > 0 \forall x \in \mathbb{R}^2 \neq 0$, i.e., $V(\cdot)$ is positive definite. Also*

$$\dot{V} \;=\; (x_1, x_2)f(x)$$
$$=\; (x_1^2 + x_2^2)(\beta^2 - x_1^2 - x_2^2).$$

*Defining the set $U$ by*

$$U = \{x \in \mathbb{R}^2 : \|x\| \leq \epsilon, \; 0 < \epsilon < \beta\}$$

*we have that $V(x) > 0 \forall x \in U, x \neq 0$, and $\dot{V} > 0 \forall x \in U, x \neq 0$. Thus the origin is unstable, by Chetaev's result.*

□

## 3.12   Exercises

(3.1) Consider the following dynamical system:

$$\dot{x}_1 \;=\; x_2$$
$$\dot{x}_2 \;=\; -x_1 + x_1^3 - x_2$$

(a) Find all of its equilibrium points.

(b) For each equilibrium point $x_e$ different from zero, perform a change of variables $y = x - x_e$, and show that the resulting system $y = g(y)$ has an equilibrium point at the origin.

(3.2) Given the systems (i) and (ii) below, proceed as follows:

(a) Find all of their equilibrium points.

(b) Find the linear approximation about each equilibrium point, find the eigenvalues of the resulting $A$ matrix and classify the stability of each equilibrium point.

(c) Using a computer package, construct the phase portrait of each nonlinear system and discuss the qualitative behavior of the system. Make sure that your analysis contains information about all the equilibrium points of these systems.

(d) Using the same computer package used in part (c), construct the phase portrait of each linear approximation found in (c) and compare it with the results in part (c). What can you conclude about the "accuracy" of the linear approximations as the trajectories deviate from the equilibrium points.

$$(i) \begin{cases} \dot{x}_1 &= x_2 \\ \dot{x}_2 &= x_1 - 2\tan^{-1}(x_1 + x_2) \end{cases}, \quad (ii) \begin{cases} \dot{x}_1 &= \frac{2}{3}x_2 \\ \dot{x}_2 &= -x_1 + x_2(1 - 3x_1^2 - 2x_2^2) \end{cases}$$

(3.3) Consider the magnetic suspension system of Section 1.9.1:

$$\dot{x}_1 = x_2$$
$$\dot{x}_2 = g - \frac{k}{m}x_2 - \frac{\lambda\mu x_3^2}{2m(1 + \mu x_1)^2}$$
$$\dot{x}_3 = \frac{1 + \mu x_1}{\lambda}\left[-Rx_3 + \frac{\lambda\mu}{(1 + \mu x_1)^2}x_2 x_3 + v\right]$$

(a) Find the input voltage $v = v_0$ necessary to keep the ball at an arbitrary position $y = y_0$ (and so $x_1 = y_0$). Find the equilibrium point $x_e = [x_{e1}, x_{e2}, x_{e3}]$ corresponding to this input.

(b) Find the linear approximation about this equilibrium point and analyze its stability.

(3.4) For each of the following systems, study the stability of the equilibrium point at the origin:

$$(i) \begin{cases} \dot{x}_1 &= -x_1 - x_1 x_2^2 \\ \dot{x}_2 &= -x_2 - x_1^2 x_2 \end{cases}, \quad (ii) \begin{cases} \dot{x}_1 &= -x_1^3 + x_1 x_2^2 \\ \dot{x}_2 &= -x_1^2 x_2 - x_2^3 \end{cases}$$

(3.5) For each of the following systems, study the stability of the equilibrium point at the origin:

$$(i) \begin{cases} \dot{x}_1 = x_2 - 2x_1(x_1^2 + x_2^2) \\ \dot{x}_2 = -x_1 - 2x_2(x_1^2 + x_2^2) \end{cases} , \quad (ii) \begin{cases} \dot{x}_1 = -x_1 + x_1 x_2 \\ \dot{x}_2 = -x_2 \end{cases}$$

(3.6) Consider the following system:

$$\begin{aligned} \dot{x}_1 &= -x_2 + \alpha x_1(x_1^2 + x_2^2) \\ \dot{x}_2 &= x_1 + \alpha x_2(x_1^2 + x_2^2) \end{aligned}$$

(a) Verify that the origin is an equilibrium point.

(b) Find the linear approximation about the origin, find the eigenvalues of the resulting $A$ matrix, and classify the stability of the equilibrium point at the origin.

(c) Assuming that the parameter $\alpha > 0$, use a computer package to study the trajectories for several initial conditions. What can you conclude about your answers from the linear analysis in part (b)?

(d) Repeat part (c) assuming $\alpha < 0$.

(3.7) It is known that a given dynamical system has an equilibrium point at the origin. For this system, a function $V(\cdot)$ has been proposed, and its derivative $\dot{V}(\cdot)$ has been computed. Assuming that $V(\cdot)$ and $\dot{V}(\cdot)$ are given below you are asked to classify the origin, in each case, as (a) stable, (b) locally asymptotically stable, (c) globally asymptotically stable, (d) unstable, and/or (e) inconclusive information. Explain your answer in each case.

(a) $V(x) = (x_1^2 + x_2^2)$, $\dot{V}(x) = -x_1^2$.

(b) $V(x) = (x_1^2 + x_2^2 - 1)$, $\dot{V}(x) = -(x_1^2 + x_2^2)$.

(c) $V(x) = (x_1^2 + x_2^2 - 1)^2$, $\dot{V}(x) = -(x_1^2 + x_2^2)$.

(d) $V(x) = (x_1^2 + x_2^2 - 1)^2$, $\dot{V}(x) = (x_1^2 + x_2^2)$.

(e) $V(x) = (x_1^2 + x_2^2 - 1)$, $\dot{V}(x) = (x_1^2 + x_2^2)$.

(f) $V(x) = (x_1^2 - x_2^2)$, $\dot{V}(x) = -(x_1^2 + x_2^2)$.

(g) $V(x) = (x_1^2 - x_2^2)$, $\dot{V}(x) = (x_1^2 + x_2^2)$.

(h) $V(x) = (x_1 + x_2)$, $\dot{V}(x) = (x_1^2 - x_2^2)$.

(i) $V(x) = (x_1 + x_2)$, $\dot{V}(x) = (x_1^2 + x_2^2)$.

(3.8) Prove the following properties of class $\mathcal{K}$ and $\mathcal{K}_\infty$ functions:

(i) If $\alpha : [0, a) \to \mathbb{R} \in \mathcal{K}$, then $\alpha^{-1} : [0, \alpha(a)) \to \mathbb{R} \in \mathcal{K}$.

(ii) If $\alpha_1$, $\alpha_2 \in \mathcal{K}$, then $\alpha_1 \circ \alpha_2 \in \mathcal{K}$.

(iii) If $\alpha \in \mathcal{K}_\infty$, then $\alpha^{-1} \in \mathcal{K}$.

(iv) If $\alpha_1$, $\alpha_2 \in \mathcal{K}_\infty$, then $\alpha_1 \circ \alpha_2 \in \mathcal{K}_\infty$.

(3.9) Consider the system defined by the following equations:

$$\dot{x}_1 = x_2 + \beta \left( \frac{x_1^3}{3} - x_1 \right)$$
$$\dot{x}_2 = -x_1$$

Study the stability of the equilibrium point $x_e = (0,0)$, in the following cases:

(i) $\beta > 0$.

(ii) $\beta = 0$.

(iii) $\beta < 0$.

(3.10) Provide a detailed proof of Theorem 3.7.

(3.11) Provide a detailed proof of Corollary 3.1.

(3.12) Provide a detailed proof of Corollary 3.2.

(3.13) Consider the system defined by the following equations:

$$\dot{x}_1 = x_2$$
$$\dot{x}_2 = -x_2 - \alpha x_1 - (2x_2 + 3x_1)^2 x_2$$

Study the stability of the equilibrium point $x_e = (0,0)$.

(3.14) Consider the system defined by the following equations:

$$\dot{x}_1 = \frac{2}{3} x_2$$
$$\dot{x}_2 = -x_1 + x_2(1 - 3x_1^2 - 2x_2^2)$$

(i) Show that the points defined by (a) $x = (0,0)$ and (b) $1 - (3x_1^2 + 2x_2^2) = 0$ are invariant sets.

(ii) Study the stability of the origin $x = (0,0)$.

(iii) Study the stability of the invariant set $1 - (3x_1^2 + 2x_2^2) = 0$.

(3.15) Given the following system, discuss the stability of the equilibrium point at the origin:

$$\dot{x}_1 = x_1^2 x_2 + 2x_1 x_2^2 + x_1^3$$
$$\dot{x}_2 = -x_1^3 + x_2^3$$

(3.16) (Lagrange stability) Consider the following notion of stability:

**Definition 3.14** *The equilibrium point $x = 0$ of the system (3.1) is said to be bounded or Lagrange stable if there exist a bound $\Delta$ such that*

$$\|x(t)\| < \Delta \qquad \forall t \geq 0.$$

Prove the following theorem

**Theorem 3.13** *[49] (Lagrange stability theorem) Let $\Omega$ be a bounded neighborhood of the origin and let $\Omega^c$ be its complement. Assume that $V(x) : \mathbb{R}^n \to \mathbb{R}$ be continuously differentiable in $\Omega^c$ and satisfying:*

*(i)* $V(x) > 0 \quad \forall x \in \Omega^c.$

*(i)* $\dot{V}(x) \leq 0 \quad \forall x \in \Omega^c.$

*(i) $V$ is radially unbounded.*

*Then the equilibrium point at the origin is Lagrange stable.*

## Notes and References

Good sources for the material of this chapter are References [48], [27], [41] [88] [68] and [95] among others. The proof of theorem 3.1 is based on Reference [32]. Section 3.7 as well as lemmas 3.4 and 3.5 follow closely the presentation in Reference [95]. Section 3.8 is based on LaSalle, [49], and Khalil, [41]. The beautiful Example 3.20 was taken from Reference [68].

# Chapter 4

# Lyapunov Stability II: Nonautonomous Systems

In this chapter we extend the results of Chapter 3 to nonautonomous system. We start by reviewing the several notions of stability. In this case, the initial time instant $t_0$ warrants special attention. This issue will originate several technicalities as well as the notion of *uniform* stability, to be defined.

## 4.1 Definitions

We now extend the several notions of stability from Chapter 3, to nonautonomous systems. For simplicity, in this chapter we state all our definitions and theorems assuming that the equilibrium point of interest is the origin, $x_e = 0$.

Consider the nonautonomous system[1]

$$\dot{x} = f(x, t) \qquad\qquad f : D \times \mathbb{R}^+ \to \mathbb{R}^n \qquad\qquad (4.1)$$

where $f : D \times [0, \infty) \to \mathbb{R}^n$ is locally Lipschitz in $x$ and piecewise continuous in $t$ on $D \times [0, \infty)$. We will say that the origin $x = 0 \in D$ is an equilibrium point of (4.1) at $t = t_0$ if

$$f(0, t) = 0 \qquad \forall t \geq t_0.$$

For autonomous systems equilibrium points are the real roots of the equation $f(x_e) = 0$. Visualizing equilibrium points for nonautonomous systems is not as simple. In general,

---

[1]Notice that, as in Chapter 3, (4.1) represents an *unforced* system.

$x_e = 0$ can be a translation of a nonzero trajectory. Indeed, consider the nonautonomous system

$$\dot{x} = f(x, t) \tag{4.2}$$

and assume that $\bar{x}(t)$ is a trajectory, or a solution of the differential equation (4.2) for $t \geq 0$. Consider the change of variable

$$y = x(t) - \bar{x}(t).$$

We have

$$\dot{y} = \dot{x} - \dot{\bar{x}}$$

$$\Rightarrow \quad \dot{y} \quad = \quad f(x, t) - \dot{\bar{x}}(t)$$
$$= \quad f(y + \bar{x}(t), t) - \dot{\bar{x}}(t)$$
$$\overset{def}{=} \quad g(y, t).$$

But $\dot{\bar{x}} = f(\bar{x}(t), t)$. Thus

$$g(y, t) \quad = \quad f(y + \bar{x}(t), t) - f(\bar{x}(t), t)$$
$$\equiv \quad 0 \quad \text{if } y = 0$$

that is, the origin $y = 0$ is an equilibrium point of the new system $\dot{y} = g(y, t)$ at $t = 0$.

**Definition 4.1** *The equilibrium point $x = 0$ of the system (4.1) is said to be*

- Stable *at $t_0$ if given $\epsilon > 0$, $\exists \delta = \delta(\epsilon, t_0) > 0$ :*

$$\|x(0)\| < \delta \quad \Rightarrow \quad \|x(t)\| < \epsilon \quad \forall t \geq t_0 > 0 \tag{4.3}$$

- Convergent *at $t_0$ if there exists $\delta_1 = \delta_1(t_0) > 0$ :*

$$\|x(0)\| < \delta_1 \quad \Rightarrow \quad \lim_{t \to \infty} x(t) = 0. \tag{4.4}$$

*Equivalently (and more precisely), $x_0$ is convergent at $t_0$ if for any given $\epsilon_1 > 0, \exists T = T(\epsilon_1, t_0)$ such that*

$$\|x(0)\| < \delta_1 \quad \Rightarrow \quad \|x(t)\| < \epsilon_1 \quad \forall t \geq t_0 + T \tag{4.5}$$

- Asymptotically stable *at $t_0$ if it is both stable and convergent.*

- Unstable *if it is not stable.*

All of these definitions are similar to their counterpart in Chapter 3. The difference is in the inclusion of the initial time $t_0$. This dependence on the initial time is not desirable and motivates the introduction of the several notions of <u>uniform</u> stability.

**Definition 4.2** *The equilibrium point $x = 0$ of the system (4.1) is said to be*

- Uniformly stable *if any given $\epsilon > 0$, $\exists \delta = \delta(\epsilon) > 0$ :*

$$\|x(0)\| < \delta \quad \Rightarrow \quad \|x(t)\| < \epsilon \quad \forall t \geq t_0 > 0 \tag{4.6}$$

- Uniformly convergent *if there is $\delta_1 > 0$, independent of $t_0$, such that*

$$\|x_0\| < \delta_1 \quad \Rightarrow \quad x(t) \to 0 \ as \ t \to \infty.$$

*Equivalently, $x = 0$ is uniformly convergent if for any given $\epsilon_1 > 0, \exists T = T(\epsilon_1)$ such that*

$$\|x(0)\| < \delta_1 \quad \Rightarrow \quad \|x(t)\| < \epsilon_1 \quad \forall t \geq t_0 + T$$

- Uniformly asymptotically stable *if it is uniformly stable and uniformly convergent.*

- Globally uniformly asymptotically stable *if it is uniformly asymptotically stable and every motion converges to the origin.*

As in the case of autonomous systems, it is often useful to restate the notions of uniform stability and uniform asymptotic stability using class $\mathcal{K}$ and class $\mathcal{KL}$ functions. The following lemmas outline the details. The proofs of both of these lemmas are almost identical to their counterpart for autonomous systems and are omitted.

**Lemma 4.1** *The equilibrium point $x = 0$ of the system (4.1) is uniformly stable if and only if there exists a class $\mathcal{K}$ function $\alpha(\cdot)$ and a constant $c > 0$, independent of $t_0$ such that*

$$\|x(0)\| < c \quad \Rightarrow \quad \|x(t)\| \leq \alpha(\|x(0)\|) \quad \forall t \geq t_0. \tag{4.7}$$

**Lemma 4.2** *The equilibrium point $x = 0$ of the system (4.1) is uniformly asymptotically stable if and only if there exists a class $\mathcal{KL}$ function $\beta(\cdot, \cdot)$ and a constant $c > 0$, independent of $t_0$ such that*

$$\|x(0)\| < c \quad \Rightarrow \quad \|x(t)\| \leq \beta(\|x(0)\|, t - t_0) \quad \forall t \geq t_0. \tag{4.8}$$

**Definition 4.3** *The equilibrium point $x = 0$ of the system (4.1) is (locally) exponentially stable if there exist positive constants $\alpha$ and $\lambda$ such that*

$$\|x(t)\| \leq \alpha \|x_0\| k e^{-\lambda t} \tag{4.9}$$

*whenever $\|x(0)\| < \delta$. It is said to be globally exponentially stable if (4.9) is satisfied for any $x \in \mathbb{R}^n$.*

## 4.2   Positive Definite Functions

As seen in Chapter 3, positive definite functions play a crucial role in the Lyapunov theory. In this section we introduce *time-dependent* positive definite functions.

In the following definitions we consider a function $W : D \times \mathbb{R}^+ \to \mathbb{R}$, i.e., a scalar function $W(x,t)$ of two variables: the vector $x \in D$ and the time variable $t$. Furthermore we assume that

(i) $0 \in D$.

(ii) $W(x,t)$ is continuous and has continuous partial derivatives with respect to all of its arguments.

**Definition 4.4** $W(\cdot,\cdot)$ *is said to be positive semi definite in $D$ if*

*(i) $W(0,t) = 0$      $\forall t \in \mathbb{R}^+$.*

*(ii) $W(x,t) \geq 0$     $\forall x \neq 0, x \in D$.*

**Definition 4.5** $W(\cdot,\cdot)$ *is said to be positive definite in $D$ if*

*(i) $W(0,t) = 0$.*

*(ii) $\exists$ a time-invariant positive definite function $V_1(x)$ such that*

$$V_1(x) \leq W(x,t) \qquad \forall x \in D.$$

**Definition 4.6** $W(\cdot,\cdot)$ *is said to be decrescent in $D$ if there exists a positive definite function $V_2(x)$ such that*

$$|W(x,t)| \leq V_2(x) \qquad \forall x \in D.$$

The essence of Definition 4.6 is to render the decay of $W(\cdot,\cdot)$ toward zero a function of $x$ *only*, and not of $t$. Equivalently, $W(x,t)$ is decrescent in $D$ if it tends to zero uniformly with respect to $t$ as $\|x\| \to 0$. It is immediate that every time-invariant positive definite function is decrescent.

**Definition 4.7** $W(\cdot,\cdot)$ *is radially unbounded if*

$$W(x,t) \to \infty \quad as \quad \|x\| \to \infty$$

*uniformly on t. Equivalently, $W(\cdot,\cdot)$ is radially unbounded if given $M, \exists N > 0$ such that*

$$W(x,t) > M$$

*for all t, provided that $\|x\| > N$.*

**Remarks:** Consider now function $W(x,t)$. By Definition 4.5, $W(\cdot,\cdot)$ is positive definite in $D$ if and only if $\exists V_1(x)$ such that

$$V_1(x) \leq W(x,t), \quad \forall x \in D \tag{4.10}$$

and by Lemma 3.1 this implies the existence of $\alpha_1(\cdot)$ such that

$$\alpha_1(\|x\|) \leq V_1(x) \leq W(x,t), \quad \forall x \in B_r \subset D. \tag{4.11}$$

If in addition $W(\cdot,\cdot)$ is decrescent, then, according to Definition 4.6 there exists $V_2$:

$$W(x,t) \leq V_2(x), \quad \forall x \in D \tag{4.12}$$

and by Lemma 3.1 this implies the existence of $\alpha_2(\cdot)$ such that

$$W(x,t) \leq V_2(x) \leq \alpha_2(\|x\|), \quad \forall x \in B_r \subset D. \tag{4.13}$$

It follows that $W(\cdot,\cdot)$ is positive definite and decrescent if and only if there exist a (time-invariant) positive definite functions $V_1(\cdot)$ and $V_2(\cdot)$, such that

$$V_1(x) \leq W(x,t) \leq V_2(x), \quad \forall x \in D \tag{4.14}$$

which in turn implies the existence of function $\alpha_1(\cdot)$ and $\alpha_2(\cdot) \in \mathcal{K}$ such that

$$\alpha_1(\|x\|) \leq W(x,t) \leq \alpha_2(\|x\|), \quad \forall x \in B_r \subset D. \tag{4.15}$$

Finally, $W(\cdot,\cdot)$ is positive definite and decrescent and radially unbounded if and only if $\alpha_1(\cdot)$ and $\alpha_2(\cdot)$ can be chosen in the class $\mathcal{K}_\infty$.

### 4.2.1 Examples

In the following examples we assume that $x = [x_1, x_2]^T$ and study several functions $W(x,t)$.

**Example 4.1** *Let $W_1(x,t) = (x_1^2 + x_2^2)e^{-\alpha t}$ $\alpha > 0$. This function satisfies*

*(i) $W_1(0,t) = 0\, e^{-\alpha t} = 0$ .*

*(ii)* $W_1(x,t) > 0 \quad \forall x \neq 0, \quad \forall t \in \mathbb{R}.$

*However,* $\lim_{t \to \infty} W_1(x,t) = 0 \quad \forall x.$ *Thus,* $W_1(\cdot, \cdot)$ *is positive semi definite, but* not *positive definite.*   □

**Example 4.2** *Let*

$$
\begin{aligned}
W_2(x,t) &= \frac{(x_1^2 + x_2^2)(t^2 + 1)}{(x_1^2 + 2)} \\
&= V_2(x)(t^2 + 1), \qquad V_2(x) \stackrel{def}{=} \frac{(x_1^2 + x_2^2)}{(x_1^2 + 2)}.
\end{aligned}
$$

*Thus,* $V_2(x) > 0 \; \forall x \in \mathbb{R}^2$ *and moreover* $W_2(x,t) \geq V_2(x) \; \forall x \in \mathbb{R}^2$, *which implies that* $W_2(\cdot, \cdot)$ *is positive definite. Also*

$$
\lim_{t \to \infty} W_2(x,t) = \infty \quad \forall x \in \mathbb{R}^2.
$$

*Thus it is not possible to find a positive definite function* $V(\cdot)$ *such that* $|W_2(x,t)| \leq V(x) \; \forall x$, *and thus* $W_2(x,t)$ *is not decrescent.* $|W_2(x,t)|$ *is not radially unbounded since it does not tend to infinity along the* $x_1$ *axis.*   □

**Example 4.3** *Let*

$$
\begin{aligned}
W_3(x,t) &= (x_1^2 + x_2^2)(t^2 + 1) \\
&= V_3(x)(t^2 + 1), \qquad V_3(x) \stackrel{def}{=} (x_1^2 + x_2^2).
\end{aligned}
$$

*Following a procedure identical to that in Example 4.2 we have that* $W_3(x)$ *is positive definite, radially unbounded and not decrescent.*   □

**Example 4.4** *Let*

$$
W_4(x,t) = \frac{(x_1^2 + x_2^2)}{(x_1^2 + 1)}.
$$

*Thus,* $W_4(\cdot, \cdot) > 0 \forall x \in \mathbb{R}^2$ *and is positive definite. It is not time-dependent, and so it is decrescent. It is not radially unbounded since it does not tend to infinity along the* $x_1$ *axis.*   □

**Example 4.5** *Let*

$$
\begin{aligned}
W_5(x,t) &= \frac{(x_1^2 + x_2^2)(t^2 + 1)}{(t^2 + 2)} \\
&= V_5(x)\frac{(t^2 + 1)}{(t^2 + 2)}, \qquad V_5(x) \stackrel{def}{=} (x_1^2 + x_2^2).
\end{aligned}
$$

Thus, $W_5(x,t) \geq k_1 V_5(x)$ for some constant $k_1$, which implies that $W_5(\cdot,\cdot)$ is positive definite. It is decrescent since

$$|W_5(x,t)| \leq k_2 V_5(x) \ \forall x \in \mathbb{R}^2$$

It is also radially unbounded since

$$W_5(x,t) \to \infty \quad as \ \|x\| \to \infty.$$

$\square$

## 4.3 Stability Theorems

We now look for a generalization of the stability results of Chapter 3 for nonautonomous systems. Consider the system (4.1) and assume that the origin is an equilibrium state:

$$f(0,t) = 0 \quad \forall t \in \mathbb{R}.$$

**Theorem 4.1** *(Lyapunov Stability Theorem) If in a neighborhood $D$ of the equilibrium state $x = 0$ there exists a differentiable function $W(\cdot,\cdot) : D \times [0,\infty) \to \mathbb{R}$ such that*

*(i) $W(x,t)$ is positive definite.*

*(ii) The derivative of $W(\cdot,\cdot)$ along any solution of (4.1) is negative semi definite in $D$, then*

*the equilibrium state is stable. Moreover, if $W(x,t)$ is also decrescent then the origin is uniformly stable.*

**Theorem 4.2** *(Lyapunov Uniform Asymptotic Stability) If in a neighborhood $D$ of the equilibrium state $x = 0$ there exists a differentiable function $W(\cdot,\cdot) : D \times [0,\infty) \to \mathbb{R}$ such that*

*(i) $W(x,t)$ is (a) positive definite, and (b) decrescent, and*

*(ii) The derivative of $\dot{W}(x,t)$ is negative definite in $D$, then*

*the equilibrium state is uniformly asymptotically stable.*

**Remarks:** There is an interesting difference between Theorems 4.1 and 4.2. Namely, in Theorem 4.1 the assumption of $W(\cdot,\cdot)$ being decrescent is optional. Indeed, if $W(\cdot,\cdot)$ is decrescent we have uniform stability, whereas if this is not the case, then we settle for stability. Theorem 4.2 is different; if the decrescent assumption is removed, the remaining conditions are not sufficient to prove asymptotic stability. This point was clarified in 1949 by Massera who found a counterexample.

Notice also that, according to inequalities (4.14)–(4.15), given a positive definite function $W(x,t)$ there exist positive definite functions $V_1(x)$ and $V_2(x)$, and class $\mathcal{K}$ functions $\alpha_1$ and $\alpha_2$ such that

$$V_1(x) \leq W(x,t) \leq V_2(x) \qquad \forall x \in D \tag{4.16}$$
$$\alpha_1(\|x\|) \leq W(x,t) \leq \alpha_2(\|x\|) \qquad \forall x \in B_r \tag{4.17}$$

with this in mind, Theorem 4.2 can be restated as follows:

**Theorem 4.2** *(Uniform Asymptotic Stability Theorem restated) If in a neighborhood $D$ of the equilibrium state $x = 0$ there exists a differentiable function $W(\cdot,\cdot) : D \times [0,\infty) \to \mathbb{R}$ such that*

(i) $V_1(x) \leq W(x,t) \leq V_2(x) \quad \forall x \in D, \forall t$

(ii) $\frac{\partial W}{\partial t} + \nabla W f(x,t) \leq -V_3(x) \quad \forall x \in D, \forall t$

*where $V_i, i = 1,2,3$ are positive definite functions in $D$, then the equilibrium state is uniformly asymptotically stable.*

**Theorem 4.3** *(Global Uniform Asymptotic Stability) If there exists a differentiable function $W(\cdot,\cdot) : \mathbb{R}^n \times [0,\infty) \to \mathbb{R}$ such that*

(i) *$W(x,t)$ is (a) positive definite, and (b) decrescent, and radially unbounded $\forall x \in \mathbb{R}^n$, and such that*

(ii) *The derivative of $\dot{W}(x,t)$ is negative definite $\forall x \in \mathbb{R}^n$, then*

*the equilibrium state at $x = 0$ is globally uniformly asymptotically stable.*

For completeness, the following theorem extends Theorem 3.4 on exponential stability to the case of nonautonomous systems. The proof is almost identical to that of Theorem 3.4 and is omitted.

**Theorem 4.4** *Suppose that all the conditions of theorem 4.2 are satisfied, and in addition assume that there exist positive constants $K_1, K_2$, and $K_3$ such that*

$$K_1\|x\|^p \leq W(x,t) \leq K_2\|x\|^p$$
$$\dot{W}(x) \leq -K_3\|x\|^p.$$

*Then the origin is exponentially stable. Moreover, if the conditions hold globally, the $x = 0$ is globally exponentially stable.*

**Example 4.6** *Consider the following system:*

$$\begin{cases} \dot{x}_1 &= -x_1 - e^{-2t}x_2 \\ \dot{x}_2 &= x_1 - x_2. \end{cases}$$

*To study the stability of the origin for this system, we consider the following Lyapunov function candidate:*

$$W(x,t) = x_1^2 + (1 + e^{-2t})x_2^2.$$

*Clearly*

$$V_1(x) = (x_1^2 + x_2^2) \leq W(x,t) \leq (x_1^2 + 2x_2^2) = V_2(x)$$

*thus, we have that*

- $W(x,t)$ *is positive definite, since $V_1(x) \leq W(x,t)$, with $V_1$ positive definite in $\mathbb{R}^2$.*

- $W(x,t)$ *is decrescent, since $W(x,t) \geq V_2(x)$, with $V_2$ also positive definite in $\mathbb{R}^2$.*

*Then*

$$\begin{aligned} \dot{W}(x,t) &= \frac{\partial W}{\partial x}f(x,t) + \frac{\partial W}{\partial t} \\ &= -2[x_1^2 - x_1x_2 + x_2^2(1 + 2e^{-2t})] \\ &\leq -2[x_1^2 - x_1x_2 + 3x_2^2]. \end{aligned}$$

*It follows that $\dot{W}(x,t)$ is negative definite and the origin is globally asymptotically stable.*

$\square$

## 4.4 Proof of the Stability Theorems

We now elaborate the proofs of theorems 4.1-4.3.

**Proof of theorem 4.1**: Choose $R > 0$ such that the closed ball

$$B_R = \{x \in \mathbb{R}^n : \|x\| \leq R\}$$

is contained in $D$. By the assumptions of the theorem $W(\cdot, \cdot)$ is positive definite and thus there exist a time-invariant positive definite function $V_1(\cdot)$ and a class $\mathcal{K}$ function $\alpha_1$ satisfying (4.11):

$$\alpha_1(\|x\|) \leq V_1(x) \leq W(x, t) \quad \forall x \in B_R. \tag{4.18}$$

Moreover, by assumption, $\dot{W}(x, t) \leq 0$, which means that $W(x, t)$ cannot increase along any motion. Thus

$$
\begin{aligned}
W(x(t), t) &\leq W(x_0, t_0) \qquad t \geq t_0 \\
\Rightarrow \quad \alpha_1(\|x\|) &\leq W(x(t), t) \leq W(x_0, t_0).
\end{aligned}
$$

Also $W$ is continuous with respect to $x$ and satisfies $W(0, t_0) = 0$. Thus, given $t_0$, we can find $\delta > 0$ such that

$$\|x_0\| < \delta \quad \Rightarrow \quad W(x_0, t_0) < \alpha(R)$$

which means that if $\|x_0\| < \delta$, then

$$\alpha_1(\|x\|) < \alpha(R) \quad \Rightarrow \quad \|x(t)\| < R \quad \forall t \geq t_0.$$

This proves stability. If in addition $W(x, t)$ is decreasing, then there exists a positive function $V_2(x)$ such that

$$\mid W(x, t) \mid \leq V_2(x)$$

and then $\exists \alpha_2$ in the class $\mathcal{K}$ such that

$$\alpha_1(\|x\|) \leq V_1(x) \leq W(x, t) \leq V_2(x) \leq \alpha_2(\|x\|) \quad \forall x \in B_R, \ \forall t \geq t_0.$$

By the properties of the function in the class $\mathcal{K}$ , for any $R > 0$, $\exists \delta = f(R)$ such that

$$\alpha_2(\delta) < \alpha_1(R)$$

$$\Rightarrow \quad \alpha_1(R) > \alpha_2(\delta) \geq W(x_0, t_0) \geq W(x, t) \geq \alpha(\|x(t)\|)$$

which implies that

$$\|x(t)\| < R \qquad \forall t \geq t_0.$$

However, this $\delta$ is a function of $R$ alone and not of $t_0$ as before. Thus, we conclude that the stability of the equilibrium is uniform. This completes the proof. $\qquad \square$

**Proof of theorem 4.2**: Choose $R > 0$ such that the closed ball

$$B_R = \{x \in \mathbb{R}^n : \|x\| \leq R\}$$

is contained in $D$. By the assumptions of the theorem there exist class $\mathcal{K}$ functions $\alpha_1$, $\alpha_2$, and $\alpha_3$ satisfying

$$\alpha_1(\|x\|) \leq W(x,t) \leq \alpha_2(\|x\|) \quad \forall t, \forall x \in B_R \tag{4.19}$$

$$\alpha_3(\|x\|) \leq -\dot{W}(x,t) \quad \forall t, \forall x \in B_R. \tag{4.20}$$

Given that the $\alpha_i$'s are strictly increasing and satisfy $\alpha_i(0) = 0, i = 1, 2, 3$, given $\epsilon > 0$, we can find $\delta_1, \delta_2 > 0$ such that

$$\alpha_2(\delta_1) < \alpha_1(R) \tag{4.21}$$

$$\alpha_2(\delta_2) < \min[\alpha_1(\epsilon), \alpha_2(\delta_1)] \tag{4.22}$$

where we notice that $\delta_1$ and $\delta_2$ are functions of $\epsilon$ but are independent of $t_0$. Notice also that inequality (4.22) implies that $\delta_1 > \delta_2$. Now define

$$T = \frac{\alpha_1(R)}{\alpha_3(\delta_2)}. \tag{4.23}$$

**Conjecture**: We claim that

$$\|x_0\| < \delta_1 \quad \Rightarrow \quad \|x(t^*)\| < \delta_2$$

for some $t = t^*$ in the interval $t_0 \leq t^* \leq t_0 + T$.

To see this, we reason by contradiction and assume that $\|x_0\| < \delta_1$ but $\|x(t^*)\| \geq \delta_2$ for all $t$ in the interval $t_0 \leq t^* \leq t_0 + T$. We have that

$$0 < \alpha_1(\delta_2) \qquad \alpha_1 \text{ is class } \mathcal{K}$$

$$\alpha_1(\delta_2) \leq W(x,t) \qquad \forall t_0 \leq t \leq t_0 + T \tag{4.24}$$

where (4.24) follows from (4.19) and the assumption $\|x(t^*)\| \geq \delta_2$. Also, since $W$ is a decreasing function of $t$ this implies that

$$0 \leq \alpha_1(\delta_2) \leq W(x(t_0 + T), t_0 + T)$$

$$\leq W(x(t_0), t_0) + \int_{t_0}^{t_0+T} \dot{W}(x(t), t) \, dt$$

$$\leq W(x(t_0), t_0) - T\alpha_3(\delta_2) \qquad \text{from (4.20) and (4.23)}.$$

But

$$\underbrace{W(x(t_0), t_0) \leq \alpha_2(\|x(t_0)\|)}_{\text{by (4.19)}} \leq \alpha_2(\delta_1), \quad \text{since } \|x_0\| < \delta_1, \text{ by assumption.}$$

Thus, we have

$$0 \le \underbrace{W(x(t_0), t_0)}_{a_2(\delta_1)} - \underbrace{T\alpha_3(\delta_2)}_{}$$
$$\Rightarrow 0 \le \quad a_2(\delta_1) \quad - \quad \alpha_1(R) \quad \text{by (4.23)}$$

which contradicts (4.21). It then follows that our conjecture is indeed correct. Now assume that $t \ge t^*$. We have

$$\alpha_1(\|x(t)\|) \le W(x(t), t) \le W(x(t^*), t^*)$$
$$\text{because } W(\cdot, \cdot) \text{ is a decreasing function of } t$$
$$\le \alpha_2(\|x(t^*)\|) \le \alpha_2(\delta_2) < \alpha_1(\epsilon).$$

This implies that $\|x(t)\| < \epsilon$. It follows that any motion with initial state $\|x_0\| < \delta_1$ is uniformly convergent to the origin. $\qquad \square$

**Proof of theorem 4.3**: Let $\alpha_1$, $\alpha_2$, and $\alpha_3$ be as in Theorem 4.2. Given that in this case $W(\cdot, \cdot)$ is radially unbounded, we must have that

$$\alpha_1(\|x\|) \to \infty \quad \text{as} \quad \|x\| \to \infty.$$

Thus for any $a > 0$ there exists $b > 0$ such that

$$\alpha_1(b) > \alpha_2(a).$$

If $\|x_0\| < a$, then

$$\alpha_1(b) > \alpha_2(a) \ge W(x_0, t_0) \ge W(x(t), t) \ge \alpha_1(\|x\|).$$

Thus, $\|x\| < a \Rightarrow \|x\| < b$, and we have that all motions are uniformly bounded. Consider $\epsilon_1$, $\delta$, and $T$ as follows:

$$\alpha_2(\delta) < \alpha_1(c) \qquad T = \frac{\alpha_1(b)}{\alpha_3(a)}$$

and using an argument similar to the one used in the proof of Theorem 4.2, we can show that for $t \ge t_0 + T$

$$\|x(t)\| < \epsilon$$

provided that $\|x_0\| < a$. This proves that all motions converge uniformly to the origin. Thus, the origin is asymptotically stable in the large. $\qquad \square$

## 4.5 Analysis of Linear Time-Varying Systems

In this section we review the stability of linear time-varying systems using Lyapunov tools. Consider the system:

$$\dot{x} = A(t)x \qquad (4.25)$$

where $A(\cdot)$ is an $n \times n$ matrix whose entries are real-valued continuous functions of $t \in \mathbb{R}$. It is a well-established result that the solution of the state equation with initial condition $x_0$ is completely characterized by the *state transition matrix* $\Phi(\cdot, \cdot)$. Indeed, the solution of (4.25) with initial condition $x_0$ is given by,

$$x(t) = \Phi(t, t_0)x_0. \qquad (4.26)$$

The following theorem, given without proof, details the necessary and sufficient conditions for stability of the origin for the nonautonomous linear system (4.25).

**Theorem 4.5** *The equilibrium $x = 0$ of the system (4.25) is exponentially stable if and only if there exist positive numbers $k_1$ and $k_2$ such that*

$$\|\Phi(t, t_0)\| \le k_1 e^{-k_2(t-t_0)} \quad \text{for any } t_0 \ge 0, \forall t \ge t_0 \qquad (4.27)$$

*where*

$$\|\Phi(t, t_0)\| = \max_{\|x\|=1} \|\Phi(t, t_0)x\|.$$

**Proof**: See Chen [15], page 404.

We now endeavor to prove the existence of a Lyapunov function that guarantees exponential stability of the origin for the system (4.25). Consider the function

$$W(x, t) = x^T P(t)x \qquad (4.28)$$

where $P(t)$ satisfies the assumptions that it is (i) continuously differentiable, (ii) symmetric, (iii) bounded, and (iv) positive definite. Under these assumptions, there exist constants $k_1, k_2 \in \mathbb{R}^+$ satisfying

$$k_1 x^T x \le x^T P(t)x \le k_2 x^T x \quad \forall t \ge 0, \forall x \in \mathbb{R}^n$$

or

$$k_1 \|x\|^2 \le W(x, t) \le k_2 \|x\|^2 \quad \forall t \ge 0, \forall x \in \mathbb{R}^n.$$

This implies that $W(x, t)$ is *positive definite, decrescent, and radially unbounded.*

$$
\begin{aligned}
\dot{W}(x, t) &= \dot{x}^T P(t)x + x^T P\dot{x} + x^T \dot{P}(t)x \\
&= x^T A^T P x + x^T P A x + x^T \dot{P}(t)x \\
&= x^T[PA + A^T P + \dot{P}(t)]x
\end{aligned}
$$

or

$$\dot{W}(x,t) = -x^T Q(t) x^T$$

where we notice that $Q(t)$ is symmetric by construction. According to theorem 4.2, if $Q(t)$ is positive definite, then the origin is uniformly asymptotically stable. This is the case if there exist $k_3, k_4 \in \mathbb{R}^+$ such that

$$k_3 x^T x \le x^T Q(t) x \le k_4 x^T x \quad \forall t \ge 0, \forall x \in \mathbb{R}^n.$$

Moreover, if these conditions are satisfied, then

$$k_3 \|x\|^2 \le Q(t) \le k_4 \|x\|^2 \quad \forall t \ge 0, \forall x \in \mathbb{R}^n$$

and then the origin is exponentially stable by Theorem 4.4.

**Theorem 4.6** *Consider the system (4.25). The equilibrium state $x = 0$ is exponentially stable if and only if for any given symmetric, positive definite, continuous, and bounded matrix $Q(t)$, there exist a symmetric, positive definite, continuously differentiable, and bounded matrix $P(t)$ such that*

$$-Q(t) = P(t)A(t) + A^T(t)P(t) + \dot{P}(t). \tag{4.29}$$

**Proof**: See the Appendix.

### 4.5.1   The Linearization Principle

Linear time-varying system often arise as a consequence of linearizing the equations of a nonlinear system. Indeed, given a nonlinear system of the form

$$\dot{x} = f(x,t) \qquad f : D \times [0, \infty) \to \mathbb{R}^n \tag{4.30}$$

with $f(x,t)$ having continuous partial derivatives of all orders with respect to $x$, then it is possible to expand $f(x,t)$ using Taylor's series about the equilibrium point $x = 0$ to obtain

$$\dot{x} = f(x,t) = f(0,t) + \left.\frac{\partial f}{\partial x}\right|_{x=0} x + \text{HOT}$$

where HOT= higher-order terms. Given that $x = 0$ is an equilibrium point, $f(0,t) = 0$, thus we can write

$$f(x,t) \approx \left.\frac{\partial f}{\partial x}\right|_{x=0}$$

or

$$\dot{x} = A(t)x + \text{HOT} \tag{4.31}$$

where

$$A(t) = \frac{\partial f}{\partial x}\Big|_{x=0} \stackrel{def}{=} \frac{\partial f(0,t)}{\partial x}$$

since the higher-order terms tend to be negligible "near" the equilibrium point, (4.31) seems to imply that "near" $x = 0$, the behavior of the nonlinear system (4.30) is similar to that of the so-called linear approximation (4.31). We now investigate this idea in more detail. More explicitly, we study under what conditions stability of the nonlinear system (4.30) can be inferred from the linear approximation (4.31). We will denote

$$g(x,t) \stackrel{def}{=} f(x,t) - A(t)x.$$

Thus

$$\dot{x} = f(x,t) \equiv A(t)x(t) + g(x,t).$$

**Theorem 4.7** *The equilibrium point $x = 0$ of the nonlinear system*

$$\dot{x} = A(t)x(t) + g(x,t)$$

*is uniformly asymptotically stable if*

(i) *The linear system $\dot{x} = A(t)x$ is exponentially stable.*

(ii) *The function $g(x,t)$ satisfies the following condition. Given $\epsilon > 0$, there exists $\delta > 0$, independent of $t$, such that*

$$\|x\| < \delta \quad \Rightarrow \quad \frac{\|g(x,t)\|}{\|x\|} < \epsilon. \tag{4.32}$$

*This means that*

$$\lim_{\|x\| \to 0} \frac{g(x,t)}{\|x\|} = 0$$

*uniformly with respect to $t$.*

**Proof:** Let $\Phi(t,t_0)$ be the transition matrix of $\dot{x} = A(t)x$, and define $P$ as in the proof of Theorem 4.6 (with $Q = I$):

$$P(t) = \int_t^\infty \Phi^T(\tau,t)\Phi(\tau,t)\,d\tau$$

according to Theorem 4.6, $P$ so defined is positive and bounded. Also, the quadratic function

$$W(x,t) = x^T P x$$

is (positive definite and) decrescent. Moreover

$$
\begin{aligned}
\dot{W}(x,t) &= \dot{x}^T P(t) x \\
&= \dot{x}^T + x^T P(t)\dot{x} + x^T \dot{P} x \\
&= (x^T A^T + g^T) P x + x^T P(Ax + g) + x^T \dot{P} x \\
&= x^T \underbrace{\left( A^T P + PA + \dot{P} \right)}_{(= -I)} x + g^T P x + x^T P g \\
&= -x^T x + g^T (x,t) P x + x^T P g(x,t).
\end{aligned}
$$

By (4.32), there exist $\delta > 0$ such that

$$
\frac{\|g(x,t)\|}{\|x\|} < \epsilon, \quad \text{provided that} \quad \|x\| < \delta.
$$

Thus, given that $P(t)$ is bounded and that condition (4.32) holds uniformly in $t$, there exist $\delta > 0$ such that

$$
\frac{\|g(x,t)\|}{\|x\|} < \frac{1}{\|P\|} \quad \Rightarrow \quad 2\|g(x,t)\|\|P\| < \|x\|
$$

thus, we can write that

$$
2\|g(x,t)\|\|P\| \stackrel{def}{=} \theta\|x\| < \|x\| \quad \text{with } 0 < \theta < 1.
$$

It then follows that

$$
\begin{aligned}
\dot{W}(x,t) &\leq -\|x\|^2 + 2\|g\|\|P\|\|x\| = -\|x\|^2 + \theta\|x\|^2 \quad 0 < \theta < 1 \\
&\leq (1-\theta)\|x\|^2.
\end{aligned}
$$

Hence, $\dot{W}(x,t)$ is negative definite in a neighborhood of the origin and satisfies the conditions of Theorem 4.4 (with $p = 2$). This completes the proof. $\qquad\square$

## 4.6   Perturbation Analysis

Inspired by the linearization principle of the previous section we now take a look at an important issue related to mathematical models of physical systems. In practice, it is not realistic to assume that a mathematical model is a true representation of a physical device. At best, a model can "approximate" a true system, and the difference between mathematical model and system is loosely referred to as *uncertainty*. How to deal with model uncertainty and related design issues is a difficult problem that has spurred a lot of research in recent years. In our first look at this problem we consider a dynamical system of the form

$$\dot{x} = f(x,t) + g(x,t) \tag{4.33}$$

where $g(x,t)$ is a perturbation term used to estimate the uncertainty between the true state of the system and its estimate given by $\dot{x} = f(x,t)$. In our first pass we will seek for an answer to the following question: Suppose that $\dot{x} = f(x,t)$ has an asymptotically stable equilibrium point; what can be said about the perturbed system $\dot{x} = f(x,t) + g(x,t)$?

**Theorem 4.8** *[41] Let $x = 0$ be an equilibrium point of the system (4.33) and assume that there exist a differentiable function $W(\cdot,\cdot) : D \times [0,\infty) \to \mathbb{R}$ such that*

*(i)* $k_1\|x\|^2 \leq W(x,t) \leq \|x\|^2.$

*(ii)* $\frac{\partial W}{\partial t} + \nabla W f(x,t) \leq -k_3\|x\|^2.$

*(iii)* $\|\nabla W\| \leq k_4\|x\|^2.$

*Then if the perturbation $g(x,t)$ satisfies the bound*

*(iv)* $\|g(x,t)\| \leq k_5\|x\|, \quad (k_3 - k_4 k_5) > 0$

*the origin is exponentially stable. Moreover, if all the assumptions hold globally, then the exponential stability is global.*

**Proof:** Notice first that assumption (i) implies that $W(x,t)$ is positive definite and decrescent, with $\alpha_1(\|x\|) = k_1\|x\|^2$ and $\alpha_2(\|x\|) = k_2\|x\|^2$. Moreover, the left-hand side of this inequality implies that $W(\cdot,\cdot)$ is radially unbounded. Assumption (ii) implies that, ignoring the perturbation term $g(x,t)$, $\dot{W}(x,t)$ is negative definite along the trajectories of the system $\dot{x} = f(x,t)$. Thus, assumptions (i) and (ii) together imply that the origin is a uniformly asymptotically stable equilibrium point *for the nominal system* $\dot{x} = f(x,t)$.

We now find $\dot{W}(\cdot,\cdot)$ along the trajectories of the perturbed system (4.33). We have

$$\dot{W} = \underbrace{\frac{\partial W}{\partial t} + \nabla W f(x,t)}_{\leq k_3\|x\|^2} + \underbrace{\nabla W g(x,t)}_{\leq k_4 k_5\|x\|^2}$$

$$\Rightarrow \quad \dot{W} \leq -(k_3 - k_4 k_5)\|x\|^2 \quad < 0$$

since $(k_3 - k_4 k_5) > 0$ by assumption. The result then follows by Theorems 4.2–4.3 (along with Theorem 4.4). $\qquad \square$

The importance of Theorem 4.8 is that it shows that the exponential stability is robust with respect to a class of perturbations. Notice also that $g(x,t)$ need not be known. Only the bound (iv) is necessary.

**Special Case:** Consider the system of the form

$$\dot{x} = Ax + g(x,t) \tag{4.34}$$

where $A \in \mathbb{R}^{n \times n}$. Denote by $\lambda_i, i = 1, 2, \cdots, n$ the eigenvalues of $A$ and assume that $\Re(\lambda_i) < 0, \forall i$. Assume also that the perturbation term $g(x,t)$ satisfies the bound

$$\|g(x,t)\|_2 \le \gamma \|x\|_2 \qquad \forall t > 0, \forall x \in \mathbb{R}^n.$$

Theorem 3.10 guarantees that for any $Q = Q^T > 0$ there exists a unique matrix $P = P^T > 0$, that is the solution of the Lyapunov equation

$$PA + A^T P = -Q.$$

The matrix $P$ has the property that defining $V(x) = x^T P x$ and denoting $\lambda_{min}(P)$ and $\lambda_{max}(P)$ respectively as the minimum and maximum eigenvalues of the matrix $P$, we have that

$$-\lambda_{min}(P)\|x\|_2^2 \le V(x) \le \lambda_{max}(P)\|x\|_2^2$$

$$-\frac{\partial V}{\partial x} Ax = -x^T Q x \le -\lambda_{min}(Q)\|x\|_2^2.$$

Thus $V(x)$ is positive definite and it is a Lyapunov function for the linear system $\dot{x} = Ax$. Also

$$\frac{\partial V}{\partial x} = Px + x^T P = 2x^T P$$

$$\|\frac{\partial V}{\partial x}\| = \|2x^T P\|_2 = 2\lambda_{max}(P)\|x\|_2.$$

It follows that

$$\begin{aligned}
\dot{V} &= \frac{\partial V}{\partial x} Ax + \frac{\partial V}{\partial x} g(x,t) \\
&\le -\lambda_{min}(Q)\|x\|_2^2 + [2\lambda_{max}(P)\|x\|_2] \, [\gamma\|x\|_2] \\
&\le -\lambda_{min}(Q)\|x\|_2^2 + 2\gamma\lambda_{max}(P)\|x\|_2 \\
\dot{V}(x) &\le [-\lambda_{min}(Q) + 2\gamma\lambda_{max}(P)] \, \|x\|_2^2.
\end{aligned}$$

It follows that the origin is asymptotically stable if

$$-\lambda_{min}(Q) + 2\gamma\lambda_{max}(P) < 0$$

that is
$$\lambda_{min}(Q) \; > \; 2\gamma\,\lambda_{max}(P)$$

or, equivalently
$$\gamma \; < \; \frac{\lambda_{min}(Q)}{2\lambda_{max}(P)}. \tag{4.35}$$

From equation (4.35) we see that the origin is exponentially stable provided that the value of $\gamma$ satisfies the bound (4.35)

## 4.7   Converse Theorems

All the stability theorems seen so far provide *sufficient* conditions for stability. Indeed, all of these theorems read more or less as follows:

> if there exists a function $V(\cdot)$ (or $W(\cdot,\cdot)$) that satisfies ..., then the equilibrium point $x_e$ satisfies one of the stability definitions.

None of these theorems, however, provides a systematic way of finding the Lyapunov function. Thus, unless one can "guess" a suitable Lyapunov function, one can never conclude anything about the stability properties of the equilibrium point.

An important question clearly arises here, namely, suppose that an equilibrium point satisfies one of the forms of stability. Does this imply the existence of a Lyapunov function that satisfies the conditions of the corresponding stability theorem? If so, then the search for the suitable Lyapunov function is not in vain. In all cases, the question above can be answered affirmatively, and the theorems related to this questions are known as *converse theorems*. The main shortcoming of these theorems is that their proof invariably relies on the construction of a Lyapunov function that is based on knowledge of the state trajectory (and thus on the solution of the nonlinear differential equation). This fact makes the converse theorems not very useful in applications since, as discussed earlier, few nonlinear equation can be solved analytically. In fact, the whole point of the Lyapunov theory is to provide an answer to the stability analysis without solving the differential equation. Nevertheless, the theorems have at least conceptual value, and we now state them for completeness.

**Theorem 4.9** *Consider the dynamical system $\dot{x} = f(x,t)$. Assume that $f$ satisfies a Lipschitz continuity condition in $D \subset \mathbb{R}^n$, and that $0 \in D$ is an equilibrium state. Then we have*

- *If the equilibrium is uniformly stable, then there exists a function $W(\cdot,\cdot) : D\times[0,\infty) \to \mathbb{R}$ that satisfies the conditions of Theorem 4.1 .*

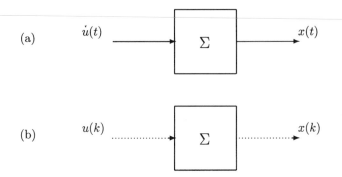

Figure 4.1: (a) Continuous-time system $\Sigma$; (b) discrete-time system $\Sigma_d$.

- *If the equilibrium is uniformly asymptotically stable, then there exists a function $W(\cdot,\cdot) : D \times [0,\infty) \to \mathbb{R}$ that satisfies the conditions of Theorem 4.2.*

- *If the equilibrium is globally uniformly stable, then there exists a function $W(\cdot,\cdot) : D \times [0,\infty) \to \mathbb{R}$ that satisfies the conditions of Theorem 4.3.*

**Proof:** The proof is omitted. See References [88], [41], or [95] for details.

## 4.8   Discrete-Time Systems

So far our attention has been limited to the stability of continuous-time systems, that is, systems defined by a vector differential equation of the form

$$\dot{x} = f(x,t) \tag{4.36}$$

where $t \in \mathbb{R}$ (i.e., is a continuous variable). In this section we consider *discrete-time systems* of the form

$$x(k+1) = f(x(k),k) \tag{4.37}$$

where $k \in \mathbb{Z}^+, x(k) \in \mathbb{R}^n$, and $f : \mathbb{R}^n \times \mathbb{Z} \to \mathbb{R}^n$, and study the stability of these systems using tools analogous to those encountered for continuous-time systems. Before doing so, however, we digress momentarily to briefly discuss discretization of continuous-time nonlinear plants.

## 4.9  Discretization

Often times discrete-time systems originate by "sampling" of a continuous-time system. This is an idealized process in which, given a continuous-time system $\Sigma : u \to x$ (a dynamical system that maps the input $u$ into the state $x$, as shown in Figure 4.1(a)) we seek a new system $\Sigma_d : u(k) \to x(k)$ (a dynamical system that maps the discrete-time signal $u(k)$ into the discrete-time state $x(k)$, as shown in Figure 4.1(b)). Both systems $\Sigma$ and $\Sigma_d$ are related in the following way. If $u(k)$ is constructed by taking "samples" every $T$ seconds of the continuous-time signal $u$, then the output $x(k)$ predicted by the model $\Sigma_d$ corresponds to the samples of the continuous-time state $x(t)$ at the same sampling instances. To develop such a model, we use the scheme shown in Figure 4.2, which consists of the cascade combination of the blocks $H$, $\Sigma$, and $S$, where each block in the figure represents the following:

- $S$ represents a *sampler*, i.e. a device that reads the continues variable $x$ every $T$ seconds and produces the discrete-time output $x(k)$, given by $x(k) = x(kT)$. Clearly, this block is an idealization of the operation of an analog-to-digital converter. For easy visualization, we have used continuous and dotted lines in Figure 4.2 to represent continuous and discrete-time signal, respectively.

- $\Sigma$ represents the plant, seen as a mapping from the input $u$ to the state $x$, that is, given $u$ the mapping $\Sigma : u \to x$ determines the trajectory $x(t)$ by solving the differential equation

$$\dot{x} = f(x, u).$$

- $H$ represents a *hold* device that converts the discrete-time signal, or sequence $u(k)$, into the continuous-time signal $u(t)$. Clearly this block is implemented by using a digital-to-analog converter. We assume the an ideal conversion process takes place in which $H$ "holds" the value of the input sequence between samples (Figure 4.3), given by

$$u(t) = u(k) \qquad \text{for } kT \leq t < (k+1)T.$$

The system $\Sigma_d$ is of the form (4.37). Finding $\Sigma_d$ is fairly straightforward when the system (4.36) is linear time-invariant (LTI), but not in the present case.

*Case (1): LTI Plants.* If the plant is LTI, then we have that

$$\dot{x} = f(x, u) = Ax + Bu \tag{4.38}$$

finding the discrete-time model of the plant reduces to solving the differential equation with initial condition $x(kT)$ at time $t(0) = kT$ and the input is the constant signal $u(kT)$. The

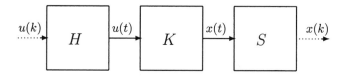

Figure 4.2: Discrete-time system $\Sigma_d$.

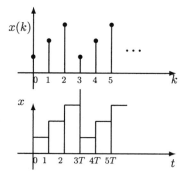

Figure 4.3: Action of the hold device $H$.

differential equation (4.38) has a well-known solution of the form

$$x(t) = e^{A(t-t_0)}x(t_0) + \int_{t_0}^{t} e^{A(t-\tau)}Bu(\tau)\, d\tau.$$

In our case we can make direct use of this solution with

$$
\begin{aligned}
t(0) &= KT \\
x(t_0) &= x(k) \\
t &= (k+1)T \\
x(t) &= x[(k+1)T] \\
u(\tau) &= u(k) \quad \text{constant for } kT \le t < k(T+1).
\end{aligned}
$$

Therefore

$$
\begin{aligned}
x(k+1) = x[(k+1)T] &= e^{AT}x(kT) + \int_{kT}^{(k+1)T} e^{A[(k+1)T-\tau]}\, d\tau\, Bu(kT) \\
&= e^{AT}x(k) + \int_{0}^{T} e^{A\tau}\, d\tau\, Bu(k).
\end{aligned}
$$

Case (2): Nonlinear Plants. In the more general case of a nonlinear plant $\Sigma$ given by (4.36) the exact model is usually impossible to find. The reason, of course, is that finding the exact solution requires solving the nonlinear differential equation (4.36), which is very difficult if not impossible. Given this fact, one is usually forced to use an approximate model. There are several methods to construct approximate models, with different degrees of accuracy and complexity. The simplest and most popular is the so-called Euler approximation, which consists of acknowledging the fact that, if $T$ is small, then

$$\dot{x} = \frac{dx}{dt} = \lim_{\Delta T \to 0} \frac{x(t+\Delta T) - x(t)}{\Delta T} \approx \frac{x(t+T) - x(t)}{T}.$$

Thus

$$\dot{x} = f(x, u)$$

can be approximated by

$$x(k+1) \approx x(k) + T\ f[x(k), u(k)].$$

## 4.10   Stability of Discrete-Time Systems

In this section we consider *discrete-time systems* of the form (4.37); that is, we assume that

$$x(k+1) = f(x(k), k)$$

where $k \in \mathbb{Z}^+, x(k) \in \mathbb{R}^n$, and $f : \mathbb{R}^n \times \mathbb{Z} \to \mathbb{R}^n$. It is refreshing to notice that, unlike the continuous case, this equation always has exactly one solution corresponding to an initial condition $x_0$. As in the continuous-time case we consider the stability of an equilibrium point $x_e$, which is defined exactly as in the continuous-time case.

### 4.10.1   Definitions

In Section 4.2 we introduced several stability definitions for continuous-time systems. For completeness, we now restate these definitions for discrete-time systems.

**Definition 4.8** *The equilibrium point $x = 0$ of the system (4.37) is said to be*

- Stable *at $k_0$ if given $\epsilon > 0$, $\exists \delta = \delta(\epsilon, k_0) > 0$ :*

$$\|x_0\| < \delta \quad \Rightarrow \quad \|x(k)\| < \epsilon \quad \forall k \geq k_0 > 0. \tag{4.39}$$

- Uniformly stable *at $k_0$ if given any given $\epsilon > 0$, $\exists \delta = \delta(\epsilon) > 0$ :*

$$\|x_0\| < \delta \quad \Rightarrow \quad \|x(k)\| < \epsilon \quad \forall k \geq k_0 > 0. \tag{4.40}$$

- Convergent *at $k_0$ if there exists $\delta_1 = \delta_1(k_0) > 0$ :*

$$\|x_0\| < \delta_1 \quad \Rightarrow \quad \lim_{t \to \infty} x(k) = 0. \tag{4.41}$$

- Uniformly convergent *if for any given $\epsilon_1 > 0, \exists M = M(\epsilon_1)$ such that*

$$\|x_0\| < \delta_1 \quad \Rightarrow \quad \|x(k)\| < \epsilon_1 \quad \forall k \geq k_0 + M.$$

- Asymptotically stable *at $k_0$ if it is both stable and convergent.*

- Uniformly asymptotically stable *if it is both stable and uniformly convergent.*

- Unstable *if it is not stable.*

## 4.10.2 Discrete-Time Positive Definite Functions

Time-independent positive functions are uninfluenced by whether the system is continuous or discrete-time. Time-dependent positive function can be defined as follows.

**Definition 4.9** *A function $W : \mathbb{R}^n \times \mathbb{Z}^+ \to \mathbb{R}$ is said to be*

- *Positive semidefinite in $D \subset \mathbb{R}^n$ if*

    *(i) $W(0, k) = 0 \quad \forall k \geq 0$.*

    *(ii) $W(x, k) \geq 0 \quad \forall x \neq 0, \ x \in D$.*

- *Positive definite in $D \subset \mathbb{R}^n$ if*

    *(i) $W(0, k) = 0 \quad \forall k \geq 0$, and*

    *(ii) $\exists$ a time invariant positive definite function $V_1(x)$ such that*

    $$V_1(x) \leq W(x, k) \quad \forall x \in D \ \forall k.$$

    *It then follows by lemma 3.1 that $W(x, k)$ is positive definite in $B_r \subset D$ if and only if there exists a class $\mathcal{K}$ function $\alpha_1(\cdot)$ such that*

    $$\alpha_1(\|x\|) \leq W(x, t), \quad \forall x \in B_r \subset D.$$

- *$W(\cdot, \cdot)$ is said to be decrescent in $D \subset \mathbb{R}^n$ if there exists a time-invariant positive definite function $V_2(x)$ such that*

    $$W(x, k) \leq V_2(x) \quad \forall x \in D, \ \forall k.$$

    *It then follows by Lemma 3.1 that $W(x, k)$ is decrescent in $B_r \subset D$ if and only if there exists a class $\mathcal{K}$ function $\alpha_2(\cdot)$ such that*

    $$W(x, t) \leq \alpha_2(\|x\|), \quad \forall x \in B_r \subset D.$$

- *$W(\cdot, \cdot)$ is said to be radially unbounded if $W(x, k) \to \infty$ as $\|x\| \to \infty$, uniformly on $k$. This means that given $M > 0$, there exists $N > 0$ such that*

    $$W(x, k) > M$$

    *provided that $\|x\| > N$.*

### 4.10.3   Stability Theorems

**Definition 4.10** *The rate of change,* $\Delta W(x, k)$*, of the function* $W(x, k)$ *along the trajectories of the system (4.37) is defined by*

$$\Delta W(x, k) = W(x(k+1), k+1) - W(x, k).$$

With these definitions, we can now state and prove several stability theorems for discrete-time system. Roughly speaking, all the theorems studied in Chapters 3 and 4 can be restated for discrete-time systems, with $\Delta W(x, k)$ replacing $\dot{W}(x, t)$. The proofs are nearly identical to their continuous-time counterparts and are omitted.

**Theorem 4.10** *(Lyapunov Stability Theorem for Discrete-Time Systems). If in a neighborhood* $D$ *of the equilibrium state* $x = 0$ *of the system (4.37) there exists a function* $W(\cdot, \cdot) : D \times \mathbb{Z}^+ \to \mathbb{R}$ *such that*

(i) $W(x, k)$ *is positive definite.*

(ii) *The rate of change* $\Delta W(x, k)$ *along any solution of (4.37) is negative semidefinite in* $D$*, then*

*the equilibrium state is stable. Moreover, if* $W(x, k)$ *is also decrescent, then the origin is uniformly stable.*

**Theorem 4.11** *(Lyapunov Uniform Asymptotic Stability for Discrete-Time Systems). If in a neighborhood* $D$ *of the equilibrium state* $x = 0$ *there exists a function* $W(\cdot, \cdot) : D \times \mathbb{Z}^+ \to \mathbb{R}$ *such that*

(i) $W(x, k)$ *is (a) positive definite, and (b) decrescent.*

(ii) *The rate of change,* $\Delta W(x, k)$ *is negative definite in* $D$*, then*

*the equilibrium state is uniformly asymptotically stable.*

**Example 4.7** *Consider the following discrete-time system:*

$$x_1(k+1) \;=\; x_1(k) + x_2(k) \tag{4.42}$$

$$x_2(k+1) \;=\; ax_1^3(k) + \frac{1}{2}x_2(k). \tag{4.43}$$

To study the stability of the origin, we consider the (time-independent) Lyapunov function candidate $V(x) = \frac{1}{2}x_1^2(k) + 2x_1(k)x_2(k) + 4x_2^2(k)$, which can be easily seen to be positive definite. We need to find $\Delta V(x) = V(x(k+1)) - V(x(k))$, we have

$$
\begin{aligned}
V(x(k+1)) &= \frac{1}{2}x_1^2(k+1) + 2x_1(k+1)x_2(k+1) + 4x_2^2(k+1) \\
&= \frac{1}{2}[x_1(k) + x_2(k)]^2 + 2[x_1(k) + x_2(k)][ax_1^3(k) + \frac{1}{2}x_2(k)] \\
&\quad + 4[ax_1^3(k) + \frac{1}{2}x_2(k)]^2
\end{aligned}
$$

$$
V(x(k)) = \frac{1}{2}x_2^2 + 2x_1x_2 + 4x_2^2.
$$

From here, after some trivial manipulations, we conclude that

$$
\Delta V(x) = V(x(k+1)) - V(x(k)) = -\frac{3}{2}x_2^2 + 2ax_1^4 + 6ax_1^3x_2 + 4a^2x_1^6.
$$

Therefore we have the following cases of interest:

- $a < 0$. In this case, $\Delta V(x)$ is negative definite in a neighborhood of the origin, and the origin is locally asymptotically stable (uniformly, since the system is autonomous).

- $a = 0$. In this case $\Delta V(x) = V(x(k+1)) - V(x(k)) = -\frac{3}{2}x_2^2 \leq 0$, and thus the origin is stable.

$\square$

## 4.11  Exercises

(4.1) Prove Lemma 4.1.

(4.2) Prove Lemma 4.2.

(4.3) Prove Theorem 4.4.

(4.4) Characterize each of the following functions $W : \mathbb{R}^2 \times \mathbb{R} \to \mathbb{R}$ as: (a) positive definite or not, (b) decreasing or not, (c) radially unbounded or not.

(i) $W_1(x, t) = (x_1^2 + x_2^2)$.
(ii) $W_2(x, t) = (x_1^2 + x_2^2)e^t$.

(iii) $W_3(x,t) = (x_1^2 + x_2^2)e^{-t}$.

(iv) $W_4(x,t) = (x_1^2 + x_2^2)(1 + e^{-t})$.

(v) $W_5(x,t) = (x_1^2 + x_2^2)\cos^2 \omega t$.

(vi) $W_6(x,t) = (x_1^2 + x_2^2)(1 + \cos^2 \omega t)$.

(vii) $W_7(x,t) = \frac{(x_1^2+x_2^2)(1+e^{-t})}{(x_1^2+1)}$.

(4.5) It is known that a given dynamical system has an equilibrium point *at the origin*. For this system, a function $W(\cdot)$ has been proposed, and its derivative $\dot{W}(\cdot)$ has been computed. Assuming that $V(\cdot)$ and $\dot{V}(\cdot)$ are given below you are asked to classify the origin, in each case, as (a) stable, (b) locally uniformly asymptotically stable, and/or (c) globally uniformly asymptotically stable. Explain you answer in each case.

(i) $W_1(x,t) = (x_1^2 + x_2^2)$, $\dot{W}_1(x,t) = -x_1^2$.

(ii) $W_2(x,t) = (x_1^2 + x_2^2)$, $\dot{W}_2(x,t) = -(x_1^2 + x_2^2)e^{-t}$.

(iii) $W_3(x,t) = (x_1^2 + x_2^2)$, $\dot{W}_3(x,t) = -(x_1^2 + x_2^2)e^{t}$.

(iv) $W_4(x,t) = (x_1^2 + x_2^2)e^{t}$, $\dot{W}_4(x,t) = -(x_1^2 + x_2^2)(1 + \sin^2 t)$.

(v) $W_5(x,t) = (x_1^2 + x_2^2)e^{-t}$, $\dot{W}_5(x,t) = -(x_1^2 + x_2^2)$.

(vi) $W_6(x,t) = (x_1^2 + x_2^2)(1 + e^{-t})$, $\dot{W}_6(x,t) = -x_1^2 e^{-t}$.

(vii) $W_7(x,t) = (x_1^2 + x_2^2)\cos^2 \omega t$, $\dot{W}_7(x,t) = -(x_1^2 + x_2^2)$.

(viii) $W_8(x,t) = (x_1^2 + x_2^2)(1 + \cos^2 \omega t)$, $\dot{W}_8(x,t) = -x_1^2$.

(ix) $W_9(x,t) = (x_1^2 + x_2^2)(1 + \cos^2 \omega t)$, $\dot{W}_9(x,t) = -(x_1^2 + x_2^2)e^{-t}$.

(x) $W_{10}(x,t) = (x_1^2 + x_2^2)(1 + \cos^2 \omega t)$, $\dot{W}_{10}(x,t) = -(x_1^2 + x_2^2)(1 + e^{-t})$.

(xi) $W_{11}(x,t) = \frac{(x_1^2+x_2^2)(1+e^{-t})}{(x_1^2+1)}$, $\dot{W}_{11}(x,t) = -(x_1^2 + x_2^2)$.

(4.6) Given the following system, study the stability of the equilibrium point at the origin:

$$\begin{aligned} \dot{x}_1 &= -x_1 - x_1 x_2^2 \cos^2 t \\ \dot{x}_2 &= -x_2 - x_1^2 x_2 \sin^2 t \end{aligned}$$

(4.7) Prove Theorem 4.10.

(4.8) Prove Theorem 4.11.

(4.9) Given the following system, study the stability of the equilibrium point at the origin:

$$\begin{aligned}
\dot{x}_1 &= x_2 + x_1^2 + x_2^2 \\
\dot{x}_2 &= -2x_1 - 3x_2
\end{aligned}$$

Hint: Notice that the given dynamical equations can be expressed in the form $\dot{x} = Ax + g(x)$.

## Notes and References

Good sources for the material of this chapter include References [27], [41], [88], [68], and [95]. Section 4.5 is based on Vidyasagar [88] and Willems [95]. Perturbation analysis has been a subject of much research in nonlinear control. Classical references on the subject are Hahn [27] and Krasovskii [44]. Section 4.6 is based on Khalil [41].

# Chapter 5

# Feedback Systems

So far, our attention has been restricted to open-loop systems. In a typical control problem, however, our interest is usually in the analysis and design of feedback control systems. Feedback systems can be analyzed using the same tools elaborated so far after incorporating the effect of the input $u$ on the system dynamics. In this chapter we look at several examples of feedback systems and introduce a simple design technique for stabilization known as *backstepping*. To start, consider the system

$$\dot{x} = f(x, u) \tag{5.1}$$

and assume that the origin $x = 0$ is an equilibrium point of the unforced system $\dot{x} = f(x, 0)$. Now suppose that $u$ is obtained using a state feedback law of the form

$$u = \phi(x). \tag{5.2}$$

To study the stability of this system, we substitute (5.2) into (5.1) to obtain

$$\dot{x} = f(x, \phi(x)). \tag{5.3}$$

According to the stability results in Chapter 3, if the origin of the unforced system (5.3) is asymptotically stable, then we can find a positive definite function $V$ whose time derivative along the trajectories of (5.3) is negative definite in a neighborhood of the origin.

It seems clear from this discussion that the stability of feedback systems can be studied using the same tool discussed in Chapters 3 and 4.

137

## 5.1   Basic Feedback Stabilization

In this section we look at several examples of stabilization via state feedback. These examples will provide valuable insight into the backstepping design of the next section.

**Example 5.1** *Consider the first order system given by*

$$\dot{x} = ax^2 + u \tag{5.4}$$

*we look for a state feedback of the form*

$$u = \phi(x)$$

*that makes the equilibrium point at the origin "asymptotically stable." One rather obvious way to approach the problem is to choose a control law u that "cancels" the nonlinear term $ax^2$. Indeed, setting*

$$u = -ax^2 - x \tag{5.5}$$

*and substituting (5.5) into (5.4) we obtain*

$$\dot{x} = -x$$

*which is linear and globally asymptotically stable, as desired.*                                 □

We mention in passing that this is a simple example of a technique known as *feedback linearization*. While the idea works quite well in our example, it does come at a certain price. We notice two things:

(i) It is based on exact cancelation of the nonlinear term $ax^2$. This is undesirable since in practice system parameters such as "$a$" in our example are never known exactly. In a more realistic scenario what we would obtain at the end of the design process with our control $u$ is a system of the form

$$\dot{x} = (a - \bar{a})x^2 - x$$

where $a$ represents the true system parameter and $\bar{a}$ the actual value used in the feedback law. In this case the true system is also asymptotically stable, but only locally because of the presence of the term $(a - \bar{a})x^2$.

(ii) Even assuming perfect modeling it may not be a good idea to follow this approach and cancel "all" nonlinear terms that appear in the dynamical system. The reason is that nonlinearities in the dynamical equation are not necessarily bad. To see this, consider the following example.

**Example 5.2** *Consider the system given by*

$$\dot{x} = ax^2 - x^3 + u$$

*following the approach in Example 5.1 we can set*

$$u \stackrel{def}{=} u_1 = -ax^2 + x^3 - x$$

*which leads to*

$$\dot{x} = -x.$$

□

Once again, $u_1$ is the feedback linearization law that renders a globally asymptotically stable linear system. Let's now examine in more detail how this result was accomplished. Notice first that our control law $u_1$ was chosen to cancel both nonlinear terms $ax^2$ and $-x^3$. These two terms are, however, quite different:

- The presence of terms of the form $x^i$ with $i$ even on a dynamical equation is never desirable. Indeed, even powers of $x$ do not discriminate sign of the variable $x$ and thus have a destabilizing effect that should be avoided whenever possible.

- Terms of the form $-x^j$, with $j$ odd, on the other hand, greatly contribute to the feedback law by providing additional damping for large values of $x$ and are usually beneficial.

- At the same time, notice that the cancellation of the term $x^3$ was achieved by incorporating the term $x^3$ *in the feedback law*. The presence of this variable in the control input $u$ can lead to very large values of the input. In practice, the physical characteristics of the actuators may place limits on the amplitude of this function, and thus the presence of the term $x^3$ on the input $u$ is not desirable.

To find an alternate solution, we proceed as follows. Given the system

$$\dot{x} = f(x, u) \qquad x \in \mathbb{R}^n, u \in \mathbb{R}, \quad f(0, 0) = 0$$

we proceed to find a feedback law of the form

$$u = \phi(x)$$

such that the feedback system

$$\dot{x} = f(x, \phi(x)) \tag{5.6}$$

has an asymptotically stable equilibrium point at the origin. To show that this is the case, we will construct a function $V_1 = V_1(x) : D \to \mathbb{R}$ satisfying

(i) $V(0) = 0$, and $V_1(x)$ is positive definite in $D - \{0\}$.

(ii) $\dot{V}_1(x)$ is negative definite along the solutions of (5.6). Moreover, there exist a positive definite function $V_2(x) : D \to \mathbb{R}^+$ such that

$$\dot{V}_1(x) = \frac{\partial V_1}{\partial x} f(x, \phi(x)) \leq -V_2(x) \quad \forall x \in D.$$

Clearly, if $D = \mathbb{R}^n$ and $V_1$ is radially unbounded, then the origin is globally asymptotically stable by Theorems 3.2 and 3.3.

**Example 5.3** *Consider again the system of example 5.2.*

$$\dot{x} = ax^2 - x^3 + u$$

*defining $V_1(x) = \frac{1}{2}x^2$ and computing $\dot{V}_1$, we obtain*

$$\begin{aligned} \dot{V}_1 &= x \cdot f(x, u) \\ &= ax^3 - x^4 + xu. \end{aligned}$$

*In Example 5.2 we chose $u = u_1 = -ax^2 + x^3 - x$. With this input function we have that*

$$\dot{V}_1 = ax^3 - x^4 + x(-ax^2 + x^3 - x) = -x^2 \overset{def}{=} -V_2(x).$$

*It then follows that this control law satisfies requirement (ii) above with $V_2(x) = x^2$. Not happy with this solution, we modify the function $V_2(x)$ as follows:*

$$\dot{V}_1 = ax^3 - x^4 + xu \leq -V_2(x) \overset{def}{=} -(x^4 + x^2).$$

*For this to be the case, we must have*

$$\begin{aligned} ax^3 - x^4 + xu &\leq -x^4 - x^2 \\ xu &\leq -x^2 - ax^3 = -x(x + ax^2) \end{aligned}$$

*which can be accomplished by choosing*

$$u = -x - ax^2.$$

*With this input function $u$, we obtain the following feedback system*

$$\begin{aligned} \dot{x} &= ax^2 - x^3 + u \\ &= -x - x^3 \end{aligned}$$

*which is asymptotically stable. The result is global since $V_1$ is radially unbounded and $D = \mathbb{R}$.*
□

## 5.2 Integrator Backstepping

Guided by the examples of the previous section, we now explore a recursive design technique known as *backstepping*. To start with, we consider a system of the form

$$\dot{x} = f(x) + g(x)\xi \tag{5.7}$$
$$\dot{\xi} = u. \tag{5.8}$$

Here $x \in \mathbb{R}^n, \xi \in \mathbb{R}$, and $[x, \xi]^T \in \mathbb{R}^{n+1}$ is the state of the system (5.7)–(5.8). The function $u \in \mathbb{R}$ is the control input and the functions $f, g : D \to \mathbb{R}^n$ are assumed to be smooth. As will be seen shortly, the importance of this structure is that can be considered as a cascade connection of the subsystems (5.7) and (5.8). We will make the following assumptions (see Figure 5.1(a)):

(i) The function $f(\cdot) : \mathbb{R}^n \to \mathbb{R}^n$ satisfies $f(0) = 0$. Thus, the origin is an equilibrium point of the subsystem $\dot{x} = f(x)$.

(ii) Consider the subsystem (5.7). Viewing the state variable $\xi$ as an independent "input" for this subsystem, we assume that there exists a state feedback control law of the form

$$\xi = \phi(x), \qquad \phi(0) = 0$$

and a Lyapunov function $V_1 : D \to \mathbb{R}^+$ such that

$$\dot{V}_1(x) = \frac{\partial V_1}{\partial x}[f(x) + g(x) \cdot \phi(x)] \leq -V_a(x) \leq 0 \quad \forall x \in D$$

where $V_a(\cdot) : D \to \mathbb{R}^+$ is a positive semidefinite function in $D$.

According to these assumptions, the system (5.7)–(5.8) consists of the subsystem (5.7), for which a known stabilizing law already exists, augmented with a pure integrator (the subsystems (5.8)). More general classes of systems are considered in the next section. We now endeavor to find a state feedback law to asymptotically stabilize the system (5.7)–(5.8). To this end we proceed as follows:

- We start by adding and subtracting $g(x)\phi(x)$ to the subsystem (5.7) (Figure 1(b)). We obtain the equivalent system

$$\dot{x} = f(x) + g(x)\phi(x) + g(x)[\xi - \phi(x)] \tag{5.9}$$
$$\dot{\xi} = u. \tag{5.10}$$

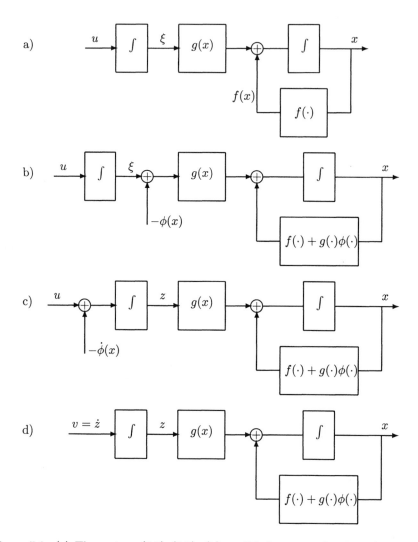

Figure 5.1: (a) The system (5.7)–(5.8); (b) modified system after introducing $-\phi(x)$; (c) "backstepping" of $-\phi(x)$; (d) the final system after the change of variables.

- Define

$$z = \xi - \phi(x) \tag{5.11}$$
$$\dot{z} = \dot{\xi} - \dot{\phi}(x) = u - \dot{\phi}(x) \tag{5.12}$$

where

$$\dot{\phi} = \frac{\partial \phi}{\partial x}\dot{x} = \frac{\partial \phi}{\partial x}[f(x) + g(x)\xi] \tag{5.13}$$

This change of variables can be seen as "backstepping" $-\phi(x)$ through the integrator, as shown in Figure 1(c). Defining

$$v = \dot{z} \tag{5.14}$$

the resulting system is

$$\dot{x} = f(x) + g(x)\phi(x) + g(x)z \tag{5.15}$$
$$\dot{z} = v \tag{5.16}$$

which is shown in Figure 1(d). These two steps are important for the following reasons:

(i) By construction, the system (5.15)–(5.16) is equivalent to the system (5.7)–(5.8).

(ii) The system (5.15)–(5.16) is, once again, the cascade connection of two subsystems, as shown in Figure 1(d). However, the subsystem (5.15) incorporates the stabilizing state feedback law $\phi(\cdot)$ and is thus asymptotically stable when the input is zero. This feature will now be exploited in the design of a stabilizing control law for the overall system (5.15)–(5.16).

- To stabilize the system (5.15)–(5.16) consider a Lyapunov function candidate of the form

$$V = V(x, \xi) = V_1(x) + \frac{1}{2}z^2. \tag{5.17}$$

We have that

$$
\begin{aligned}
\dot{V} &= \frac{\partial V_1}{\partial x}[f(x) + g(x)\phi(x) + g(x)z] + z\dot{z} \\
&= \frac{\partial V_1}{\partial x}f(x) + \frac{\partial V_1}{\partial x}g(x)\phi(x) + \frac{\partial V_1}{\partial x}g(x)z + zv.
\end{aligned}
$$

We can choose

$$v = -\left(\frac{\partial V_1}{\partial x}g(x) + kz\right), \quad k > 0 \tag{5.18}$$

Thus

$$\begin{aligned}
\dot{V} &= \frac{\partial V_1}{\partial x} f(x) + \frac{\partial V_1}{\partial x} g(x)\phi(x) - kz^2 \\
&= \frac{\partial V_1}{\partial x}[f(x) + g(x)\phi(x)] - kz^2 \\
&\leq -V_a(x) - kz^2.
\end{aligned} \qquad (5.19)$$

It then follows by (5.19) that the origin $x = 0, z = 0$ is asymptotically stable. Moreover, since $z = \xi - \phi(x)$ and $\phi(0) = 0$ by assumption, the result also implies that the origin of the original system $x = 0, \xi = 0$ is also asymptotically stable. If all the conditions hold globally and $V_1$ is radially unbounded, then the origin is *globally* asymptotically stable. Finally, notice that, according to (5.12), the stabilizing state feedback law is given by

$$u = \dot{z} + \dot{\phi} \qquad (5.20)$$

and using (5.13), (5.18), and (5.11), we obtain

$$u = \frac{\partial \phi}{\partial x}[f(x) + g(x)\xi] - \frac{\partial V_1}{\partial x}g(x) - k[\xi - \phi(x)]. \qquad (5.21)$$

**Example 5.4** *Consider the following system, which is a modified version of the one in Example 5.3:*

$$\begin{aligned}
\dot{x}_1 &= ax_1^2 - x_1^3 + x_2 \qquad &(5.22) \\
\dot{x}_2 &= u. \qquad &(5.23)
\end{aligned}$$

*Clearly this system is of the form (5.15)–(5.16) with*

$$\begin{aligned}
x &= x_1 \\
\xi &= x_2 \\
f(x) = f(x_1) &= ax_1^2 - x_1^3 \\
g(x) &= 1
\end{aligned}$$

Step 1: *Viewing the "state" $\xi$ as an independent input for the subsystem (5.15), find a state feedback control law $\xi = \phi(x)$ to stabilize the origin $x = 0$. In our case we have $\dot{x}_1 = ax_1^2 - x_1^3 + x_2$. Proceeding as in Example 5.3, we define*

$$\begin{aligned}
V_1(x_1) &= \frac{1}{2}x_1^2 \\
\Rightarrow \dot{V}_1(x_1) &= ax_1^3 - x_1^4 + x_1x_2 \leq -V_a(x_1) \overset{def}{=} -(x_1^4 + x_1^2)
\end{aligned}$$

*which can be accomplished by choosing*

$$x_2 = \phi(x_1) = -x_1 - ax_1^2$$

*leading to*

$$\dot{x}_1 = -x_1 - x_1^3.$$

*Step 2: To stabilize the original system (5.22)– (5.23), we make use of the control law (5.21). We have*

$$
\begin{aligned}
u &= \frac{\partial \phi}{\partial x}[f(x) + g(x)\xi] - \frac{\partial V_1}{\partial x}g(x) - k[\xi - \phi(x)] \\
&= -(1 + 2ax_1)[ax_1^2 - x_1^3 + x_2] - x_1 - k[x_2 + x_1 + ax_1^2].
\end{aligned}
$$

*With this control law the origin is globally asymptotically stable (notice that $V_1$ is radially unbounded). The composite Lyapunov function is*

$$
\begin{aligned}
V = V_1 + \frac{1}{2}z^2 &= \frac{1}{2}x_1^2 + \frac{1}{2}[x_2 - \phi(x_1)]^2 \\
&= \frac{1}{2}x_1^2 + \frac{1}{2}[x_2 - x_1 + ax_1^2]^2.
\end{aligned}
$$

$\square$

## 5.3 Backstepping: More General Cases

In the previous section we discussed integrator backstepping for systems with a state of the form $[x, \xi]$, $x \in \mathbb{R}^n$, $\xi \in \mathbb{R}$, under the assumption that a stabilizing law $\phi$. We now look at more general classes of systems.

### 5.3.1 Chain of Integrators

A simple but useful extension of this case is that of a "chain" of integrators, specifically a system of the form

$$
\begin{aligned}
\dot{x} &= f(x) + g(x)\xi_1 \\
\dot{\xi}_1 &= \xi_2
\end{aligned}
$$

$$\vdots$$

$$\dot{\xi}_{k-1} = \xi_k$$
$$\dot{\xi}_k = u$$

Backstepping design for this class of systems can be approached using successive iterations of the procedure used in the previous section. To simplify our notation, we consider, without loss of generality, the third order system

$$\dot{x} = f(x) + g(x)\xi_1 \tag{5.24}$$
$$\dot{\xi}_1 = \xi_2 \tag{5.25}$$
$$\dot{\xi}_2 = u \tag{5.26}$$

and proceed to design a stabilizing control law. We first consider the first two "subsystems"

$$\dot{x} = f(x) + g(x)\xi_1 \tag{5.27}$$
$$\dot{\xi}_1 = \xi_2 \tag{5.28}$$

and assume that $\xi_1 = \phi(x_1)$ is a stabilizing control law for the system

$$\dot{x} = f(x) + g(x)\phi(x).$$

Moreover, we also assume that $V_1$ is the corresponding Lyapunov function for this subsystem. The second-order system (5.27)–(5.28) can be seen as having the form (5.7)-(5.8) with $\xi_2$ considered as an independent input. We can asymptotically stabilize this system using the control law (5.21) and associated Lyapunov function $V_2$:

$$\xi_2 = \phi(x, \xi_1) = \frac{\partial \phi(x)}{\partial x}[f(x) + g(x)\xi_1] - \frac{\partial V_1}{\partial x}g(x) - k[\xi_1 - \phi(x)], \quad k > 0$$
$$V_2 = V_1 + \frac{1}{2}[\xi_1 - \phi(x)]^2$$

We now iterate this process and view the third-order system given by the first three equations as a more general version of (5.7)–(5.8) with

$$x = \begin{bmatrix} x \\ \xi_1 \end{bmatrix}, \quad \xi = \xi_2, \quad f = \begin{bmatrix} f(x) + g(x)\xi_1 \\ 0 \end{bmatrix}, \quad g = \begin{bmatrix} 0 \\ 1 \end{bmatrix}$$

Applying the backstepping algorithm once more, we obtain the stabilizing control law:

$$u = \frac{\partial \phi(x)}{\partial x}\dot{x} - \frac{\partial V_2}{\partial x}g(x) - k[\xi_2 - \phi(x)], \quad k > 0$$
$$= \left[\frac{\partial \phi(x, \xi_1)}{\partial x}, \frac{\partial \phi(x, \xi_1)}{\partial \xi_1}\right][\dot{x}, \dot{\xi}_1]^T - \left[\frac{\partial V_2}{\partial x}, \frac{\partial V_2}{\partial \xi_1}\right][0, 1]^T - k[\xi_2 - \phi(x, \xi_1)], \quad k > 0$$

or

$$u = \frac{\partial \phi(x, \xi_1)}{\partial x}[f(x) + g(x)\xi_1] + \frac{\partial \phi(x, \xi_1)}{\partial \xi_1}\xi_2 - \frac{\partial V_2}{\partial \xi_1} - k[\xi_2 - \phi(x, \xi_1)], \quad k > 0.$$

The composite Lyapunov function is

$$\begin{aligned}
V &= V_2 + \frac{1}{2}[\xi_2 - \phi(x, \xi_1)]^2 \\
&= V_1 + \frac{1}{2}[\xi_1 - \phi(x)]^2 + \frac{1}{2}[\xi_2 - \phi(x, \xi_1)]^2.
\end{aligned}$$

We finally point out that while, for simplicity, we have focused attention on third-order systems, the procedure for $n$th-order systems is entirely analogous.

**Example 5.5** *Consider the following system, which is a modified version of the one in Example 5.4:*

$$\begin{aligned}
\dot{x}_1 &= ax_1^2 + x_2 \\
\dot{x}_2 &= x_3 \\
\dot{x}_3 &= u.
\end{aligned}$$

*We proceed to stabilize this system using the backstepping approach. To start, we consider the first equation treating $x_2$ is an independent "input" and proceed to find a state feedback law $\phi(x_1)$ that stabilizes this subsystem. In other words, we consider the system $\dot{x}_1 = ax_1^2 + \phi(x_1)$ and find a stabilizing law $u = \phi(x_1)$. Using the Lyapunov function $V_1 = 1/2x_1^2$, it is immediate that $\phi(x_1) = -x_1 - ax_1^2$ is one such law.*

*We can now proceed to the first step of backstepping and consider the first two subsystems, assuming at this point that $x_3$ is an independent input. Using the result in the previous section, we propose the stabilizing law*

$$\phi(x_1, x_2)(= x_3) = \frac{\partial \phi(x_1)}{\partial x_1}[f(x_1) + g(x_1)x_2] - \frac{\partial V_1}{\partial x_1}g(x_1) - k[x_2 - \phi(x_1)], \quad k > 0$$

*with associated Lyapunov function*

$$\begin{aligned}
V_2 &= V_1 + \frac{1}{2}z^2 \\
&= V_1 + \frac{1}{2}[x_2 - \phi(x_1)] \\
&= V_1 + \frac{1}{2}[x_2 + x_1 + ax_1^2]^2.
\end{aligned}$$

*In our case*

$$\frac{\partial \phi(x_1)}{\partial x_1} = -(1 + 2ax_1)$$

$$\frac{\partial V_1}{\partial x_1} = x_1$$

$$\Rightarrow \quad \phi(x_1, x_2) = -(1 + 2ax_1)[ax_1^2 + x_2] - x_1 - [x_2 + x_1 + ax_1^2]$$

*where we have chosen $k = 1$. We now move on to the final step, in which we consider the third order system as a special case of (5.7)–(5.8) with*

$$x = \begin{bmatrix} x_1 \\ x_2 \end{bmatrix}, \quad \xi = x_3, \quad f = \begin{bmatrix} f(x_1) + g(x_1)x_2 \\ 0 \end{bmatrix}, \quad g = \begin{bmatrix} 0 \\ 1 \end{bmatrix}$$

*From the results in the previous section we have that*

$$u = \frac{\partial \phi(x_1, x_2)}{\partial x_1}[f(x_1) + g(x_1)x_2] + \frac{\partial \phi(x_1, x_2)}{\partial x_2}x_3 - \frac{\partial V_2}{\partial x_2} - k[x_3 - \phi(x_1, x_2)], \quad k > 0$$

*is a stabilizing control law with associated Lyapunov function*

$$V = V_2 + \frac{1}{2}[x_3 + \phi(x_1, x_2)]^2$$

□

## 5.3.2   Strict Feedback Systems

We now consider systems of the form

$$\dot{x} = f(x) + g(x)\xi_1$$
$$\dot{\xi}_1 = f_1(x, \xi_1) + g_1(x, \xi_1)\xi_2$$
$$\dot{\xi}_2 = f_2(x, \xi_1, \xi_2) + g_2(x, \xi_1, \xi_2)\xi_3$$
$$\vdots$$
$$\dot{\xi}_{k-1} = f_{k-1}(x, \xi_1, \xi_2, \cdots, \xi_{k-1}) + g_{k-1}(x, \xi_1, \xi_2, \cdots, \xi_{k-1})\xi_k$$
$$\dot{\xi}_k = f_k(x, \xi_1, \xi_2, \cdots, \xi_k) + g_k(x, \xi_1, \xi_2, \cdots, \xi_k)u$$

where $x \in \mathbb{R}^n$, $\xi_i \in \mathbb{R}$, and $f_i$, $g_i$ are smooth, for all $i = 1, \cdots, k$. Systems of this form are called *strict feedback systems* because the nonlinearities $f$, $f_i$, and $g_i$ depend only on the variables $x, \xi_1, \cdots$ that are fed back. Strict feedback systems are also called *triangular systems.* We begin our discussion considering the special case where the $\xi$ system is of order one (equivalently, $k = 1$ in the system defined above):

$$\dot{x} = f(x) + g(x)\xi \tag{5.29}$$
$$\dot{\xi} = f_a(x, \xi) + g_a(x, \xi)u. \tag{5.30}$$

Assuming that the $x$ subsystem (5.29) satisfies assumptions (i) and (ii) of the backstepping procedure in Section 5.2, we now endeavor to stabilize (5.29)–(5.30). This system reduces to the integrator backstepping of Section 5.2 in the special case where $f_a(x, \xi) \equiv 0, g_a(x, \xi) \equiv 1$. To avoid trivialities we assume that this is not the case. If $g_a(x, \xi) \neq 0$ over the domain of interest, then we can define

$$u = \phi(x, \xi) \stackrel{def}{=} \frac{1}{g_a(x, \xi)}[u_1 - f_a(x, \xi)]. \tag{5.31}$$

Substituting (5.31) into (5.30) we obtain the modified system

$$\dot{x} = f(x) + g(x)\xi \tag{5.32}$$
$$\dot{\xi}_1 = u_1 \tag{5.33}$$

which is of the form (5.7)–(5.8). It then follows that, using (5.21), (5.17), and (5.31), the stabilizing control law and associated Lyapunov function are

$$\phi_1(x, \xi) = \frac{1}{g_a(x, \xi)} \left\{ \frac{\partial \phi}{\partial x}[f(x) + g(x)\xi] - \frac{\partial V_1}{\partial x}g(x) - k_1[\xi - \phi(x)] - f_a(x, \xi) \right\}, \quad k_1 > 0 \tag{5.34}$$

$$V_2 = V_2(x, \xi) = V_1(x) + \frac{1}{2}[\xi - \phi(x)]^2. \tag{5.35}$$

We now generalize these ideas by moving one step further and considering the system

$$\dot{x} = f(x) + g(x)\xi_1$$
$$\dot{\xi}_1 = f_1(x, \xi_1) + g_1(x, \xi_1)\xi_2$$
$$\dot{\xi}_2 = f_2(x, \xi_1, \xi_2) + g_2(x, \xi_1, \xi_2)\xi_3$$

which can be seen as a special case of (5.29)–(5.30) with

$$x = \begin{bmatrix} x \\ \xi_1 \end{bmatrix}, \quad \xi = \xi_2, \quad u = \xi_3, f = \begin{bmatrix} f + g\,\xi_1 \\ f_1 \end{bmatrix}, g = \begin{bmatrix} 0 \\ g_1 \end{bmatrix}, f_a = f_2, \, g_a = g_2.$$

With these definitions, and using the control law and associated Lyapunov function (5.34)–(5.35), we have that a stabilizing control law and associated Lyapunov function for this systems are as follows:

$$\phi_2(x, \xi_1, \xi_2) = \frac{1}{g_2}\left\{\frac{\partial\phi_1}{\partial x}(f + g\xi_1) + \frac{\partial\phi_1}{\partial\xi_1}(f_1(x) + g_1(x)\xi_2) - \frac{\partial V_2}{\partial\xi_1}g_1 - k_2[\xi_2 - \phi_1] - f_2\right\},$$
$$k_2 > 0 \qquad (5.36)$$

$$V_3(x, \xi_1, \xi_2) = V_2(x) + \frac{1}{2}[\xi_2 - \phi_1(x, \xi_1)]^2. \qquad (5.37)$$

The general case can be solved iterating this process.

**Example 5.6** *Consider the following systems:*

$$\dot{x}_1 = ax_1^2 - x_1 + x_1^2 x_2$$
$$\dot{x}_2 = x_1 + x_2 + (1 + x_2^2)u.$$

*We begin by stabilizing the x subsystem. Using the Lyapunov function candidate $V_1 = 1/2x_1^2$ we have that*

$$\dot{V}_1 = x_1[ax_1^2 - x_1 + x_1^2 x_2]$$
$$= ax_1^3 - x_1^2 + x_1^3 x_2.$$

*Thus, the control law $x_2 = \phi(x_1) = -(x_1 + a)$ results in*

$$\dot{V}_1 = -(x_1^2 + x_1^4)$$

*which shows that the x system is asymptotically stable. It then follows by (5.34)–(5.35) that a stabilizing control law for the second-order system and the corresponding Lyapunov function are given by*

$$\phi_1(x_1, x_2) = \frac{1}{(1 + x_2^2)}\{-(1 + a)[ax_1^2 - x_1^3 + x_1^2 x_2] - x_1^3 - k_1[x_2 + x_1 + a] - (x_1 + x_2)\},$$
$$k_1 > 0$$

$$V_2 = \frac{1}{2}x_1^2 + \frac{1}{2}[x_1 + x_2 + a]^2.$$

□

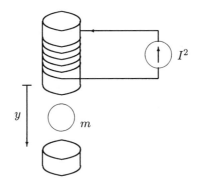

Figure 5.2: Current-driven magnetic suspension system.

## 5.4   Example

Consider the magnetic suspension system of Section 1.9.1, but assume that to simplify matters, the electromagnet is driven by a current source $I^2$, as shown in Figure 5.2. The equation of the motion of the ball remain the same as (1.31):

$$m\ddot{y} = -k\dot{y} + mg - \frac{\lambda\mu I^2}{2(1 + \mu y)^2}.$$

Defining states $\bar{x}_1 = y, \bar{x}_2 = \dot{y}$, we obtain the following state space realization:

$$\dot{\bar{x}}_1 = \bar{x}_2 \tag{5.38}$$

$$\dot{\bar{x}}_2 = g - \frac{k}{m}\bar{x}_2 - \frac{\lambda I^2}{2m(1 + \mu\bar{x}_1)^2}. \tag{5.39}$$

A quick look at this model reveals that it is "almost" in strict feedback form. It is not in the proper form of Section 5.3.2 because of the square term in $I^2$. In this case, however, we can ignore this matter and proceed with the design without change. Notice that $I$ represents the direct current flowing through the electromagnet, and therefore negative values of the input cannot occur.

We are interested in a control law that maintains the ball at an arbitrary position $y = y_0$. We can easily obtain the current necessary to achieve this objective. Setting $\dot{\bar{x}}_1 = 0$

along with $\dot{\bar{x}}_1 = \dot{\bar{x}}_2 = 0$ in equations (5.38)–(5.39), we obtain

$$I_0^2 = \frac{2mg}{\lambda\mu}(1 + \mu y_0)^2.$$

It is straightforward to show that this equilibrium point is unstable, and so we look for a state feedback control law to stabilize the closed loop around the equilibrium point. We start by applying a coordinate transformation to translate the equilibrium point $\bar{x}_e = (y_0, 0)^T$ to the origin. To this end we define new coordinates:

$$
\begin{aligned}
x_1 &= \bar{x}_1 - y_0, &\Rightarrow& \quad \dot{x}_1 = \dot{\bar{x}}_1 \\
x_2 &= \bar{x}_2 &\Rightarrow& \quad \dot{x}_2 = \dot{\bar{x}}_2 \\
u &= I^2 - \frac{2mg(1 + \mu y_0)^2}{\lambda\mu}.
\end{aligned}
$$

In the new coordinates the model takes the form:

$$\dot{x}_1 = x_2 \tag{5.40}$$

$$\dot{x}_2 = g - \frac{k}{m}x_2 - \frac{g(1 + \mu y_0)^2}{[1 + \mu(x_1 + y_0)]^2} - \frac{\lambda\mu}{2m[1 + \mu(x_1 + y_0)]^2}u \tag{5.41}$$

which has an equilibrium point at the origin with $u = 0$. The new model is in the form (5.29)–(5.30) with

$$x = x_1, \quad \xi = x_2, \quad f(x) = 0, \quad g(x) = 1$$

$$f_a(x, \xi) = g - \frac{k}{m}x_2 - \frac{g(1 + \mu y_0)^2}{[1 + \mu(x_1 + y_0)]^2}, \quad \text{and} \quad g_a(x, \xi) = -\frac{\lambda\mu}{[1 + \mu(x_1 + y_0)]^2}.$$

*Step 1:* We begin our design by stabilizing the $x_1$-subsystem. Using the Lyapunov function candidate $V_1 = \frac{1}{2}x_1^2$ we have

$$\dot{V}_1 = x_1 x_2$$

and setting $x_2 = -x_1$ we obtain $\phi(x_1) = -x_1$, which stabilizes the first equation.

*Step 2:* We now proceed to find a stabilizing control law for the two-state system using backsteppping. To this end we use the control law (5.34) with

$$g_a = -\frac{\lambda\mu}{2m[1 + \mu(x_1 + y_0)]^2}, \quad \frac{\partial\phi}{\partial x} = -1, \quad [f(x) + g(x)\xi] = x_2$$

$$V_1 = \frac{1}{2}x_1^2, \quad \frac{\partial V_1}{\partial x} = x_1, \quad g(x) = 1$$

$$[\xi - \phi(x)] = x_1 + x_2, \quad f_a(x, \xi) = g - \frac{k}{m}x_2 - \frac{g(1 + \mu y_0)^2}{[1 + \mu(x_1 + y_0)]^2}.$$

Substituting values, we obtain

$$
\begin{aligned}
u = \phi_1(x_1, x_2) \quad = \quad & -\frac{2m[1 + \mu(x_1 + y_0)]^2}{\lambda \mu}\left[-(1 + k_1)(x_1 + x_2) + \frac{k}{m}x_2 - g\right. \\
& \left. + g\frac{(1 + \mu y_0)^2}{[1 + \mu(x_1 + y_0)]^2}\right]. \tag{5.42}
\end{aligned}
$$

The corresponding Lyapunov function is

$$
\begin{aligned}
V_2 = V_2(x_1, x_2) \quad &= \quad V_1 + \frac{1}{2}[x_2 - \phi(x)]^2 \\
&= \quad \frac{1}{2}x_1^2 + \frac{1}{2}(x_1 + x_2)^2.
\end{aligned}
$$

Straightforward manipulations show that with this control law the closed-loop system reduces to the following:

$$
\begin{aligned}
\dot{x}_1 \quad &= \quad -x_1 \\
\dot{x}_2 \quad &= \quad -(1 + k)(x_1 + x_2)
\end{aligned}
$$

## 5.5 Exercises

(5.1) Consider the following system:

$$
\begin{cases}
\dot{x}_1 = x_1 + \cos x_1 - 1 + x_2 \\
\dot{x}_2 = u
\end{cases}
$$

Using backstepping, design a state feedback control law to stabilize the equilibrium point at the origin.

(5.2) Consider the following system:

$$
\begin{cases}
\dot{x}_1 = x_2 \\
\dot{x}_2 = x_1 + x_1^3 + x_3 \\
\dot{x}_2 = u
\end{cases}
$$

Using backstepping, design a tate feedback control law to stabilize the equilibrium point at the origin.

(5.3) Consider the following system, consisting of a chain of integrators:

$$\begin{cases} \dot{x}_1 = x_1 + e^{x_1} - 1 + x_2 \\ \dot{x}_2 = x_3 \\ \dot{x}_2 = u \end{cases}$$

Using backstepping, design a state feedback control law to stabilize the equilibrium point at the origin.

(5.4) Consider the following system:

$$\begin{cases} \dot{x}_1 = x_1 + x_1^2 + x_1 x_2 \\ \dot{x}_2 = x_1 + (1 + x_2^2)u \end{cases}$$

Using backstepping, design a state feedback control law to stabilize the equilibrium point at the origin.

(5.5) Consider the following system:

$$\begin{cases} \dot{x}_1 = x_1^3 + x_2 \\ \dot{x}_2 = x_1^2 x_2 - x_1^4 + u \end{cases}$$

Using backstepping, design a state feedback control law to stabilize the equilibrium point at the origin.

## Notes and References

This chapter is based heavily on Reference [47], with help from chapter 13 of Khalil [41]. The reader interested in backstepping should consult Reference [47], which contains a lot of additional material on backstepping, including interesting applications as well as the extension of the backstepping approach to adaptive control of nonlinear plants.

# Chapter 6

# Input–Output Stability

So far we have explored the notion of stability in the sense of Lyapunov, which, as discussed in Chapters 3 and 4, corresponds to stability of equilibrium points for the free or unforced system. This notion is then characterized by the lack of external excitations and is certainly *not* the only way of defining stability.

In this chapter we explore the notion of input–output stability, as an alternative to the stability in the sense of Lyapunov. The input–output theory of systems was initiated in the 1960s by G. Zames and I. Sandberg, and it departs from a conceptually very different approach. Namely, it considers systems as mappings from inputs to outputs and defines stability in terms of whether the system output is bounded whenever the input is bounded. Thus, roughly speaking, a system is viewed as a black box and can be represented graphically as shown in Figure 6.1.

To define the notion of mathematical model of physical systems, we first need to choose a suitable space of functions, which we will denote by $\mathcal{X}$. The space $\mathcal{X}$ must be

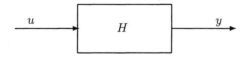

Figure 6.1: The system $H$.

sufficiently rich to contain all input functions of interest as well as the corresponding outputs. Mathematically this is a challenging problem since we would like to be able to consider systems that are not well behaved, that is, where the output to an input in the space $\mathcal{X}$ may not belong to $\mathcal{X}$. The classical solution to this dilemma consists of making use of the so-called extended spaces, which we now introduce.

## 6.1   Function Spaces

In Chapter 2 we introduced the notion of *vector space*, and so far our interest has been limited to the $n$th-dimensional space $\mathbb{R}^n$. In this chapter we need to consider "function spaces," specifically, spaces where the "vectors," or "elements," of the space are functions of time. By far, the most important spaces of this kind in control applications are the so-called $\mathcal{L}_p$ spaces which we now introduce.

In the following definition, we consider a function $u : \mathbb{R}^+ \to \mathbb{R}^q$, i.e., $u$ is of the form:

$$u(t) = \begin{bmatrix} u_1(t) \\ u_2(t) \\ \cdots \\ u_q(t) \end{bmatrix}$$

**Definition 6.1** *(The Space $\mathcal{L}_2$) The space $\mathcal{L}_2$ consists of all piecewise continuous functions* $u : \mathbb{R}^+ \to \mathbb{R}^q$ *satisfying*

$$\|u\|_{\mathcal{L}_2} \overset{def}{=} \sqrt{\int_0^\infty [|u_1|^2 + |u_2|^2 + \cdots + |u_q|^2]\, dt} \quad < \quad \infty. \tag{6.1}$$

The norm $\|u\|_{\mathcal{L}_2}$ defined in this equation is the so-called $\mathcal{L}_2$ norm of the function $u$.

**Definition 6.2** *(The Space $\mathcal{L}_\infty$) The space $\mathcal{L}_\infty$ consists of all piecewise continuous functions* $u : \mathbb{R}^+ \to \mathbb{R}^q$ *satisfying*

$$\|u\|_{\mathcal{L}_\infty} \overset{def}{=} \sup_{t \in \mathbb{R}^+} \|u(t)\|_\infty < \infty. \tag{6.2}$$

The reader should not confuse the two different norms used in equation (6.2). Indeed, the norm $\|u\|_{\mathcal{L}_\infty}$ is the $\mathcal{L}_\infty$ norm of the function $u$, whereas $\|u(t)\|_\infty$ represents the infinity norm of the vector $u(t)$ in $\mathbb{R}^q$, defined in section 2.3.1. In other words

$$\|u\|_{\mathcal{L}_\infty} \overset{def}{=} \sup_{t \in \mathbb{R}^+} (\max_i |u_i|) < \infty \quad 1 \le i \le q.$$

Both $\mathcal{L}_2$ and $\mathcal{L}_\infty$ are special cases of the so-called $\mathcal{L}_p$ spaces. Given $p : 1 \leq p < \infty$, the space $\mathcal{L}_p$ consists of all piecewise continuous functions $u : \mathbb{R}^+ \to \mathbb{R}^q$ satisfying

$$\|u\|_{\mathcal{L}_p} \stackrel{def}{=} \left( \int_0^\infty [|u_1|^p + |u_2|^p + \cdots + |u_q|^p]\, dt \right)^{1/p} < \infty. \tag{6.3}$$

Another useful space is the so-called $\mathcal{L}_1$. From (6.3), $\mathcal{L}_1$ is the space of all piecewise continuous functions $u : \mathbb{R}^+ \to \mathbb{R}^q$ satisfying:

$$\|u\|_{\mathcal{L}_1} \stackrel{def}{=} \left( \int_0^\infty [|u_1| + |u_2| + \cdots + |u_q|]\, dt \right) < \infty. \tag{6.4}$$

**Property:** (Hölder's inequality in $\mathcal{L}_p$ spaces). If $p$ and $q$ are such that $\frac{1}{p} + \frac{1}{q} = 1$ with $1 \leq p \leq \infty$ and if $f \in \mathcal{L}_p$ and $g \in \mathcal{L}_q$, then $fg \in \mathcal{L}_1$, and

$$\|(fg)_T\|_{\mathcal{L}_1} = \int_0^T |f(t)g(t)|\, dt \leq \left( \int_0^T |f(t)|^p\, dt \right)^{1/p} \left( \int_0^T |g(t)|\, dt \right)^{1/q}. \tag{6.5}$$

For the most part, we will focus our attention on the space $\mathcal{L}_2$, with occasional reference to the space $\mathcal{L}_\infty$. However, most of the stability theorems that we will encounter in the sequel, as well as all the stability definitions, are valid in a much more general setting. To add generality to our presentation, we will state all of our definitions and most of the main theorems referring to a generic space of functions, denoted by $\mathcal{X}$.

### 6.1.1 Extended Spaces

We are now in a position to introduce the notion of *extended spaces*.

**Definition 6.3** *Let $u \in \mathcal{X}$. We define the truncation operator $P_T : \mathcal{X} \to \mathcal{X}$ by*

$$(P_T u)(t) \equiv u_T(t) \stackrel{def}{=} \begin{cases} u(t), & t \leq T \\ \\ 0, & t > T \end{cases} \qquad t, T \in \mathbb{R}^+ \tag{6.6}$$

**Example 6.1** *Consider the function $u : [0, \infty) \to [0, \infty)$ defined by $u(t) = t^2$. The truncation of $u(t)$ is the following function:*

$$u_T(t) = \begin{cases} t^2, & 0 \leq t \leq T \\ 0, & t > T \end{cases}$$

□

Notice that according to definition 6.3, $P_T$ satisfies

(i) $[P_T(u + v)](t) = u_T(t) + v_T(t) \quad \forall u, v \in \mathcal{X}_e$ .

(ii) $[P_T(\alpha u)](t) = \alpha u_T(t) \quad \forall u \in \mathcal{X}_e, \alpha \in \mathbb{R}.$

Thus, the truncation operator is a *linear* operator.

**Definition 6.4** *The extension of the space $\mathcal{X}$, denoted $\mathcal{X}_e$ is defined as follows:*

$$\mathcal{X}_e = \{u : \mathbb{R}^+ \to \mathbb{R}^q, \ such \ that \ x_T \in \mathcal{X} \ \forall T \in \mathbb{R}^+\}. \tag{6.7}$$

In other words, $\mathcal{X}_e$ is the space consisting of all functions *whose truncation* belongs to $\mathcal{X}$, regardless of whether $u$ itself belongs to $\mathcal{X}$. In the sequel, the space $\mathcal{X}$ is referred to as the "parent" space of $\mathcal{X}_e$. We will assume that the space $\mathcal{X}$ satisfy the following properties:

(i) $\mathcal{X}$ is a normed linear space of piecewise continuous functions of the form $u : \mathbb{R}^+ \to \mathbb{R}^q$. The norm of functions in the space $\mathcal{X}$ will be denoted $\| \cdot \|_{\mathcal{X}}$ .

(ii) $\mathcal{X}$ is such that if $u \in \mathcal{X}$, then $u_T \in \mathcal{X} \ \forall T \in \mathbb{R}^+$, and moreover, $\mathcal{X}$ is such that $u = \lim_{T \to \infty} u_T$. Equivalently, $\mathcal{X}$ is closed under the family of projections $\{P_T\}$.

(iii) If $u \in \mathcal{X}$ and $T \in \mathbb{R}^+$, then $\|u_T\|_{\mathcal{X}} \leq \|x\|_{\mathcal{X}}$ ; that is, $\|x_T\|_{\mathcal{X}}$ is a nondecreasing function of $T \in \mathbb{R}^+$.

(iv) If $u \in \mathcal{X}_e$, then $u \in \mathcal{X}$ if and only if $\lim_{T \to \infty} \|x_T\|_{\mathcal{X}} < \infty.$

It can be easily seen that all the $\mathcal{L}_p$ spaces satisfy these properties. Notice that although $\mathcal{X}$ is a normed space, $\mathcal{X}_e$ is a linear (not normed) space. It is not normed because in general, the norm of a function $u \in \mathcal{X}_e$ is not defined. Given a function $u \in \mathcal{X}_e$, however, using property (iv) above, it is possible to check whether $u \in \mathcal{X}$ by studying the limit $\lim_{T \to \infty} \|u_T\|$.

**Example 6.2** *Let the space of functions $\mathcal{X}$ be defined by*

$$\mathcal{X} = \{x \mid x : \mathbb{R}^+ \to \mathbb{R}, x(t) \ integrable \ and \ \|x\| = \int_0^\infty | \, x(t) \, | \, dt < \infty\}.$$

*In other words, $\mathcal{X}$ is the space of real-valued function in $\mathcal{L}_1$. Consider the function $x(t) = t$. We have*

$$x_T(t) = \begin{cases} t, & 0 \leq t \leq T \\ 0, & t > T \end{cases}$$

$$\|x_T\| = \int_0^\infty | \, x_T(t) \, | \, dt = \int_0^T t \, dt = \frac{T^2}{2}$$

*Thus $x_T \in \mathcal{X}_e \ \forall T \in \mathbb{R}^+$. However $x \notin \mathcal{X}$   since $\lim_{T \to \infty} |x_T| = \infty$.*                □

**Remarks**: In our study of feedback systems we will encounter unstable systems, i.e., systems whose output grows without bound as time increases. Those systems cannot be described with any of the $\mathcal{L}_p$ spaces introduced before, or even in any other space of functions used in mathematics. Thus, the extended spaces are the right setting for our problem. As mentioned earlier, our primary interest is in the spaces $\mathcal{L}_2$ and $\mathcal{L}_\infty$. The extension of the space $\mathcal{L}_p$ , $1 \le p \le \infty$, will be denoted $\mathcal{L}_{pe}$. It consists of all the functions $u(t)$ whose truncation belongs to $\mathcal{L}_p$ .

## 6.2   Input–Output Stability

We start with a precise definition of the notion of system.

**Definition 6.5** *A system, or more precisely, the mathematical representation of a physical system, is defined to be a mapping $H : \mathcal{X}_e \to \mathcal{X}_e$ that satisfies the so-called causality condition:*

$$[Hu(\cdot)]_T = [Hu_T(\cdot)]_T \quad \forall u \in \mathcal{X}_e \ and \ \forall T \in \mathbb{R}. \tag{6.8}$$

Condition (6.8) is important in that it formalizes the notion, satisfied by all physical systems, that the past and present outputs do not depend on future inputs. To see this, imagine that we perform the following experiments (Figures 6.2 and 6.3):

(1) First we apply an arbitrary input $u(t)$, we find the output $y(t) = Hu(t)$, and from here the truncated output $y_T(t) = [Hu(t)]_T$. Clearly $y_T = [Hu(t)]_T(t)$ represents the left-hand side of equation (6.8). See Figure 6.3(a)-(c).

(2) In the second experiment we start by computing the truncation $\bar{u} = u_T(t)$ of the input $u(t)$ used above, and repeat the procedure used in the first experiment. Namely, we compute the output $\bar{y}(t) = H\bar{u}(t) = Hu_T(t)$ to the input $\bar{u}(t) = u_T(t)$, and finally we take the truncation $\bar{y}_T = [Hu_T(t)]_T$ of the function $\bar{y}$. Notice that this corresponds to the right-hand side of equation (6.8). See Figure 6.3(d)-(f).

The difference in these two experiments is the truncated input used in part (2). Thus, if the outputs $[Hu(t)]_T$ and $[Hu_T(t)]_T$ are identical, the system output in the interval $0 \le t \le T$ does not depend on values of the input *outside* this interval (i.e., $u(t)$ for $t > T$). All physical systems share this property, but care must be exercised with mathematical models since <u>not</u> all functions behave like this.

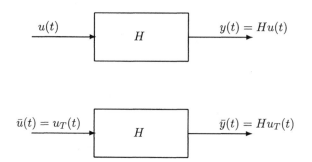

Figure 6.2: *Experiment 1*: input $u(t)$ applied to system $H$. *Experiment 2*: input $\bar{u}(t) = u_T(t)$ applied to system $H$.

We end this discussion by pointing out that the causality condition (6.8) is frequently expressed using the projection operator $P_T$ as follows:

$$P_T H = P_T H P_T \quad (i.e., P_T(Hx) = P_T[H(P_T x)] \; \forall x \in \mathcal{X}_e \text{ and } \forall T \in \mathbb{R}^+.$$

It is important to notice that the notion of input–output stability, and in fact, the notion of input–output system itself, *does not* depend in any way on the notion of *state*. In fact, the internal description given by the state is unnecessary in this framework. The essence of the input–output theory is that only the relationship between inputs and outputs is relevant. Strictly speaking, there is no room in the input–output theory for the existence of nonzero (variable) initial conditions. In this sense, the input–output theory of systems in general, and the notion of input–output stability in particular, are complementary to the Lyapunov theory. Notice that the Lyapunov theory deal with equilibrium points of systems with zero inputs and nonzero initial conditions, while the input–output theory considers *relaxed* systems with non-zero inputs. In later chapters, we review these concepts and consider input–output systems with an internal description given by a state space realization.

We may now state the definition of input–output stability.

**Definition 6.6** *A system $H : \mathcal{X}_e \to \mathcal{X}_e$ is said to be input–output $\mathcal{X}$-stable if whenever the input belongs to the parent space $\mathcal{X}$, the output is once again in $\mathcal{X}$. In other words, $H$ is $\mathcal{X}$-stable if $Hx$ is in $\mathcal{X}$ whenever $u \in \mathcal{X}$.*

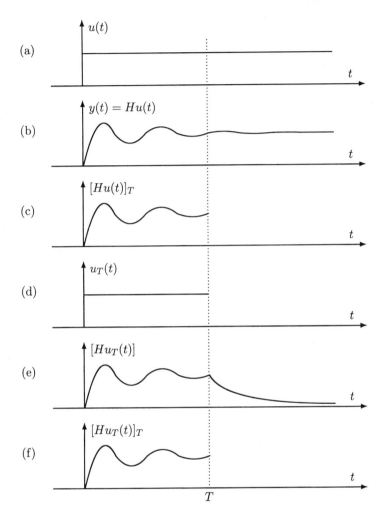

Figure 6.3: Causal systems: (a) input $u(t)$; (b) the response $y(t) = Hu(t)$; (c) truncation of the response $y(t)$. Notice that this figure corresponds to the left-hand side of equation (6.8); (d) truncation of the function $u(t)$; (e) response of the system when the input is the truncated input $u_T(t)$; (f) truncation of the system response in part (e). Notice that this figure corresponds to the right-hand side of equation (6.8).

For simplicity we will usually say that $H$ is input–output stable instead of input–output $\mathcal{X}$-stable whenever confusion is unlikely. It is clear that input–output stability is a notion that depends both on the system and the space of functions.

One of the most useful concepts associated with systems is the notion of gain.

**Definition 6.7** *A system $H : \mathcal{X}_e \to \mathcal{X}_e$ is said to have a finite gain if there exists a constant $\gamma(H) < \infty$ called the gain of $H$, and a constant $\beta \in \mathbb{R}^+$ such that*

$$\|(Hu)_T\|_{\mathcal{X}} \ \leq \gamma(H)\, \|u_T\|_{\mathcal{X}} \ + \beta. \tag{6.9}$$

Systems with finite gain are said to be *finite-gain-stable*. The constant $\beta$ in (6.9) is called the bias term and is included in this definition to allow the case where $Hu \neq 0$ when $u = 0$. A different, and perhaps more important interpretation of this constant will be discussed in connection with input–output properties of state space realizations. If the system $H$ satisfies the condition

$$Hu = 0 \quad \text{whenever} \quad u = 0$$

then the gain $\gamma(H)$ can be calculated as follows

$$\gamma(H) = \sup \frac{\|(Hu)_T\|_{\mathcal{X}}}{\|x_T\|_{\mathcal{X}}} \tag{6.10}$$

where the supremum is taken over all $u \in \mathcal{X}_e$ and all $T$ in $\mathbb{R}^+$ for which $u_T \neq 0$.

**Example 6.3** *Let $\mathcal{X} = \mathcal{L}_\infty$, and consider the nonlinear operator $N(\cdot)$ defined by the graph in the plane shown in Figure 6.4, and notice that $N(0) = 0$. The gain $\gamma(H)$ is easily determined from the slope of the graph of $N$.*

$$\gamma(H) = \sup \frac{\|(Hu)_T\|_{\mathcal{L}_\infty}}{\|u_T\|_{\mathcal{L}_\infty}} \ = \ 1.$$

$\square$

Systems such as $N(\cdot)$ in example 6.3 have no "dynamics" and are called *static* or *memoryless* systems, given that the response is an instantaneous function of the input.

We conclude this section by making the following observation. It is immediately obvious that if a system has finite gain, then it is input–output stable. The converse is, however, not true. For instance; any static nonlinearity without a bounded slope *does not* have a finite gain. As an example, the memoryless systems $H_i : \mathcal{L}_{\infty e} \to \mathcal{L}_{\infty e}$, $i = 1, 2$ shown in Figure 6.5 are input–output-stable, but do not have a finite gain.

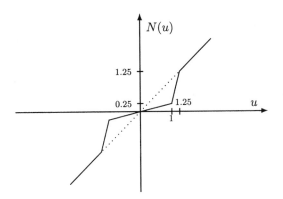

Figure 6.4: Static nonlinearity $N(\cdot)$.

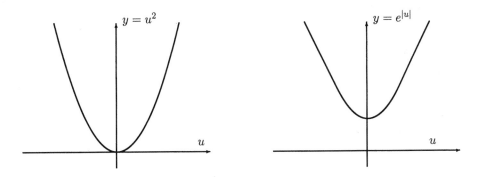

Figure 6.5: The systems $H_1 u = u^2$, and $H_2 u = e^{|u|}$.

## 6.3   Linear Time-Invariant Systems

So far, whenever dealing with LTI systems we focused our attention on state space real-izations. In this section we include a brief discussion of LTI system in the context of the input–output theory of systems. For simplicity, we limit our discussion to single-input–single-output systems.

**Definition 6.8** *We denote by* $\mathcal{A}$ *the set of distributions (or generalized functions) of the form*

$$f(t) = \begin{cases} f_0\delta(t) + f_a(t), & t \geq 0 \\ 0, & t < 0 \end{cases}$$

*where* $f_0 \in \mathbb{R}$, $\delta(\cdot)$ *denotes the unit impulse, and* $f_a(\cdot)$ *is such that*

$$\int_0^\infty |f_a(\tau)| \, d\tau < \infty$$

*namely,* $f_a \in \mathcal{L}_1$. *The norm of* $f \in \mathcal{A}$ *is defined by*

$$\|f\|_{\mathcal{A}} = |f_0| + \int_0^\infty |f_a(t)| \; dt. \tag{6.11}$$

We will also denote by $\widehat{\mathcal{A}}$ the set consisting of all functions that are Laplace transforms of elements of $\mathcal{A}$. Note that, with the norm of $f$ defined in (6.11), if $f \in \mathcal{L}_1$ (i.e., if $f_0 = 0$), then $\|f\|_{\mathcal{A}} = \|f\|_1$. The extension of the algebra $\mathcal{A}$, denoted $\mathcal{A}_e$, is defined to be the set of all functions whose truncation belongs to $\mathcal{A}$.

We now introduce the following notation:

- $\mathcal{R}[s]$: set of polynomials in the variable s.

- $\mathcal{R}(s)$: field of fractions associated with $\mathcal{R}[s]$, i.e. $\mathcal{R}(s)$ consists of all rational functions in s with real polynomials. A rational function $\widehat{M} \in \mathcal{R}(s)$ will be said to be proper if it satisfies

$$\lim_{s \to \infty} \widehat{M} < \infty.$$

It is said to be strictly proper if

$$\lim_{s \to \infty} \widehat{M} < 0.$$

Notice that, according to this, if $\widehat{H}(s)$ is strictly proper, then it is also proper. The converse is, however, not true.

**Theorem 6.1** *Consider a function $\widehat{F}(s) \in \mathcal{R}(s)$. Then $\widehat{F}(s) \in \widehat{\mathcal{A}}$ if and only if (i) $\widehat{F}(s)$ is proper, and (ii) all poles of $\widehat{F}(s)$ lie in the left half of the complex plane.*

**Proof**: See the Appendix. □

**Definition 6.9** *The convolution of $f$ and $g$ in $\mathcal{A}$, denoted by $f * g$, is defined by*

$$(f * g)(t) = \int_0^\infty f(\tau)g(t-\tau)\, d\tau = \int_0^\infty f(t-\tau)g(\tau)\, d\tau. \tag{6.12}$$

It is not difficult to show that if $f, g \in \mathcal{A}$, then $f * g \in \mathcal{A}$ and $g * f \in \mathcal{A}$.

We can now define what will be understood by a linear time-invariant system.

**Definition 6.10** *A linear time-invariant system $H$ is defined to be a convolution operator of the form*

$$(Hu)(t) = h(t) * u(t) = \int_{-\infty}^\infty h(\tau)u(t-\tau)\, d\tau = \int_{-\infty}^\infty h(t-\tau)u(\tau)\, d\tau \tag{6.13}$$

*where $h(\cdot) \in \mathcal{A}$. The function $h(\cdot)$ is called the "kernel" of the operator $H$.*

Given the causality assumption $h(\tau) = 0$ for $\tau < 0$, we have that, denoting $y(t) \overset{def}{=} Hu(t)$

$$y(t) = h_0 u(0) + \int_0^\infty h(\tau)u(t-\tau)\, d\tau = h_0 u(0) + \int_{-\infty}^t h(t-\tau)u(\tau)\, d\tau \tag{6.14}$$

If in addition, $u = 0$ for $t < 0$, then

$$y(t) = h_0 u(0) + \int_0^t h(\tau)u(t-\tau)\, d\tau = h_0 u(0) + \int_0^t h(t-\tau)u(\tau)\, d\tau \tag{6.15}$$

Definition 6.10 includes the possible case of infinite-dimensional systems, such as systems with a time delay. In the special case of finite-dimensional LTI systems, Theorem 6.1 implies that a (finite-dimensional) LTI system is stable if and only if the roots of the polynomial denominator lie in the left half of the complex plane. It is interesting, however, to consider a more general class of LTI systems. We conclude this section with a theorem that gives necessary and sufficient conditions for the $\mathcal{L}_p$ stability of a (possibly infinite-dimensional) linear time-invariant system.

**Theorem 6.2** *Consider a linear time-invariant system $H$, and let $h(\cdot)$ represent its impulse response. Then $H$ is $\mathcal{L}_p$ stable if and only if $h(\cdot) = h_0 \delta(t) + h_a(t) \in \mathcal{A}$ and moreover, if $H$ is $\mathcal{L}_p$ stable, then $\|Hu\|_{\mathcal{L}_p} \leq \|h\|_{\mathcal{A}}\|u\|_{\mathcal{L}_p}$.*

**Proof**: See the Appendix.

## 6.4   $\mathcal{L}_p$ Gains for LTI Systems

Having settled the question of input–output stability in $\mathcal{L}_p$ spaces, we focus our attention on the study of *gain*. Once again, for simplicity we restrict our attention to single-input-single-output (SISO) systems. It is clear that the notion of gain depends on the space of input functions in an essential manner.

### 6.4.1   $\mathcal{L}_\infty$ Gain

By definition

$$\gamma(H)_\infty = \sup_u \frac{\|Hu\|_{\mathcal{L}_\infty}}{\|u\|_{\mathcal{L}_\infty}} \quad = \quad \sup_{\|u\|_{\mathcal{L}_\infty}=1} \|Hu\|_{\mathcal{L}_\infty} \tag{6.16}$$

This is a very important case. The space $\mathcal{L}_\infty$ consists of all the functions of $t$ whose absolute value is bounded, and constitutes perhaps the most natural choice for the space of functions $\mathcal{X}$. We will show that in this case

$$\gamma(H)_\infty = \|h(t)\|_{\mathcal{A}} \tag{6.17}$$

Consider an input $u(t)$ applied to the system $H$ with impulse response $h(\cdot) = h_0\delta(t)+h_a(t) \in \mathcal{A}$. We have

$$
\begin{aligned}
y(t) &= (h * u)(t) = h_0 u(t) + \int_0^t h_a(\tau)u(t-\tau)\mathrm{d}\tau \\
|y(t)| &\leq |h_0||u(t)| + \int_0^t |h_a(\tau)||u(t-\tau)|\mathrm{d}\tau \\
&\leq \sup_t |u(t)| \left\{ |h_0| + \int_0^t |h_a(\tau)|\mathrm{d}\tau \right\} \\
&= \|u\|_{\mathcal{L}_\infty}\|h\|_{\mathcal{A}}
\end{aligned}
$$

Thus

$$
\begin{aligned}
\|y\|_\infty &\leq \|u\|_\infty\|h\|_{\mathcal{A}} \quad \text{or} \\
\|h\|_{\mathcal{A}} &\geq \frac{\|Hu\|_{\mathcal{L}_\infty}}{\|u\|_{\mathcal{L}_\infty}}.
\end{aligned}
$$

This shows that $\|h\|_{\mathcal{A}} \geq \gamma(H)_\infty$. To show that $\|h\|_{\mathcal{A}} = \gamma(H)$. we must show that equality can actually occur. We do this by constructing a suitable input. For each fixed $t$, let

$$u(t-\tau) = \mathrm{sgn}[h(\tau)] \quad \forall \tau$$

where

$$\text{sgn}[h(t)] = \begin{cases} 1 & \text{if } h(\tau) \geq 0 \\ 0 & \text{if } h(\tau) < 0 \end{cases}$$

It follows that $\|u\|_{\mathcal{L}_\infty} = 1$, and

$$
\begin{aligned}
y(t) &= (h * u)(t) = h_0 u(t) + \int_0^t h_a(\tau) u(t - \tau) d\tau \\
&= |h_0| + \int_0^t |h_a(\tau)| d\tau = \|h\|_{\mathcal{A}}
\end{aligned}
$$

and the result follows.                                                                                      □

## 6.4.2   $\mathcal{L}_2$ Gain

This space consists of all the functions of $t$ that are square integrable or, to state this in different words, functions that have finite energy. Although from the input–output point of view this class of functions is not as important as the previous case (e.g. sinusoids and step functions are not in this class), the space $\mathcal{L}_2$ is the most widely used in control theory because of its connection with the frequency domain that we study next. We consider a linear time-invariant system $H$, and let $h(\cdot) \in \mathcal{A}$ be the kernel of $H$., i.e., $(Hu)(t) = h(t) * u(t), \forall u \in \mathcal{L}_2$. We have that

$$\gamma_2(H) = \sup_x \frac{\|Hx\|_{\mathcal{L}_2}}{\|x\|_{\mathcal{L}_2}} \tag{6.18}$$

where

$$\|x\|_{\mathcal{L}_2} = \left\{ \int_{-\infty}^{\infty} |x(t)|^2 \, dt \right\}^{1/2} \tag{6.19}$$

We will show that in this case, the gain of the system $H$ is given by

$$\gamma_2(H) = \sup_\omega |\widehat{H}(\jmath\omega)| \overset{def}{=} \|H\|_\infty \tag{6.20}$$

where $\widehat{H}(\jmath\omega) = \mathcal{F}[h(t)]$, the Fourier transform of $h(t)$. The norm (6.20) is the so-called $H$-infinity norm of the system $H$. To see this, consider the output $y$ of the system to an input $u$

$$\|y\|_{\mathcal{L}_2}^2 = \|Hu\|_{\mathcal{L}_2}^2 = \int_{-\infty}^{\infty} [h(t) * u(t)] dt = \frac{1}{2\pi} \int_{-\infty}^{\infty} |\widehat{H}(\jmath\omega)|^2 |\widehat{U}(\jmath\omega)|^2 d\omega$$

where the last identity follows from Parseval's equality. From here we conclude that

$$\Rightarrow \ \|y\|_{\mathcal{L}_2}^2 \leq \{\sup_\omega |\widehat{H}(\jmath\omega)|\}^2 \left\{ \frac{1}{2\pi} \int_{-\infty}^{\infty} |\widehat{U}(\jmath\omega)|^2 d\omega \right\}$$

but

$$\sup_{\omega} |\widehat{H}(\jmath\omega)| \stackrel{def}{=} \|H\|_{\infty}$$

$$\Rightarrow \quad \|y\|_{\mathcal{L}_2}^2 \quad \leq \quad \|H\|_{\infty}^2 \|u\|_2^2 \tag{6.21}$$

$$\Rightarrow \quad \gamma_2(H) \quad \leq \quad \sup_{\omega} |\widehat{H}(\jmath\omega)| \, \|u\|_2^2. \tag{6.22}$$

Equation (6.21) or (6.22) proves that the $H$-infinity norm of $H$ is an upper bound for the gain $\gamma(H)$. As with the $\mathcal{L}_{\infty}$ case, to show that it is the least upper bound, we proceed to construct a suitable input. Let $u(t)$ be such that its Fourier transform, $\mathcal{F}[u(t)] = U(\jmath\omega)$, has the following properties:

$$|U(\jmath\omega)| = \begin{cases} A & \text{if } |\omega - \omega_0| < \Delta\omega \text{ , or } |\omega + \omega_0| < \Delta\omega \\ 0 & \text{otherwise} \end{cases}$$

In this case

$$\|y\|_{\mathcal{L}_2}^2 = \frac{1}{2\pi} \left\{ \int_{-\omega-\Delta\omega_0}^{-\omega+\Delta\omega_0} A^2 |\widehat{H}(\jmath\omega)|^2 d\omega \;+\; \int_{\omega-\Delta\omega_0}^{\omega+\Delta\omega_0} A^2 |\widehat{H}(\jmath\omega)|^2 d\omega \right\}$$

Therefore, as $\Delta\omega \to 0$, $\widehat{H}(\jmath\omega) \to \widehat{H}(\jmath\omega)$ and

$$\|y\|_{\mathcal{L}_2}^2 \to \frac{1}{2\pi} A^2 |\widehat{H}(\jmath\omega)|^2 \, 4 \, \Delta\omega$$

Thus, defining $A = \{\pi/2\Delta\omega\}^{1/2}$, we have that $\|y\|_2^2 \to \|H\|_{\infty}^2$ as $\Delta\omega \to \infty$, which completes the proof.　　　　　　　　　　　　　　　　　　　　　　　　　　　　　　　□

It is useful to visualize the $\mathcal{L}_2$ gain (or $H$-infinity norm) using Bode plots, as shown in Figure 6.6.

## 6.5   Closed-Loop Input–Output Stability

Until now we have concentrated on open-loop systems. One of the main features of the input–output theory of systems, however, is the wealth of theorems and results concerning stability of feedback systems. To study feedback systems, we first define what is understood by closed loop input–output stability, and then prove the so-called *small gain theorem*. As we will see, the beauty of the input–output approach is that it permits us to draw conclusions about stability of feedback interconnections based on the properties of the several subsystems encountered around the feedback loop. The following model is general enough to encompass most cases of interest.

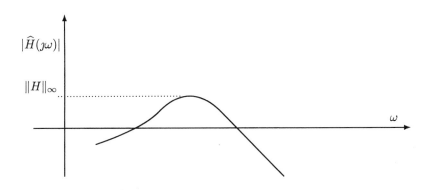

Figure 6.6: Bode plot of $|\widehat{H}(\jmath\omega)|$, indicating the $\|H\|_\infty$ norm of $H$.

**Definition 6.11** *We will denote by feedback system to the interconnection of the subsystems $H_1$ and $H_2 : \mathcal{X}_e \to \mathcal{X}_e$ that satisfies the following assumptions:*

*(i) $e_1$, $e_2$, $y_1$, and $y_2 \in \mathcal{X}_e$ for all pairs of inputs $u_1, u_2 \in \mathcal{X}_e$.*

*(ii) The following equations are satisfied for all $u_1$, $u_2 \in \mathcal{X}_e$:*

$$e_1 = u_1 - H_2 e_2 \qquad (6.23)$$
$$e_2 = u_2 + H_1 e_1. \qquad (6.24)$$

Here $u_1$ and $u_2$ are input functions and may represent different signals of interest such as commands, disturbances, and sensor noise. $e_1$ and $e_2$ are outputs, usually referred to as *error signals* and $y_1 = H_1 e_1, y_2 = H_2 e_2$ are respectively the outputs of the subsystems $H_1$ and $H_2$. Assumptions (i) and (ii) ensure that equations (6.23) and (6.24) can be solved for all inputs $u_1, u_2 \in \mathcal{X}_e$. If this assumptions are not satisfied, the operators $H_1$ and $H_2$ do not adequately describe the physical systems they model and should be modified. It is immediate that equations (6.23) and (6.24) can be represented graphically as shown in Figure 6.7.

In general, the subsystems $H_1$ and $H_2$ can have several inputs and several outputs. For compatibility, it is implicitly assumed that the number of inputs of $H_1$ equals the number of outputs of $H_2$, and the number of outputs of $H_1$ equals the number of inputs of $H_2$. We also notice that we do not make explicit the system dimension in our notation. For example, if $\mathcal{X} = \mathcal{L}_\infty$, the space of functions with bounded absolute value, we write $H : \mathcal{X}_e \to \mathcal{X}_e$ (or $H : \mathcal{L}_{\infty e} \to \mathcal{L}_{\infty e}$) regardless of the number of inputs and outputs of the system.

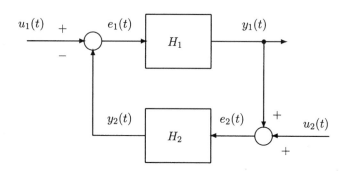

Figure 6.7: The Feedback System $S$.

In the following definition, we introduce vectors $u$, $e$, and $y$, defined as follows:

$$u(t) = \begin{bmatrix} u_1(t) \\ u_2(t) \end{bmatrix}, \quad e(t) = \begin{bmatrix} e_1(t) \\ e_2(t) \end{bmatrix}, \quad y(t) = \begin{bmatrix} y_1(t) \\ y_2(t) \end{bmatrix} \qquad (6.25)$$

**Definition 6.12** *Consider the feedback interconnection of subsystems $H_1$ and $H_2$. For this system we introduce the following input–output relations. Given $u_1, u_2 \in X_e$, $i,j = 1,2$, and with $u$, $e$, and $y$ given by (6.25) we define*

$$E = \{(u,e) \in X_e \times X_e \text{ and } e \text{ satisfies (6.23) and (6.24)}\} \qquad (6.26)$$
$$F = \{(u,y) \in X_e \times X_e \text{ and } y \text{ satisfies (6.23) and (6.24)}\}. \qquad (6.27)$$

In words, $E$ and $F$ are the relations that relate the inputs $u_i$ with $e_i$ and $y_i$, respectively. Notice that questions related to the existence and uniqueness of the solution of equations (6.23) and (6.24) are taken for granted. That is the main reason why we have chosen to work with relations rather than functions.

**Definition 6.13** *A relation $P$ on $X_e$ is said to be bounded if the image under $P$ of every bounded subset of $X_e \in dom(P)$ is a bounded subset of $X_e$.*

In other words, $P$ is bounded if $Pu \in X$ for every $u \in X$, $u \in dom(P)$.

**Definition 6.14** *The feedback system of equations (6.23) and (6.24) is said to be bounded or input–output–stable if the closed-loop relations $E$ and $F$ are bounded for all possible $u_1, u_2$ in the domain of $E$ and $F$.*

In other words, a feedback system is input–output–stable if whenever the inputs $u_1$ and $u_2$ are in the parent space $\mathcal{X}$, the outputs $y_1$ and $y_2$ and errors $e_1$ and $e_2$ are also in $\mathcal{X}$. In the sequel, input–output–stable systems will be referred to simply as "stable" systems.

**Remarks:** Notice that, as with open-loop systems, the definition of boundedness depends strongly on the selection of the space $\mathcal{X}$. To emphasize this dependence, we will sometimes denote $\mathcal{X}$-stable to a system that is bounded in the space $\mathcal{X}$.

## 6.6 The Small Gain Theorem

In this section we study the so-called small gain theorem, one of the most important results in the theory of input–output systems. The main goal of the theorem is to provide open-loop conditions for closed-loop stability. Theorem 6.3 given next, is the most popular and, in some sense, the most important version of the small gain theorem. It says that if the product of the gains of two systems, $H_1$ and $H_2$, is less than 1, then its feedback interconnection is stable, in a sense to be made precise below. See also remark (b) following Theorem 6.3.

**Theorem 6.3** *Consider the feedback interconnection of the systems $H_1$ and $H_2 : \mathcal{X}_e \to \mathcal{X}_e$. Then, if $\gamma(H_1)\gamma(H_2) < 1$, the feedback system is input–output–stable.*

**Proof:** To simplify our proof, we assume that the bias term $\beta$ in Definition 6.7 is identically zero (see Exercise 6.5). According to Definition 6.14 we must show that $u_1, u_2 \in \mathcal{X}$ imply that $e_1$, $e_2$, $y_1$ and $y_2$ are also in $\mathcal{X}$. Consider a pair of elements $(u_1, e_1), (u_2, e_2)$ that belong to the relation $E$. Then, $u_1$, $u_2$, $e_1$, $e_2$ must satisfy equations (6.23) and (6.24) and, after truncating these equations, we have

$$
\begin{align}
e_{1T} &= u_{1T} - (H_2 e_2)_T \tag{6.28}\\
e_{2T} &= u_{2T} + (H_1 e_1)_T. \tag{6.29}
\end{align}
$$

Thus,

$$
\begin{align}
\|e_{1T}\| &\leq \|u_{1T}\| + \|(H_2 e_2)_T\| \leq \|u_{1T}\| + \gamma(H_2)\|e_{2T}\| \tag{6.30}\\
\|e_{2T}\| &\leq \|u_{2T}\| + \|(H_1 e_1)_T\| \leq \|u_{2T}\| + \gamma(H_1)\|e_{1T}\|. \tag{6.31}
\end{align}
$$

Substituting (6.31) in (6.30) we obtain

$$
\begin{align}
\|e_{1T}\| &\leq \|u_{1T}\| + \gamma(H_2)\{\|u_{2T}\| + \gamma(H_1)\|e_{1T}\|\}\\
&\leq \|u_{1T} + \gamma(H_2)\|u_{2T}\| + \gamma(H_1)\gamma(H_2)\|e_{1T}\|\\
\Rightarrow \quad &[1 - \gamma(H_1)\gamma(H_2)]\|e_{1T}\| \leq \|u_{1T}\| + \gamma(H_2)\|u_{2T}\| \tag{6.32}
\end{align}
$$

and since, by assumption, $\gamma(H_1)\gamma(H_2) < 1$, we have that $1 - \gamma(H_1)\gamma(H_2) \neq 0$ and then

$$\|e_{1T}\| \leq [1 - \gamma(H_1)\gamma(H_2)]^{-1}\{\|u_{1T}\| + \gamma(H_2)\|u_{2T}\|\}. \tag{6.33}$$

Similarly

$$\|e_{2T}\| \leq [1 - \gamma(H_1)\gamma(H_2)]^{-1}\{\|u_{2T}\| + \gamma(H_1)\|u_{1T}\|\}. \tag{6.34}$$

Thus, the norms of $e_{1T}$ and $e_{2T}$ are bounded by the right-hand side of (6.34) and (6.33). If, in addition, $u_1$ and $u_2$ are in $\mathcal{X}$ (i.e., $\|u_i\| < \infty$ for $i = 1, 2$), then (6.34) and (6.33) must also be satisfied if we let $T \to \infty$. We have

$$\|e_1\| \leq [1 - \gamma(H_1)\gamma(H_2)]^{-1}\{\|u_1\| + \gamma(H_2)\|u_2\|\} \tag{6.35}$$
$$\|e_2\| \leq [1 - \gamma(H_1)\gamma(H_2)]^{-1}\{\|u_2\| + \gamma(H_1)\|u_1\|\}. \tag{6.36}$$

It follows that $e_1$ and $e_2$ are also in $\mathcal{X}$ (see the assumptions about the space $\mathcal{X}$) and the closed-loop relation $E$ is bounded. That $F$ is also bounded follows from (6.35) and (6.36) and the equation

$$\|(H_i e_i)_T\| \leq \gamma(H_i)\|e_{iT}\|, \quad i = 1, 2 \tag{6.37}$$

evaluated as $T \to \infty$.                                                                                  $\square$

**Remarks:**

(a) Theorem 6.3 says exactly that if the product of the gains of the open-loop systems $H_1$ and $H_2$ is less than 1, then each bounded input in the domain of the relations $E$ and $F$ produces a bounded output. It does not, however, imply that the solution of equations (6.23) and (6.24) is unique. Moreover, it does not follow from Theorem 6.3 that for every pair of functions $u_1$ and $u_2 \in \mathcal{X}$ the outputs $e_1$, $e_2$, $y_1$ $y_2$ are bounded, because the existence of a solution of equations (6.23) and (6.24) was not proved for every pair of inputs $u_1$ and $u_2$. In practice, the question of existence of a solution can be studied separately from the question of stability. If the only thing that is known about the system is that it satisfies the condition $\gamma(H_1)\gamma(H_2) < 1$, then Theorem 6.3 guarantees that, if a solution of equations (6.23) and (6.24) exists, then it is bounded. Notice that we were able to ignore the question of existence of a solution by making use of *relations*. An alternative approach was used by Desoer and Vidyasagar [21], who assume that $e_1$ and $e_2$ belong to $\mathcal{X}_e$ and *define* $u_1$ and $u_2$ to satisfy equations (6.23) and (6.24).

(b) Theorem 6.3 provides sufficient but not necessary conditions for input–output stability. In other words, it is possible, and indeed usual, to find a system that does not satisfy the small gain condition $\gamma(H_1)\gamma(H_2) < 1$ and is nevertheless input–output-stable.

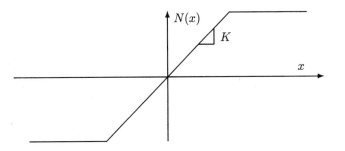

Figure 6.8: The nonlinearity $N(\cdot)$.

Examples 6.4 and 6.5 contain very simple applications of the small gain theorem. We assume that $\mathcal{X} = \mathcal{L}_2$.

**Example 6.4** *Let $H_1$ be a linear time-invariant system with transfer function*

$$\widehat{G}(s) = \frac{1}{s^2 + 2s + 4},$$

*and let $H_2$ be the nonlinearity $N(\cdot)$ defined by a graph in the plane, as shown in Figure 6.8.*

*We apply the small gain theorem to find the maximum slope of the nonlinearity $N(\cdot)$ that guarantees input–output stability of the feedback loop. First we find the gain of $H_1$ and $H_2$. For a linear time-invariant system, we have*

$$\gamma(H) = \sup_{\omega} |\widehat{G}(\jmath\omega)|$$

*In this case, we have*

$$
\begin{aligned}
|\widehat{G}(\jmath\omega)| &= \frac{1}{|(4 - \omega^2) + 2\jmath\omega|} \\
&= \frac{1}{\sqrt{[(4 - \omega^2)^2 + 4\omega^2]}}.
\end{aligned}
$$

*Since $|\widehat{G}(\jmath\omega)| \to 0$ as $\omega \to \infty$, the supremum must be located at some finite frequency, and since $|\widehat{G}(\jmath\omega)|$ is a continuous function of $\omega$, the maximum values of $|\widehat{G}(\jmath\omega)|$ exists and satisfies*

$$\gamma(H_1) = \max_{\omega} |\widehat{G}(\jmath\omega)|$$

$$= |\widehat{G}(\jmath\omega^*)| : \left.\frac{d|\widehat{G}(\jmath\omega)|}{d\omega}\right|_{\omega=\omega^*} = 0 , \quad and$$

$$\left.\frac{d^2|\widehat{G}(\jmath\omega)|}{d\omega^2}\right|_{\omega=\omega^*} < 0.$$

*Differentiating* $|\widehat{G}(\jmath\omega)|$ *twice with respect to* $\omega$ *we obtain* $\omega = \sqrt{2}$, *and* $|\widehat{G}(\jmath\omega^*)| = 1/\sqrt{12}$. *It follows that* $\gamma(H_1) = 1/\sqrt{12}$. *The calculation of* $\gamma(H_2)$ *is straightforward. We have* $\gamma(H_2) = |K|$. *Applying the small gain condition* $\gamma(H_1)\gamma(H_2) < 1$, *we obtain*

$$\gamma(H_1)\gamma(H_2) < 1 \qquad \Rightarrow \qquad |K| < \sqrt{12}.$$

*Thus, if the absolute value of the slope of the nonlinearity* $N(\cdot)$ *is less than* $\sqrt{12}$ *the system is closed loop stable.*                                                                                   □

**Example 6.5** *Let* $H_1$ *be as in Example 6.4 and let* $H_2$ *be a constant gain (i.e.,* $H_2$ *is linear time-invariant and* $H_2 = k$). *Since the gain of* $H_2$ *is* $\gamma(H_2) = |k|$, *application of the small gain theorem produces the same result obtained in Example 6.4, namely, if* $|k| < \sqrt{12}$, *the system is closed loop stable. However, in this simple example we can find the closed loop transfer function, denoted* $\widehat{H}(s)$, *and check stability by obtaining the poles of* $\widehat{H}(s)$. *We have*

$$\widehat{H}(s) = \frac{\widehat{G}(s)}{1 + k\widehat{G}(s)} = \frac{1}{s^2 + 2s + (4+k)}$$

*The system is closed-loop-stable if and only if the roots of the polynomial denominator of* $\widehat{H}(s)$ *lie in the open left half of the complex plane. For a second-order polynomial this is satisfied if and only if all its coefficients have the same sign. It follows that the system is closed-loop-stable if and only if* $(4+k) > 0$, *or equivalently,* $k > -4$. *Comparing the results obtained using this two methods, we have*

- *Small gain theorem:* $-\sqrt{12} < k < \sqrt{12}$.

- *Pole analysis:* $-4 < k < \infty$.

*We conclude that, in this case, the small gain theorem provides a poor estimate of the stability region.*                                                                                   □

## 6.7   Loop Transformations

As we have seen, the small gain theorem provides sufficient conditions for the stability of a feedback loop, and very often results in conservative estimates of the system stability.

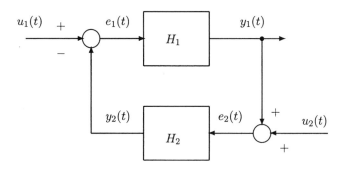

Figure 6.9: The Feedback System $S$.

The same occurs with any other sufficient but not necessary stability condition, such as the passivity theorem (to be discussed in Chapter 8). One way to obtain improved stability conditions (i.e. less conservative) is to apply the theorems to a modified feedback loop that satisfies the following two properties: (1) it guarantees stability of the original feedback loop, and (2) it lessens the overall requirements over $H_1$ and $H_2$. In other words, it is possible that a modified system satisfies the stability conditions imposed by the theorem in use whereas the original system does not. The are two basic transformations of feedback loops that will be used throughout the book, and will be referred to as transformations of Types I and Type II.

**Definition 6.15** *(Type I Loop Transformation) Consider the feedback system $S$ of Figure 6.9. Let $H_1, H_2, K$ and $(I + KH_1)^{-1}$ be causal maps from $X_e$ into $X_e$ and assume that $K$ is linear. A loop transformation of Type I is defined to be the modified system, denoted $S_K$, formed by the feedback interconnection of the subsystems $H_1' = H_1(I + KH_1)^{-1}$ and $H_2' = H_2 - K$, with inputs $u_1' = u_1 - Ku_2$ and $u_2' = u_2$, as shown in Figure 6.10. The closed-loop relations of $S_K$ will be denoted $E_K$ and $F_K$.*

The following theorem shows that the system $S$ is stable if and only if the system $S_K$ is stable. In other words, for stability analysis, the system $S$ can always be replaced by the system $S_K$.

**Theorem 6.4** *Consider the system $S$ of Figure 6.9 and let $S_K$ be the modified system obtained after a type I loop transformation and assume the $K$ and $(I + KH_1)^{-1} : X \rightarrow X$. Then (i) The system $S$ is stable if and only if the system $S_K$ is stable.*

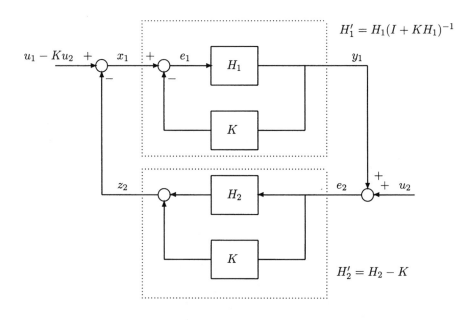

Figure 6.10: The Feedback System $S_K$.

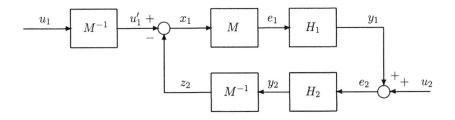

Figure 6.11: The Feedback System $S_M$.

**Proof:** The proof is straightforward, although laborious, and is omitted . Notice, however, that the transformation consists essentially of adding and subtracting the term $Ky_1$ at the same point in the loop (in the bubble in front of $H_1$), thus leading to the result.                    □

**Definition 6.16** *(Type II Loop Transformation) Consider the feedback system $S$ of Figure 6.9. Let $H_1, H_2$ be causal maps of $X_e$ into $X_e$  and let $M$ be a causal linear operator satisfying*

*(i) $M : X  \to X$ .*

*(ii) $\exists M^{-1} : X  \to X  : MM^{-1} = I$, $M^{-1}$ causal.*

*(iii) Both $M$ and $M^{-1}$ have finite gain.*

*    A type II loop transformation is defined to be the modified system $S_M$, formed by the feedback interconnection of the subsystem $H_1' = H_1M$ and $H_2' = M^{-1}H_2$, with inputs $u_1' = M^{-1}u_1$ and $u_2' = u_2$, as shown in Figure 6.11. The closed-loop relation of this modified system will be denoted $E_M$ and $F_M$.*

**Theorem 6.5** *Consider the system $S$ of Figure 6.9 and let $S_M$ be the modified system obtained after a type II loop transformation. Then the system $S$ is stable if and only if the system $S_M$  is stable.*

**Proof:** The proof is straightforward and is omitted (see Exercise 6.9).

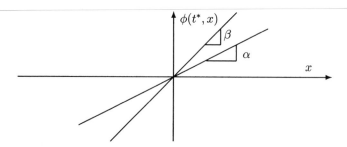

Figure 6.12: The nonlinearity $\phi(t^*, x)$ in the sector $[\alpha, \beta]$.

## 6.8   The Circle Criterion

Historically, one of the first applications of the small gain theorem was in the derivation of the celebrated circle criterion for the $\mathcal{L}_2$ stability of a class of nonlinear systems. We first define the nonlinearities to be considered.

**Definition 6.17** *A function $\phi : \mathbb{R}^+ \times \mathbb{R} \to \mathbb{R}$ is said to belong to the sector $[\alpha, \beta]$ where $\alpha \le \beta$ if*

$$\alpha x^2 \le x\phi(t, x) \le \beta x^2 \qquad \forall t \ge 0, \forall x \in \mathbb{R}. \tag{6.38}$$

According to this definition, if $\phi$ satisfies a sector condition then, in general, it is time-varying and for each fixed $t = t^*$, $\phi(t^*, x)$ is confined to a graph on the plane, as shown in Figure 6.12.

We assume that the reader is familiar with the Nyquist stability criterion. The Nyquist criterion provides necessary and sufficient conditions for closed-loop stability of lumped linear time-invariant systems. Given a proper transfer function

$$\widehat{G}(s) = \frac{p(s)}{q(s)}$$

where $p(s)$ and $q(s)$ are polynomials in $s$ with no common zeros, expanding $\widehat{G}(s)$ in partial fractions it is possible to express this transfer function in the following form:

$$\widehat{G}(s) = \widehat{g}(s) + \frac{n(s)}{d(s)}$$

where

(i) $\widehat{g}(s)$ has no poles in the open left-half plane (i.e., $\widehat{g}(s)$ is the transfer function of an exponentially stable system).

(ii) $n(s)$ and $d(s)$ are polynomials, and

$$\frac{n(s)}{d(s)}$$

is a proper transfer function.

(iii) All zeros of $d(s)$ are in the closed right half plane. Thus,

$$\frac{n(s)}{d(s)}$$

contains the unstable part of $\widehat{G}$. The number of open right half plane zeros of $d(s)$ will be denoted by $\nu$.

With this notation, the Nyquist criteria can be stated as in the following lemma.

**Lemma 6.1** *(Nyquist) Consider the feedback interconnection of the systems $H_1$ and $H_2$. Let $H_1$ be linear time-invariant with a proper transfer function $\widehat{G}(s)$ satisfying (i)–(iii) above, and let $H_2$ be a constant gain $K$. Under these conditions the feedback system is closed-loop-stable in $\mathcal{L}_p$, $1 \leq p \leq \infty$, if and only if the Nyquist plot of $\widehat{G}(s)$ [i.e., the polar plot of $\widehat{G}(\jmath\omega)$ with the standard indentations at each $\jmath\omega$-axis pole of $\widehat{G}(\jmath\omega)$ if required] is bounded away from the critical point $(-1/K + \jmath 0)$, $\forall \omega \in \mathbb{R}$ and encircles it exactly $\nu$ times in the counterclockwise direction as $\omega$ increases from $-\infty$ to $\infty$.*

The circle criterion of Theorem 6.6 analyzes the $\mathcal{L}_2$ stability of a feedback system formed by the interconnection of a linear time-invariant system in the forward path and a nonlinearity in the sector $[\alpha, \beta]$ in the feedback path. This is sometimes referred to as the *absolute stability* problem, because it encompasses not a particular system but an entire class of systems.

In the following theorem, whenever we refer to the gain $\gamma(H)$ of a system $H$, it will be understood in the $\mathcal{L}_2$ sense.

**Theorem 6.6** *Consider the feedback interconnection of the subsystems $H_1$ and $H_2 : \mathcal{L}_{2e} \to \mathcal{L}_{2e}$. Assume $H_2$ is a nonlinearity $\phi$ in the sector $[\alpha, \beta]$, and let $H_1$ be a linear time-invariant system with a proper transfer function $\widehat{G}(s)$ that satisfies assumptions (i)–(iii) above. Under these assumptions, if one of the following conditions is satisfied, then the system is $\mathcal{L}_2$-stable:*

(a) **If**  $0 < \alpha < \beta$: *The Nyquist plot of $\widehat{G}(s)$ is bounded away from the critical circle $C^*$, centered on the real line and passing through the points $(-\alpha^{-1} + j0)$ and $(-\beta^{-1} + j0)$ and encircles it $\nu$ times in the counterclockwise direction, where $\nu$ is the number of poles of $\widehat{G}(s)$ in the open right half plane.*

(b) **If**  $0 = \alpha < \beta$: $\widehat{G}(s)$ *has no poles in the open right half plane and the Nyquist plot of $\widehat{G}(s)$ remains to the right of the vertical line with abscissa $-\beta^{-1}$ for all $\omega \in \mathbb{R}$.*

(c) **If**  $\alpha < 0 < \beta$: $\widehat{G}(s)$ *has no poles in the closed right half of the complex plane and the Nyquist plot of $\widehat{G}(s)$ is entirely contained within the interior of the circle $C^*$.*

**Proof**: See the Appendix.

## 6.9   Exercises

(6.1) Very often physical systems are combined to form a new system (e.g. by adding their outputs or by cascading two systems). Because physical systems are represented using causal operators, it is important to determine whether the addition of two causal operators [with addition defined by $(A+B)x = Ax + Bx$ and the composition product (defined by $(AB)x = A(Bx)$], are again causal. With this introduction, you are asked the following questions:

  (a) Let $A : \mathcal{X}_e \to \mathcal{X}_e$ and $B : \mathcal{X}_e \to \mathcal{X}_e$ be causal operators. Show that the *sum* operator $C : \mathcal{X}_e \to \mathcal{X}_e$ defined by $C(x) = (A + B)(x)$ is also causal.

  (b) Show that the *cascade* operator $D : \mathcal{X}_e \to \mathcal{X}_e$ defined by $D(x) = (AB)(x)$ is also causal.

(6.2) Consider the following alternative definition of causality

  **Definition 6.18** *An operator $H : \mathcal{X}_e \to \mathcal{X}_e$ is said to be causal if*

  $$P_T u_1 = P_T u_2 \quad \Rightarrow \quad P_T H u_1 = P_T H u_2 \quad \forall u_1, u_2 \in \mathcal{X}_e \text{ and } \forall T \in \mathbb{R}^+ \quad (6.39)$$

  According to this definition, if the truncations of $u_1$ and $u_2$ are identical (i.e., if $u_1 = u_2, \forall t \leq T$), then the truncated *outputs* are also identical.

  You are asked to prove the following theorem, which states that the two definitions are equivalent.

  **Theorem 6.7** *Consider an operator $H : \mathcal{X}_e \to \mathcal{X}_e$. Then $H$ is causal according to Definition 6.5 if and only if it is causal according to Definition 6.18.*

(6.3) Prove the following theorem, which states that, for bounded causal operators, all of the truncations in the definition of gain may be dropped if the input space is restricted to the space $\mathcal{X}$.

**Theorem 6.8** *Consider a causal operator* $H : \mathcal{X}_e \to \mathcal{X}_e$ *satisfying* $H0 = 0$, *and assume that* $\exists \gamma^*(H) = \sup \|Hx\|/\|x\| < \infty \forall x \in \mathcal{X}$ , $x \neq 0$. *Then* $H$ *has a finite gain, according to Definition 6.7, and* $\gamma^*(H) = \gamma(H)$.

(6.4) Prove the following theorem.

**Theorem 6.9** *Let* $H_1, H_2 : \mathcal{X} \to \mathcal{X}$ *be causal bounded operator satisfying* $H_i 0 = 0$, $i = 1, 2$. *Then*

$$\gamma(H_1 H_2) \leq \gamma(H_1)\gamma(H_2) \tag{6.40}$$

(6.5) Prove the small gain theorem (Theorem 6.3) in the more general case when $\beta \neq 0$.

(6.6) Prove Theorem 6.4.

(6.7) Prove Theorem 6.5.

(6.8) In Section 6.2 we introduced system gains and the notion of finite gain stable system. We now introduce a stronger form of input–output stability:

**Definition 6.19** *A system* $H : \mathcal{X}_e \to \mathcal{X}_e$ *is said to be Lipschitz continuous, or simply continuous, if there exists a constant* $\Gamma(H) < \infty$, *called its incremental gain, satisfying*

$$\Gamma(H) = \sup \frac{\|(Hu_1)_T - (Hu_2)_T\|}{\|(u_1)_T - u_2)_T\|} \tag{6.41}$$

*where the supremum is taken over all* $u_1, u_2$ *in* $\mathcal{X}_e$ *and all* $T$ *in* $\mathbb{R}^+$ *for which* $u_{1T} \neq u_{2T}$.

Show that if $H : \mathcal{X}_e \to \mathcal{X}_e$ is Lipschitz continuous, then it is finite-gain-stable.

(6.9) Find the incremental gain of the system in Example 6.3.

(6.10) For each of the following transfer functions, find the sector $[\alpha, \beta]$ for which the closed-loop system is absolutely stable:

$$(i) \quad \widehat{H}(s) = \frac{1}{(s+1)(s+3)}, \quad (ii) \quad \widehat{H}(s) = \frac{s+1}{(s+2)(s-3)}$$

**Notes and References**

The classical input/output theory was initiated by Sandberg, [65], [66], (see also the more complete list of Sandberg papers in [21]), and Zames [97], [98]. Excellent general references for the material of this chapter are [21], [88] and [92]. Our presentation follows [21]) as well as [98].

# Chapter 7

# Input-to-State Stability

So far we have seen two different notions of stability: (1) stability in the sense of Lyapunov and (2) input–output stability. These two concepts are at opposite ends of the spectrum. On one hand, Lyapunov stability applies to the equilibrium points of unforced state space realizations. On the other hand, input–output stability deals with systems as mappings between inputs and outputs, and ignores the internal system description, which may or may not be given by a state space realization.

In this chapter we begin to close the gap between these two notions and introduce the concept of *input-to-state-stability*. We assume that systems are described by a state space realization that includes a variable input function, and discuss stability of these systems in a way to be defined.

## 7.1  Motivation

Throughout this chapter we consider the nonlinear system

$$\dot{x} = f(x, u) \tag{7.1}$$

where $f : D \times D_u \to \mathbb{R}^n$ is locally Lipschitz in $x$ and $u$. The sets $D$ and $D_u$ are defined by $D = \{x \in \mathbb{R}^n : \|x\| < r\}$, $D_u = \{u \in \mathbb{R}^m : \sup_{t>0} \|u(t)\| = \|u\|_{\mathcal{L}_\infty} < r_u\}$.

These assumptions guarantee the local existence and uniqueness of the solutions of the differential equation (7.1). We also assume that the unforced system

$$\dot{x} = f(x, 0)$$

has a uniformly asymptotically stable equilibrium point at the origin $x = 0$.

Under these conditions, the problem to be studied in this chapter is as follows. Given that in the absence of external inputs (i.e., $u = 0$) the equilibrium point $x = 0$ is asymptotically stable, does it follow that in the presence of nonzero external input either

(a) $\lim_{t\to\infty} u(t) = 0 \quad \Rightarrow \quad \lim_{t\to\infty} x(t) = 0$ ?

(b) Or perhaps that *bounded inputs* result in *bounded states* ?, specifically, $\|u_T(t)\|_{\mathcal{L}_\infty} < \delta,\ 0 \le T \le t \quad \Rightarrow \quad \sup_t \|x(t)\| < \epsilon$ ?

Drawing inspiration from the linear time invariant (LTI) case, where *all* notions of stability coincide, the answer to both questions above seems to be affirmative. Indeed, for LTI systems the solution of the state equation is well known. Given an LTI system of the form

$$\dot{x} = Ax + Bu$$

the trajectories with initial condition $x_0$ and nontrivial input $u(t)$ are given by

$$x(t) = e^{At}x_0 + \int_0^t e^{A(t-\tau)} Bu(\tau)\, d\tau \qquad (7.2)$$

If the origin is asymptotically stable, then all of the eigenvalues of $A$ have negative real parts and we have that $\|e^{At}\|$ is bounded for all $t$ and satisfies a bound of the form

$$\|e^{At}\| \le k e^{\lambda t}, \quad \lambda < 0$$

It then follows that

$$
\begin{aligned}
\|x(t)\| &\le k e^{\lambda t}\|x_0\| + \int_0^t k e^{\lambda(t-\tau)}\|B\|\|u(\tau)\|\, d\tau \\
&\le k e^{\lambda t}\|x_0\| + \frac{k\|B\|}{\lambda} \sup_{0 \le \tau \le t} \|u(t)\| \\
&= k e^{\lambda t}\|x_0\| + \frac{k\|B\|}{\lambda}\|u_T(t)\|_{\mathcal{L}_\infty}, \quad 0 \le T \le t
\end{aligned}
$$

Thus, it follows trivially that bounded inputs give rise to bounded states, and that if $\lim_{t\to\infty} u(t) = 0$, then $\lim_{t\to\infty} x(t) = 0$.

The nonlinear case is, however a lot more subtle. Indeed, it is easy to find counterexamples showing that, in general, these implications fail. To see this, consider the following simple example.

**Example 7.1** *Consider the following first-order nonlinear system:*

$$\dot{x} = -x + (x + x^3)u.$$

*Setting $u = 0$, we obtain the autonomous LTI system $\dot{x} = -x$, which clearly has an asymptotically stable equilibrium point. However, when the bounded input $u(t) = 1$ is applied, the forced system becomes $\dot{x} = x^3$, which results in an unbounded trajectory for any initial condition, however small, as can be easily verified using the graphic technique introduced in Chapter 1.* □

## 7.2 Definitions

In an attempt to rescue the notion of "bounded input–bounded state", we now introduce the concept of input-to-state stability (ISS).

**Definition 7.1** *The system (7.1) is said to be locally input-to-state-stable (ISS) if there exist a $\mathcal{KL}$ function $\beta$, a class $\mathcal{K}$ function $\gamma$ and constants $k_1, k_2 \in \mathbb{R}^+$ such that*

$$\|x(t)\| \leq \beta(\|x_0\|, t) + \gamma(\|u_T(\cdot)\|_{\mathcal{L}_\infty}), \qquad \forall t \geq 0, \quad 0 \leq T \leq t \qquad (7.3)$$

*for all $x_0 \in D$ and $u \in D_u$ satisfying : $\|x_0\| < k_1$ and $\sup_{t>0} \|u_T(t)\| = \|u_T\|_{\mathcal{L}_\infty} < k_2$, $0 \leq T \leq t$. It is said to be input-to-state stable, or globally ISS if $D = \mathbb{R}^n$, $D_u = \mathbb{R}^m$ and (7.3) is satisfied for any initial state and any bounded input $u$.*

Definition 7.1 has several implications, which we now discuss.

- *Unforced systems*: Assume that $\dot{x} = f(x, u)$ is ISS and consider the unforced system $\dot{x} = f(x, 0)$. Given that $\gamma(0) = 0$ (by virtue of the assumption that $\gamma$ is a class $\mathcal{K}$ function), we see that the response of (7.1) with initial state $x_0$ satisfies.

$$\|x(t)\| \leq \beta(\|x_0\|, t) \qquad \forall t \geq 0, \|x_0\| < k_1,$$

  which implies that the origin is uniformly asymptotically stable.

- Interpretation: For bounded inputs $u(t)$ satisfying $\|u\|_\infty < \delta$, trajectories remain bounded by the ball of radius $\beta(\|x_0\|, t) + \gamma(\delta)$, i.e.,

$$\|x(t)\| \leq \beta(\|x_0\|, t) + \gamma(\delta).$$

As $t$ increases, the term $\beta(\|x_0\|, t) \to 0$ as $t \to \infty$, and the trajectories approach the ball of radius $\gamma(\delta)$, i.e.,

$$\lim_{t \to \infty} \|x(t)\| \leq \gamma(\delta).$$

For this reason, $\gamma(\cdot)$ is called the <u>ultimate bound</u> of the system 7.1.

- *Alternative Definition:* A variation of Definition 7.1 is to replace equation (7.3) with the following equation:

$$\|x(t)\| \leq \max\{\beta(\|x_0\|, t), \gamma(\|u_T(\cdot)\|_{\mathcal{L}_\infty})\}, \quad \forall t \geq 0, \ 0 \leq T \leq t. \tag{7.4}$$

The equivalence between (7.4) and (7.3) follows from the fact that given $\beta > 0$ and $\gamma > 0$, $\max\{\beta, \gamma\} \leq \beta + \gamma \leq \{2\beta, 2\gamma\}$. On occasions, (7.4) might be preferable to (7.3), especially in the proof of some results.

It seems clear that the concept of input-to-state stability is quite different from that of stability in the sense of Lyapunov. Nevertheless, we will show in the next section that ISS can be investigated using Lyapunov-like methods. To this end, we now introduce the concept of input-to-state Lyapunov function (ISS Lyapunov function).

**Definition 7.2** *A continuously differentiable function $V : D \to \mathbb{R}$ is said to be an ISS Lyapunov function on $D$ for the system (7.1) if there exist class $\mathcal{K}$ functions $\alpha_1$, $\alpha_2$, $\alpha_3$, and $\mathcal{X}$ such that the following two conditions are satisfied:*

$$\alpha_1(\|x\|) \leq V(x(t)) \leq \alpha_2(\|x\|) \qquad \forall x \in D, \ t > 0 \tag{7.5}$$

$$\frac{\partial V(x)}{\partial x} f(x, u) \leq -\alpha_3(\|x\|) \qquad \forall x \in D, u \in D_u : \|x\| \geq \mathcal{X}(\|u\|). \tag{7.6}$$

*$V$ is said to be an ISS Lyapunov function if $D = \mathbb{R}^n$, $D_u = \mathbb{R}^m$, and $\alpha_1$, $\alpha_2$, $\alpha_3 \in \mathcal{K}_\infty$.*

**Remarks:** According to Definition 7.2, $V$ is an ISS Lyapunov function for the system (7.1) if it has the following properties:

(a) It is positive definite in $D$. Notice that according to the property of Lemma 3.1, given a positive definite function $V$, there exist class $\mathcal{K}$ functions $\alpha_1$ and $\alpha_2$ satisfying equation (7.5).

(b) It is negative definite in along the trajectories of (7.1) *whenever the trajectories are outside of the ball defined by $\|x^*\| = \mathcal{X}(\|u\|)$.*

## 7.3   Input-to-State Stability (ISS) Theorems

**Theorem 7.1** *(Local ISS Theorem) Consider the system (7.1) and let $V : D \to \mathbb{R}$ be an ISS Lyapunov function for this system. Then (7.1) is input-to-state-stable according to*

*Definition 7.1 with*

$$\gamma = \alpha_1^{-1} \circ \alpha_2 \circ \mathcal{X} \tag{7.7}$$

$$k_1 = \alpha_2^{-1}(\alpha_1(r)) \tag{7.8}$$

$$k_2 = \mathcal{X}^{-1}(\min\{k_1, \mathcal{X}(r_u)\}). \tag{7.9}$$

**Theorem 7.2** *(Global ISS Theorem) If the preceeding conditions are satisfied with $D = \mathbb{R}^n$ and $D_u = \mathbb{R}^m$, and if $\alpha_1, \alpha_2, \alpha_3 \in \mathcal{K}_\infty$, then the system (7.1) is globally input-to-state stable.*

**Proof of Theorem 7.1:** Notice first that if $u = 0$, then defining conditions (7.5) and (7.6) of the ISS Lyapunov function guarantee that the origin is asymptotically stable. Now consider a nonzero input $u$, and let

$$r_u \stackrel{def}{=} \sup_{t>0} \|u_T\| = \|u\|_{\mathcal{L}_\infty}, \quad 0 \le T \le t.$$

Also define

$$c \stackrel{def}{=} \alpha_2(\mathcal{X}(r_u))$$

$$\Omega_c \stackrel{def}{=} \{x \in D : V(x) \le c\}$$

and notice that $\Omega_c$, so defined, is *bounded* and *closed* (i.e., $\Omega_c$ is a compact set), and thus it includes its boundary, denoted $\partial(\Omega_c)$. It then follows by the right-hand side of (7.5) that the open set of points

$$\{x \in \mathbb{R}^n : \|x\| < \mathcal{X}(r_u) \le c\} \subset \Omega_c \subset D.$$

Moreover, this also implies that $\|x\| \ge \mathcal{X}(\|u(t)\|)$ at each point $x$ in the boundary of $\Omega_c$. Notice also that condition (7.6) implies that $\dot{V}(x(t)) < 0 \; \forall t > 0$ whenever $x(t) \notin \Omega_c$. We now consider two cases of interest: (1) $x_0$ is inside $\Omega_c$ and (2) $x_0$ is outside $\Omega_c$.

**Case (1)** $(x_0 \in \Omega_c)$: The previous argument shows that the closed set $\Omega_c$ is surrounded by $\partial(\Omega_c)$ along which $\dot{V}(x(t))$ is negative definite. Therefore, whenever $x_0 \in \Omega_c$, $x(t)$ is locked inside $\Omega_c$ for all $t > 0$. Trajectories $x(t)$ are such that

$$\alpha_1(\|x\|) \le V(x(t))$$

$$\Rightarrow \quad \|x(t)\| \le \alpha_1^{-1}(V(x(t))) \le \alpha_1^{-1}(c) = \alpha_1^{-1}(\alpha_2(\mathcal{X}(r_u)))$$

defining

$$\gamma \stackrel{def}{=} \alpha_1^{-1} \circ \alpha_2 \circ \mathcal{X}(\|u_T\|_{\mathcal{L}_\infty}) \quad 0 \le T \le t$$

we conclude that whenever $x_0 \in \Omega_0$,

$$\|x(t)\|_\infty \leq \gamma(\|u_T\|_{\mathcal{L}_\infty}) \qquad \forall t \geq 0, \quad 0 \leq T \leq t. \tag{7.10}$$

**Case (ii)** $(x_0 \notin \Omega_c)$: Assume that $x_0 \in \Omega_c$, or equivalently, $V(x_0) > c$. By condition (7.6), $\|x\| \geq \mathcal{X}(\|u\|)$ and thus $\dot{V}(x(t)) < 0$. It then follows that for some $t_1 > 0$ we must have

$$
\begin{aligned}
V(x(t)) &> 0 \qquad \text{for } 0 \leq t < t_1 \\
V(x(t_1)) &= c.
\end{aligned}
$$

But this implies that when $t = t_1$, then $x(t) \in \partial(\Omega_c)$, and we are back in case (1). This argument also shows that $x(t)$ is bounded and that there exist a class $\mathcal{KL}$ function $\beta$ such that

$$\|x(t)\|_\infty \leq \beta(\|x_o\|, t) \qquad \text{for } 0 \leq t \leq t_1 \tag{7.11}$$

Combining (7.10) and (7.11), we obtain

$$
\begin{aligned}
\|x(t)\|_\infty &\leq \gamma(\|u\|_\infty) \qquad \text{for } t \geq t_1 \\
\|x(t)\|_\infty &\leq \beta(\|x_o\|, t) \quad \text{for } 0 \leq t < t_1
\end{aligned}
$$

and thus we conclude that

$$\|x(t)\| \leq \max\{\beta(\|x_o\|, t), \gamma(\|u\|_\infty)\} \quad \forall t \geq 0$$

or that

$$\|x(t)\| \leq \beta(\|x_o\|, t) + \gamma(\|u\|_\infty) \quad \forall t \geq 0.$$

To complete the proof, it remains to show that $k_1$ and $k_2$ are given by (7.8) and (7.9), respectively. To see (7.8), notice that, by definition

$$
\begin{aligned}
c &= \alpha_2(\mathcal{X}(r_u)), \text{ and} \\
\Omega_c &= \{x \in D : V(x) \leq c\}.
\end{aligned}
$$

Assuming that $x_0 \notin \Omega_c$ we must have that $V(x) > c$. Thus

$$
\begin{aligned}
\alpha_1(\|x\|) &\leq V(x) \leq \alpha_2(\|x\|) \\
\alpha_1(r) &\leq V(x_0) \leq \alpha_2(\|x_0\|)
\end{aligned}
$$

and

$$\|x_0\| \leq \alpha_1^{-1}(\alpha_1(r)).$$

Thus

$$k_1 \overset{def}{=} \|x_0\| \leq \alpha_1^{-1}(\alpha_1(r))$$

which is (7.8). To see (7.9) notice that

$$\|x(t)\| \leq \max\{\beta(\|x_0\|, t), \gamma(\|u_t\|_\infty)\} \quad \forall t \geq 0, \quad 0 \leq T \leq t$$
$$\Rightarrow \quad \mathcal{X}(k_2) \leq \|x\| \leq \min\{k_1, \mathcal{X}(r_u))\}.$$

Thus

$$k_2 = \mathcal{X}^{-1}(\min\{k_1, \mathcal{X}(r_u)\}).$$

$\square$

### 7.3.1   Examples

We now present several examples in which we investigate the ISS property of a system. It should be mentioned that application of Theorems 7.1 and 7.2 is not an easy task, something that should not come as a surprise since even proving asymptotic stability has proved to be quite a headache. Thus, our examples are deliberately simple and consist of various alterations of the same system.

**Example 7.2** *Consider the following system:*

$$\dot{x} = -ax^3 + u \quad a > 0.$$

*To check for input-to-state stability, we propose the ISS Lyapunov function candidate* $V(x) = \frac{1}{2}x^2$. *This function is positive definite and satisfies (7.5) with* $\alpha_1(\|x\|) = \alpha_2(\|x\|) = \frac{1}{2}x^2$. *We have*

$$\dot{V} = -x(ax^3 - u). \tag{7.12}$$

*We need to find* $\alpha_3(\cdot)$ *and* $\mathcal{X}(\cdot) \in \mathcal{K}$ *such that* $\dot{V}(x) \leq -\alpha_3(\|x\|)$, *whenever* $\|x\| \geq \mathcal{X}(\|u\|)$. *To this end we proceed by adding and subtracting* $a\theta x^4$ *to the right-hand side of (7.12), where the parameter* $\theta$ *is such that* $0 < \theta < 1$

$$\begin{aligned}
\dot{V} &= -ax^4 + a\theta x^4 - a\theta x^4 + xu \\
&= -a(1-\theta)x^4 - x(a\theta x^3 - u) \\
&\leq -a(1-\theta)x^4 = \alpha_3(\|x\|)
\end{aligned}$$

*provided that*

$$x(a\theta x^3 - u) > 0.$$

*This will be the case, provided that*

$$a\theta|x|^3 > |u|$$

*or, equivalently*

$$|x| > \left( \frac{|u|}{a\theta} \right)^{1/3}$$

*It follows that the system is globally input-to-state-stable with* $\gamma(u) = \left( \frac{|u|}{a\theta} \right)^{1/3}$.

□

**Example 7.3** *Now consider the following system, which is a slightly modified version of the one in Example 7.2:*

$$\dot{x} = -ax^3 + x^2 u \qquad a > 0$$

*Using the same ISS Lyapunov function candidate used in Example 7.2, we have that*

$$
\begin{aligned}
\dot{V} &= -ax^4 + x^3 u \\
&= -ax^4 + a\theta x^4 - a\theta x^4 + x^3 u \quad 0 < \theta < 1 \\
&= -a(1-\theta)x^4 - x^3(a\theta x - u) \\
&\leq -a(1-\theta)x^4, \quad provided
\end{aligned}
$$

$$
\begin{aligned}
x^3(a\theta x - u) &> 0 \quad or, \\
|x| &> \frac{|u|}{a\theta}.
\end{aligned}
$$

*Thus, the system is globally input-to-state stable with* $\gamma(u) = \frac{|u|}{a\theta}$.

□

**Example 7.4** *Now consider the following system, which is yet another modified version of the one in Examples 7.2 and 7.3:*

$$\dot{x} = -ax^3 + x(1+x^2)u \qquad a > 0$$

*Using the same ISS Lyapunov function candidate in Example 7.2 we have that*

$$
\begin{aligned}
\dot{V} &= -ax^4 + x(1+x^2)u \\
&= -ax^4 + a\theta x^4 - a\theta x^4 + x(1+x^2)u \\
&= -a(1-\theta)x^4 - x[a\theta x^3 - (1+x^2)u] \\
&\leq -a(1-\theta)x^4, \quad provided
\end{aligned}
$$

$$x[a\theta x^3 - (1 + x^2)u] > 0. \tag{7.13}$$

*Now assume now that the sets $D$ and $D_u$ are the following: $D = \{x \in \mathbb{R}^n : \|x\| < r\}$, $D_u = \mathbb{R}$. Equation (7.13) is satisfied if*

$$|x| > \left( \frac{(1+r^2)|u|}{a\theta} \right)^{1/3}.$$

*Therefore, the system is input-to-state stable with $k_1 = r$:*

$$\begin{aligned}
\gamma(u) &= \left( \frac{(1+r^2)|u|}{a\theta} \right)^{1/3}, \\
k_2 &= \mathcal{X}^{-1}[\min\{k_1, \mathcal{X}(r_u)\}] \\
&= \mathcal{X}^{-1}(k_1) = \mathcal{X}^{-1}(r) \\
\Rightarrow k_2 &= \frac{a\theta r}{(1+r^2)}.
\end{aligned}$$

□

## 7.4   Input-to-State Stability Revisited

In this section we provide several remarks and further results related to input-to-state stability. We begin by stating that, as with all the Lyapunov theorems in chapters 3 and 4, Theorems 7.1 and 7.2 state that the existence of an ISS Lyapunov function is a sufficient condition for input-to-state stability. As with all the Lyapunov theorems in Chapters 3 and 4, there is a converse result that guarantees the existence of an ISS Lyapunov function whenever a system is input-to-state-stable. For completeness, we now state this theorem without proof.

**Theorem 7.3** *The system (7.1) is input-to-state-stable if and only if there exist an ISS Lyapunov function $V : D \to \mathbb{R}$ satisfying the conditions of Theorems 7.1 or 7.2.*

**Proof:** See Reference [73].

Theorems 7.4 and 7.5 are important in that they provide conditions for local and global input-to-state stability, respectively, using only "classical" Lyapunov stability theory.

**Theorem 7.4** *Consider the system (7.1). Assume that the origin is an asymptotically stable equilibrium point for the autonomous system $\dot{x} = f(x, 0)$, and that the function $f(x, u)$ is continuously differentiable. Under these conditions (7.1) is locally input-to-state-stable.*

**Theorem 7.5** *Consider the system (7.1). Assume that the origin is an exponentially stable equilibrium point for the autonomous system $\dot{x} = f(x, 0)$, and that the function $f(x, u)$ is continuously differentiable and globally Lipschitz in $(x, u)$. Under these conditions (7.1) is input-to-state-stable.*

**Proof of theorems 7.4 and 7.5**: See the Appendix. □

We now state and prove the main result of this section. Theorem 7.6 gives an alternative characterization of ISS Lyapunov functions that will be useful in later sections.

**Theorem 7.6** *A continuous function $V : D \rightarrow \mathbb{R}$ is an ISS Lyapunov function on $D$ for the system (7.1) if and only if there exist class $\mathcal{K}$ functions $\alpha_1$, $\alpha_2$, $\alpha_3$, and $\sigma$ such that the following two conditions are satisfied:*

$$\alpha_1(\|x\|) \leq V(x(t)) \leq \alpha_2(\|x\|) \qquad\qquad \forall x \in D,\ t > 0 \qquad\qquad (7.14)$$

$$\frac{\partial V(x)}{\partial x} f(x, u) \leq -\alpha_3(\|x\|) + \sigma(\|u\|) \qquad\qquad \forall x \in D, u \in D_u \qquad (7.15)$$

*$V$ is an ISS Lyapunov function if $D = \mathbb{R}^n$, $D_u = \mathbb{R}^m$, and $\alpha_1, \alpha_2, \alpha_3$, and $\sigma \in \mathcal{K}_\infty$.*

**Proof:** Assume first that (7.15) is satisfied. Then we have that

$$
\begin{aligned}
\frac{\partial V}{\partial x} f(x, u) &\leq -\alpha_3(\|x\|) + \sigma(\|u\|) & \forall x \in D, \forall u \in D_u \\
&\leq -(1 - \theta)\alpha_3(\|x\|) - \theta\alpha_3(\|x\|) + \sigma(\|u\|) & \theta < 1 \\
&\leq -(1 - \theta)\alpha_3(\|x\|)
\end{aligned}
$$

$$\forall \|x\| \geq \alpha_3^{-1}\left[\frac{\sigma(\|u\|)}{\theta}\right]$$

which shows that (7.6) is satisfied. For the converse, assume that (7.6) holds:

$$\frac{\partial V}{\partial x} f(x, u) \leq -\alpha_3(\|x\|) \qquad \forall x \in D, u \in D_u : \|x\| \geq \mathcal{X}(\|u\|).$$

To see that (7.15) is satisfied, we consider two different scenarios:

(a) $\|x\| \geq \mathcal{X}(\|u\|)$: This case is trivial. Indeed, under these conditions, (7.6) holds for any $\sigma(\cdot)$.

(b) $\|x\| < \mathcal{X}(\|u\|)$: Define

$$\phi(r) = \max \left[ \frac{\partial V}{\partial x} f(x, u) + \alpha_3(\mathcal{X}(\|u\|)) \right] \qquad \|u\| = r, \quad \|x\| \leq \mathcal{X}(r)$$

then

$$\frac{\partial V}{\partial x} f(x, u) \leq -\alpha_3(\|x\|) + \phi(r).$$

Now, defining

$$\overline{\sigma} = \max\{0, \phi(r)\}$$

we have that $\overline{\sigma}$, so defined, satisfies the following: (i) $\overline{\sigma}(0) = 0$, (ii) it is nonnegative, and (iii) it is continuous. It may not be in the class $\mathcal{K}$ since it may not be strictly increasing. We can always, however, find $\sigma \in \mathcal{K}$ such that $\sigma(r) \geq \overline{\sigma}$.

This completes the proof.                                                             □

The only difference between Theorem 7.6 and definition 7.2 is that condition (7.6) in Definition 7.2 has been replaced by condition (7.15). For reasons that will become clear in Chapter 9, (7.15) is called the *dissipation inequality*.

**Remarks:** According to Theorems 7.6 and 7.3, we can claim that the system (7.1) is input-to-state-stable if and only if there exists a continuous function $V : D \to \mathbb{R}$ satisfying conditions (7.14) and (7.15). Inequality (7.15) permits a more transparent view of the concept and implications of input-to-state stability in terms of ISS Lyapunov functions. To see this, notice that, given $r_u > 0$, there exist points $x \in \mathbb{R}^n$ such that

$$\alpha_3(\|x\|) = \sigma(r_u).$$

This implies that $\exists d \in \mathbb{R}^+$ such that

$$\alpha_3(d) = \sigma(r_u), \quad \text{or}$$

$$d = \alpha^{-1}(\sigma(r_u)).$$

Denoting $B_d = \{x \in \mathbb{R}^n : \|x\| \leq d\}$, we have that for any $\|x\| > d$ and any $u : \|u\|_{\mathcal{L}_\infty} < r_u$:

$$\frac{\partial V}{\partial x} f(x, u) \leq -\alpha(\|x\|) + \sigma(\|u\|)$$
$$\leq -\alpha(\|d\|) + \sigma(\|u\|_{\mathcal{L}_\infty}).$$

This means that the trajectory $x(t)$ resulting from an input $u(t) : \|u\|_{\mathcal{L}_\infty} < r_u$ will eventually (i.e., at some $t = t^*$) enter the region

$$\Omega_d = \max_{\|x\| \leq d} V(x).$$

Once inside this region, the trajectory is trapped inside $\Omega_d$, because of the condition on $\dot{V}$.

According to the discussion above, the region $\Omega_d$ seems to depend on the composition of $\alpha^{-1}(\cdot)$ and $\sigma(\cdot)$. It would then appear that it is the composition of these two functions that determines the correspondence between a bound on the input function $u$ and a bound on the state $x$. The functions $\alpha(\cdot)$ and $\sigma(\cdot)$ constitute a pair $[\alpha(\cdot), \sigma(\cdot)]$ referred to as an ISS pair for the system (7.1). The fact that $\Omega_d$ depends on the composition of $\alpha^{-1}(\cdot)$ and $\sigma(\cdot)$ suggests that, for a given system, the pair $[\alpha(\cdot), \sigma(\cdot)]$ is perhaps nonunique. Our next theorem shows that this is in fact the case.

In the following theorem we consider the system (7.1) and we assume that it is globally input-to-state-stable with an ISS pair $[\alpha, \sigma]$.

$$\underline{\alpha}(\|x\|) \;\leq\; V(x) \;\leq\; \overline{\alpha}(\|x\|) \qquad\qquad \forall x \in \mathbb{R}^n, \qquad\qquad (7.16)$$
$$\nabla V(x) \cdot f(x, u) \;\leq\; -\alpha(\|x\|) + \sigma(\|u\|) \qquad \forall x \in \mathbb{R}^n, u \in \mathbb{R}^m \qquad (7.17)$$

for some $\alpha, \underline{\alpha}, \overline{\alpha} \in \mathcal{K}_\infty$, and $\sigma \in \mathcal{K}$.

We will need the following notation. Given functions $x(\cdot), y(\cdot) : \mathbb{R} \to \mathbb{R}$, we say that

$$x(s) = O(y(s)) \qquad \text{as } s \to \infty+$$

if

$$\lim_{s \to \infty} \frac{|x(s)|}{|y(s)|} < \infty.$$

Similarly

$$x(s) = O(y(s)) \quad \cdot \text{ as } s \to 0+$$

if

$$\lim_{s \to 0} \frac{|x(s)|}{|y(s)|} < \infty.$$

**Theorem 7.7** *Let $(\alpha, \sigma)$ be a supply pair for the system (7.1). Assume that $\tilde{\sigma}$ is a $\mathcal{K}_\infty$ function satisfying $\sigma(r) = O(\tilde{\sigma}(r))$ as $r \to \infty+$. Then, there exist $\tilde{\alpha} \in \mathcal{K}_\infty$ such that $(\tilde{\alpha}, \tilde{\sigma})$ is a supply pair.*

**Theorem 7.8** *Let $(\alpha, \sigma)$ be a supply pair for the system (7.1). Assume that $\tilde{\alpha}$ is a $\mathcal{K}_\infty$ function satisfying $\alpha(r) = O(\tilde{\alpha}(r))$ as $r \to 0+$. Then, there exist $\tilde{\sigma} \in \mathcal{K}_\infty$ such that $(\tilde{\alpha}, \tilde{\sigma})$ is a supply pair.*

**Proof of theorems 7.7 and 7.8:** See the Appendix.

**Remarks:** Theorems 7.7 and 7.8 have theoretical importance and will be used in the next section is connection with the stability of cascade connections if ISS systems. The proof of theorems 7.7 and 7.8 shows how to construct the new ISS pairs using only the bounds $\underline{\alpha}$ and $\overline{\alpha}$, but not $V$ itself.

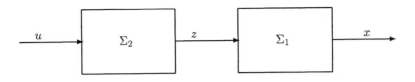

Figure 7.1: Cascade connection of ISS systems.

## 7.5  Cascade-Connected Systems

Throughout this section we consider the composite system shown in Figure 7.1, where $\Sigma_1$ and $\Sigma_2$ are given by

$$\Sigma_1 : \quad \dot{x} = f(x, z) \tag{7.18}$$
$$\Sigma_2 : \quad \dot{z} = g(z, u) \tag{7.19}$$

where $\Sigma_2$ is the system with input $u$ and state $z$. The state of $\Sigma_2$ serves as input to the system $\Sigma_1$.

In the following lemma we assume that both systems $\Sigma_1$ and $\Sigma_2$ are input-to-state-stable with ISS pairs $[\alpha_1, \sigma_1]$ and $[\alpha_2, \sigma_2]$, respectively. This means that there exist positive definite functions $V_1$ and $V_2$ such that

$$\nabla V_1 \, f(x, z) \leq -\alpha_1(\|x\|) + \sigma_1(\|z\|) \tag{7.20}$$
$$\nabla V_2 \, g(z, u) \leq -\alpha_2(\|z\|) + \sigma_2(\|u\|). \tag{7.21}$$

The lemma follows our discussion at the end of the previous section and guarantees the existence of alternative ISS pairs $[\tilde{\alpha}_1, \tilde{\sigma}_1]$ and $[\tilde{\alpha}_2, \tilde{\sigma}_2]$ for the two systems. As it turns out, the new ISS-pairs will be useful in the proof of further results.

**Lemma 7.1**  *Given the systems $\Sigma_1$ and $\Sigma_2$, we have that*

*(i) Defining*

$$\tilde{\alpha}_2 = \begin{cases} \alpha_2(s) & \text{for } s \text{ ``small''} \\ \sigma_2(s) & \text{for } s \text{ ``large''} \end{cases}$$

*then there exist $\tilde{\sigma}_2$ such that $(\tilde{\alpha}_2, \tilde{\sigma}_2)$ is an ISS pair for the system $\Sigma_2$.*

*(ii) Defining*

$$\tilde{\sigma}_1(s) = \frac{1}{2}\tilde{\alpha}_2$$

*there exist* $\tilde{\sigma}_1 : [\tilde{\alpha}_1, \tilde{\sigma}_1] = [\tilde{\alpha}_1, \frac{1}{2}\tilde{\alpha}_2]$ *is an ISS pair for the system* $\Sigma_1$.

**Proof**: A direct application of Theorems 7.7 and 7.8.                              □

We now state and prove the main result of this section.

**Theorem 7.9** *Consider the cascade interconnection of the systems* $\Sigma_1$ *and* $\Sigma_2$. *If both systems are input-to-state-stable, then the composite system* $\Sigma$

$$\Sigma : u \to \begin{bmatrix} x \\ z \end{bmatrix}$$

*is input-to-state-stable.*

**Proof**: By (7.20)–(7.21) and Lemma 7.1, the functions $V_1$ and $V_2$ satisfy

$$\nabla V_1 \, f(x,z) \;\le\; -\tilde{\alpha}_1(\|x\|) + \frac{1}{2}\tilde{\alpha}_2(\|z\|)$$
$$\nabla V_2 \, g(z,u) \;\le\; -\tilde{\alpha}_2(\|z\|) + \tilde{\sigma}_2(\|u\|).$$

Define the ISS Lyapunov function candidate

$$V = V_1 + V_2$$

for the composite system. We have

$$
\begin{aligned}
\dot{V}((x,z),u) &= \nabla V_1 \, f(x,z) + \nabla V_2 \, f(z,u) \\
&= -\tilde{\alpha}_1(\|x\|) - \frac{1}{2}\tilde{\alpha}_2(\|z\|) + \tilde{\sigma}_2(\|u\|).
\end{aligned}
$$

It the follows that $V$ is an ISS Lyapunov function for the composite system, and the theorem is proved.                                                                                      □

This theorem is somewhat obvious and can be proved in several ways (see exercise 7.3). As the reader might have guessed, a local version of this result can also be proved. For completeness, we now state this result without proof.

**Theorem 7.10** *Consider the cascade interconnection of the systems* $\Sigma_1$ *and* $\Sigma_2$. *If both systems are locally input-to-state-stable, then the composite system* $\Sigma$

$$\Sigma : u \to \begin{bmatrix} x \\ z \end{bmatrix}$$

*is locally input-to-state-stable.*

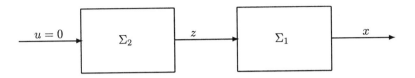

Figure 7.2: Cascade connection of ISS systems with input $u = 0$.

The following two corollaries, which are direct consequences of Theorems 7.9 and 7.10 are also important and perhaps less obvious. Here we consider the following special case of the interconnection of Figure 7.1, shown in Figure 2:

$$\Sigma_1 : \quad \dot{x} = f(x, z) \qquad (7.22)$$
$$\Sigma_2 : \quad \dot{z} = g(z) = g(z, 0) \qquad (7.23)$$

and study the Lyapunov stability of the origin of the interconnected system with state

$$\begin{bmatrix} x \\ z \end{bmatrix}$$

**Corollary 7.1** *If the system $\Sigma_1$ with input $z$ is locally input-to-state-stable and the origin $x = 0$ of the system $\Sigma_2$ is asymptotically stable, then the origin of the interconnected system (7.22)–(7.23)*

$$\begin{bmatrix} x \\ z \end{bmatrix} = \begin{bmatrix} 0 \\ 0 \end{bmatrix}$$

*is locally asymptotically stable.*

**Corollary 7.2** *Under the conditions of Corollary 7.1, if $\Sigma_1$ is input-to-state-stable and the origin $x = 0$ of the system $\Sigma_2$ is globally asymptotically stable, then the origin of the interconnected system (7.22)–(7.23).*

$$\begin{bmatrix} x \\ z \end{bmatrix} = \begin{bmatrix} 0 \\ 0 \end{bmatrix}$$

*is globally asymptotically stable.*

**Proof of Corollaries 7.1 and 7.2:** The proofs of both corollaries follow from the fact that, under the assumptions, the system $\Sigma_2$ is trivially (locally or globally) input-to-state-stable, and then so is the interconnection by application of Theorem 7.9 or 7.10. □

## 7.6   Exercises

(7.1) Consider the following 2-input system ([70]):

$$\dot{x} = -x^3 + x^2 u_1 - x u_2 + u_1 u_2$$

Is it input-to-state stable?

(7.2) Sketch the proof of theorem 7.5.

(7.3) Provide an alternative proof of Theorem 7.9, using only the definition of input-to-state-stability, Definition 7.1.

(7.4) Sketch a proof of Theorem 7.10.

(7.5) Consider the following system:

$$\begin{cases} \dot{x}_1 & = & -x_1 + x_1^2 x_2 \\ \dot{x}_2 & = & -x_2 + x_1 x_2^2 + u \end{cases}$$

   (i) Is it locally input-to-state-stable?
   (ii) Is it input-to-state-stable?

(7.6) Consider the following system:

$$\begin{cases} \dot{x}_1 & = & -x_1 - x_2 \\ \dot{x}_2 & = & x_1 - x_2^3 + u \end{cases}$$

   (i) Is it locally input-to-state-stable?
   (ii) Is it input-to-state-stable?

(7.7) Consider the following cascade connection of systems

$$\begin{cases} \dot{x} & = & -x^3 + x^2 u \\ \dot{z} & = & -z^3 + z(1 + z^2)x \end{cases}$$

   (i) Is it locally input-to-state-stable?
   (ii) Is it input-to-state-stable?

(7.8) Consider the following cascade connection of systems:

$$\begin{cases} \dot{x} & = & -x^3 + x^2 u_1 - x u_2 + u_1 u_2 \\ \dot{z} & = & -z^3 + z^2 x \end{cases}$$

(i) Is it locally input-to-state-stable?

(ii) Is it input-to-state-stable?

(7.9) Consider the following cascade connection of systems:

$$\begin{cases} \dot{x}_1 &= -x_1 + 2x_2 + u_1 \\ \dot{x}_2 &= -x_2 + u_2 \\ \dot{z} &= -z^3 + z^2 x_1 - zx_2 + x_1 x_2 \end{cases}$$

(i) Is it locally input-to-state-stable?

(ii) Is it input-to-state-stable?

## Notes and References

The concept of input-to-state stability, as presented here, was introduced by Sontag [69]. Theorems 7.3 and 7.6 were taken from reference [73]. The literature on input-to-state stability is now very extensive. See also References [70], [74], [71], and [72] for a thorough introduction to the subject containing the fundamental results. See also chapter 10 of Reference [37] for a good survey of results in this area. ISS pairs were introduced in Reference [75]. Section 7.5 on cascade connection of ISS systems, as well as Theorems 7.7 and 7.8 are based on this reference.

# Chapter 8

# Passivity

The objective of this chapter is to introduce the concept of passivity and to present some of the stability results that can be obtained using this framework. Throughout this chapter we focus on the classical input–output definition. State space realizations are considered in Chapter 9 in the context of the theory of dissipative systems. As with the small gain theorem, we look for open-loop conditions for closed loop stability of feedback interconnections.

## 8.1   Power and Energy: Passive Systems

Before we introduce the notion of passivity for abstract systems, it is convenient to motivate this concept with some examples from circuit theory. We begin by recalling from basic physics that power is the time rate at which energy is absorbed or spent,

$$p(t) = \frac{\mathrm{d}w(t)}{\mathrm{d}t} \tag{8.1}$$

where

$$
\begin{aligned}
p(\cdot) &: \quad \text{power} \\
w(\cdot) &: \quad \text{energy} \\
t &: \quad \text{time}
\end{aligned}
$$

Then

$$w(t) = \int_{t_0}^{t} p(t)\,\mathrm{d}t. \tag{8.2}$$

Now consider a basic circuit element, represented in Figure 8.1 using a black box.

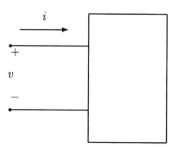

Figure 8.1: Passive network.

In Figure 8.1 the voltage across the terminals of the box is denoted by $v$, and the current in the circuit element is denoted by $i$. The assignment of the reference polarity for voltage, and reference direction for current is completely arbitrary. We have

$$p(t) = v(t)i(t) \tag{8.3}$$

thus, the energy *absorbed* by the circuit at time "$t$" is

$$w(t) = \int_{-\infty}^{t} v(t)i(t)\ dt\ =\ \int_{-\infty}^{0} v(t)i(t)\ dt + \int_{0}^{t} v(t)i(t)\ dt. \tag{8.4}$$

The first term on the right hand side of equation (8.4) represents the effect of initial conditions different from zero in the circuit elements. With the indicated sign convention, we have

(i) If $w(t) > 0$, the box absorbs energy (this is the case, for example, for a resistor).

(ii) If $w(t) < 0$, the box delivers energy (this is the case, for example, for a battery, with negative voltage with respect to the polarity indicated in Figure 8.1).

In circuit theory, elements that do not generate their own energy are called *passive*, i.e., a circuit element is passive if

$$\int_{-\infty}^{t} v(t)i(t)\ dt \geq 0. \tag{8.5}$$

Resistors, capacitors and inductors indeed satisfy this condition, and are therefore called *passive elements*. Passive networks, in general, are *well behaved*, in an admittedly ambiguous

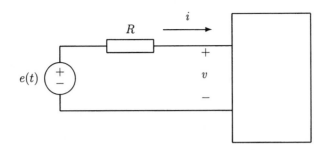

Figure 8.2: Passive network.

sense. It is not straightforward to capture the notion of *good behavior* within the context of a theory of networks, or *systems* in a more general sense. Stability, in its many forms, is a concept that has been used to describe a desirable property of a physical system, and it is intended to capture precisely the notion of a system that is well behaved, in a certain precise sense. If the notion of passivity in networks is to be of any productive use, then we should be able to infer some general statements about the behavior of a passive network.

To study this proposition, we consider the circuit shown in Figure 8.2, where we assume that the black box contains a passive (linear or not) circuit element. Assuming that the network is initially relaxed, and using Kirchhoff voltage law, we have

$$e(t) = i(t)R + v(t)$$

Assume now that the electromotive force (emf) source $e(\cdot)$ is such that

$$\int_0^T e^2(t) \, dt < \infty$$

we have,

$$\int_0^T e^2(t) \, dt = \int_0^T (i(t)R + v(t))^2 \, dt$$

$$= R^2 \int_0^T i^2(t) \, dt + 2R \int_0^T i(t)v(t) \, dt + \int_0^T v^2(t) \, dt$$

and since the black box is passive, $\int_0^T i(t)v(t) \, dt > 0$. It follows that

$$\int_0^T e^2(t) \, dt \geq R^2 \int_0^T i^2(t) \, dt + \int_0^T v^2(t) \, dt.$$

Moreover, since the applied voltage is such that $\int_0^\infty e^2(t)\,dt < \infty$, we can take limits as $T \to \infty$ on both sides of this inequality, and we have

$$R^2 \int_0^\infty i(t)^2\,dt + \int_0^\infty v^2(t)\,dt \le \int_0^\infty e^2(t)\,dt < \infty$$

which implies that both $i$ and $v$ have *finite energy*. This, in turn, implies that the energy in these two quantities can be controlled from the input source $e(\cdot)$, and in this sense we can say that the network is *well behaved*. In the next section, we formalize these ideas in the context of the theory of *input–output systems* and generalize these concepts to more general classes of systems. In particular, we want to draw conclusions about the feedback interconnection of systems based on the properties of the individual components.

## 8.2   Definitions

Before we can define the concept of passivity and study some of its properties we need to introduce our notation and lay down the mathematical machinery. The essential tool needed in the passivity definition is that of an *inner product space*.

**Definition 8.1** *A real vector space $\mathcal{X}$ is said to be a real inner product space, if for every 2 vectors $x, y \in \mathcal{X}$, there exists a real number $\langle x, y \rangle$ that satisfies the following properties:*

(i) $\langle x, y \rangle = \langle y, x \rangle$.

(ii) $\langle x + y, z \rangle = \langle x, z \rangle + \langle y + z \rangle \quad \forall x, y, z \in \mathcal{X}$.

(iii) $\langle \alpha x, y \rangle = \alpha \langle y, x \rangle \quad \forall x, y \in \mathcal{X}, \forall \alpha \in \mathbb{R}$.

(iv) $\langle x, x \rangle \ge 0$.

(v) $\langle x, x \rangle = 0$ *if and only if $x = 0$.*

The function $\langle \cdot, \cdot \rangle : \mathcal{X} \times \mathcal{X} \to \mathbb{R}$ is called the *inner product* of the space $\mathcal{X}$. If the space $\mathcal{X}$ is complete, then the inner product space is said to be a *Hilbert space*. Using these properties, we can define a norm for each element of the space $\mathcal{X}$ as follows:

$$\|x\|_{\mathcal{X}}^2 = \langle x, x \rangle.$$

An important property of inner product space is the so-called Schwarz inequality:

$$|\langle x, y \rangle| \le \|x\|_{\mathcal{X}}\ \|y\|_{\mathcal{X}} \quad \forall x, y \in \mathcal{X}. \tag{8.6}$$

Throughout the rest of this chapter we will assume that $\mathcal{X}$ is a real *inner product space*.

**Example 8.1** *Let $\mathcal{X}$ be $R^n$. Then the usual "dot product" in $\mathbb{R}^n$, defined by*

$$x \cdot y = x^T y = x_1 y_1 + x_2 y_2 + \cdots + x_n y_n$$

*defines an inner product in $\mathbb{R}^n$. It is straightforward to verify that defining properties (i)–(v) are satisfied.* □

For the most part, our attention will be centered on continuous-time systems. Virtually all of the literature dealing with this type of system makes use of the following inner product

$$\langle x, y \rangle = \int_0^\infty x(t) \cdot y(t) \, dt \tag{8.7}$$

where $x \cdot y$ indicates the usual dot product in $\mathbb{R}^n$. This inner product is usually referred to as the *natural* inner product in $\mathcal{L}_2$. Indeed, with this inner product, we have that $\mathcal{X} = \mathcal{L}_2$, and moreover

$$\|x\|_{\mathcal{L}_2}^2 = \langle x, x \rangle = \int_0^\infty \|x(t)\|_2^2 \, dt. \tag{8.8}$$

We have chosen, however, to state our definitions in more general terms, given that our discussion will not be restricted to continuous-time systems. Notice also that, even for continuous time systems, there is no a priori reason to assume that this is the only interesting inner product that one can find.

In the sequel, we will need the extension of the space $\mathcal{X}$ (defined as usual as the space of all functions whose truncation belongs to $\mathcal{X}$), and assume that the inner product satisfies the following:

$$\langle x_T, y \rangle = \langle x, y_T \rangle = \langle x_T, y_T \rangle \stackrel{def}{=} \langle x, y \rangle_T. \tag{8.9}$$

**Example 8.2** : *Let $\mathcal{X} = \mathcal{L}_2$, the space of finite energy functions:*

$$\mathcal{X} = \{x : \mathbb{R} \to \mathbb{R}, \text{and satisfy}\}$$

$$\|x\|_{\mathcal{L}_2}^2 = \langle x, x \rangle = \int_0^\infty x^2(t) \, dt < \infty$$

*Thus, $\mathcal{X}_e = \mathcal{L}_{2e}$ is the space of all functions whose truncation $x_T$ belongs to $\mathcal{L}_2$, regardless of whether $x(t)$ itself belongs to $\mathcal{L}_2$. For instance, the function $x(t) = e^t$ belongs to $\mathcal{L}_{2e}$ even though it is not in $\mathcal{L}_2$.* □

**Definition 8.2** : *(Passivity) A system $H : \mathcal{X}_e \to \mathcal{X}_e$ is said to be passive if*

$$\langle u, Hu \rangle_T \geq \beta \qquad \forall u \in \mathcal{X}_e, \forall T \in \mathbb{R}^+. \tag{8.10}$$

**Definition 8.3** : *(Strict Passivity)*  *A system $H : \mathcal{X}_e \to \mathcal{X}_e$ is said to be strictly passive if there exists $\delta > 0$ such that*

$$\langle u, Hu \rangle_T \geq \delta \|u_T\|_{\mathcal{X}}^2 + \beta \qquad \forall u \in \mathcal{X}_e, \forall T \in \mathbb{R}^+. \qquad (8.11)$$

The constant $\beta$ in Definitions 8.2 and 8.3 is a bias term included to account for the possible effect of energy initially stored in the system at $t = 0$. Definition 8.2 states that only a finite amount of energy, initially stored at time $t = 0$, can be extracted from a passive system. To emphasize these ideas, we go back to our network example:

**Example 8.3** *Consider again the network of Figure 8.1.  To analyze this network as an abstract system with input $u$ and output $y = Hu$, we define*

$$u = v(t)$$
$$y = Hu = i(t).$$

*According to Definition 8.2, the network is passive if and only if*

$$\langle x, Hx \rangle_T = \langle v(t), i(t) \rangle_T \geq \beta.$$

*Choosing the inner product to be the inner product in $\mathcal{L}_2$, the last inequality is equivalent to the following:*

$$\int_0^T x(t)y(t)dt = \int_0^T v(t)i(t)\ dt \geq \beta \quad \forall v(t) \in \mathcal{X}_e, \forall T \in \mathbb{R}.$$

*From equation (8.4) we know that the total energy absorbed by the network at time $t$ is*

$$\int_{-\infty}^t v(t)i(t)\ dt = \int_0^t v(t)i(t)\ dt + \int_{-\infty}^0 v(t)i(t)\ dt$$
$$= \langle v(t), i(t) \rangle_T + \int_{-\infty}^0 v(t)i(t)\ dt.$$

*Therefore, according to definition 8.2 the network is passive if and only if*

$$\langle v(t), i(t) \rangle_T \geq \beta \overset{def}{=} -\int_{-\infty}^0 v(t)i(t)\ dt.$$

□

Closely related to the notions of passivity and strict passivity are the concepts of *positivity* and *strict positivity*, introduced next.

**Definition 8.4** : *A system $H : \mathcal{X} \to \mathcal{X}$ is said to be strictly positive if there exists $\delta > 0$ such that*

$$\langle u, Hu \rangle \geq \delta \|u\|_{\mathcal{X}}^2 + \beta \qquad \forall u \in \mathcal{X}. \tag{8.12}$$

*$H$ is said to be positive if it satisfies (8.12) with $\delta = 0$.*

The only difference between the notions of passivity and positivity (strict passivity and strict positivity) is the lack of truncations in (8.12). As a consequence, the notions of positivity and strict positivity apply to input-output stable systems exclusively. Notice that, if the system $H$ is not input-output stable, then the left-hand side of (8.12) is unbounded.

The following theorem shows that if a system is (i) causal and (ii) stable, then the notions of positivity and passivity are entirely equivalent.

**Theorem 8.1** *Consider a system $H : \mathcal{X} \to \mathcal{X}$ , and let $H$ be causal. We have that*

*(i) $H$ positive $\iff$ $H$ passive.*

*(ii) $H$ strictly positive $\iff$ $H$ strictly passive.*

**Proof:** First assume that $H$ satisfies (8.12) and consider an arbitrary input $u \in \mathcal{X}_e$. It follows that $u_T \in \mathcal{X}$, and by (8.12) we have that

$$\langle u_T, Hu_T \rangle \geq \delta \|u_T\|_{\mathcal{X}}^2 + \beta$$

but

$$
\begin{aligned}
\langle u_T, Hu_T \rangle &= \langle u_T, (Hu_T)_T \rangle && \text{by (8.9)} \\
&= \langle u_T, (Hu)_T \rangle && \text{since } H \text{ is causal} \\
&= \langle u, Hu \rangle_T && \text{by (8.9).}
\end{aligned}
$$

It follows that $\langle u, Hu \rangle_T \geq \delta \|u_T\|^2$, and since $u \in \mathcal{X}_e$ is arbitrary, we conclude that (8.12) implies (8.11). For the converse, assume that $H$ satisfies (8.11) and consider an arbitrary input $u \in \mathcal{X}$. By (8.11), we have that

$$\langle u, Hu \rangle_T \geq \delta \|u_T\|_{\mathcal{X}}^2 + \beta$$

but

$$
\begin{aligned}
\langle u, Hu \rangle_T &= \langle u_T, (Hu)_T \rangle \\
&= \langle u_T, (Hu_T)_T \rangle \\
&= \langle u_T, Hu_T \rangle
\end{aligned}
$$

Thus $\langle u_T, H u_T \rangle \geq \|u_T\|^2 + \beta$, which is valid for all $T \in \mathbb{R}^+$. Moreover, since $u \in \mathcal{X}$ and $H : \mathcal{X} \to \mathcal{X}$, we can take limits as $T \to \infty$ to obtain

$$\langle u, H u \rangle \geq \delta \|u\|_{\mathcal{X}}^2 + \beta.$$

Thus we conclude that (8.11) implies (8.12) and the second assertion of the theorem is proved. Part (i) is immediately obvious assuming that $\delta = 0$. This completes the proof. $\square$

## 8.3   Interconnections of Passivity Systems

In many occasions it is important to study the properties of combinations of passive systems. The following Theorem considers two important cases

**Theorem 8.2** *Consider a finite number of systems $H_i : \mathcal{X}_e \to \mathcal{X}_e$, $i = 1, \cdots, n$. We have*

(i) *If all of the systems $H_i, i = 1, \cdots, n$ are passive, then the system $H : \mathcal{X}_e \to \mathcal{X}_e$, defined by (see Figure 8.3)*
$$H = H_1 + \cdots + H_n \tag{8.13}$$
   *is passive.*

(ii) *If all the systems $H_i, i = 1, \cdots, n$ are passive, and at least one of them is strictly passive, then the system $H$ defined by equation (8.13) is strictly passive.*

(iii) *If the systems $H_i, i = 1, 2$ are passive and the feedback interconnection defined by the equations (Figure 8.4)*
$$e = u - H_2 y \tag{8.14}$$
$$y = H_1 e \tag{8.15}$$
   *is well defined (i.e., $e(t) \in \mathcal{X}_e$ and is uniquely determined for each $u(t) \in \mathcal{X}_e$), then, the mapping from $u$ into $y$ defined by equations (8.14)–(8.15) is passive.*

**Proof of Theorem 8.2**

Proof of (i): We have
$$\begin{aligned}
\langle x, (H_1 + \cdots + H_n)x \rangle_T &= \langle x, H_1 x + \cdots + H_n x \rangle_T \\
&= \langle x, H_1 x \rangle_T + \cdots + \langle x, H_n x \rangle_T \\
&\geq \beta_1 + \cdots + \beta_n \overset{def}{=} \beta.
\end{aligned}$$

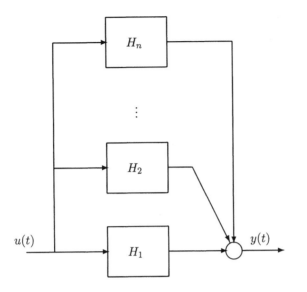

Figure 8.3: $H = H_1 + H_2 + \cdots + H_n$.

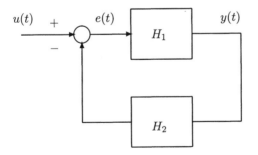

Figure 8.4: The Feedback System $S_1$.

Thus, $H \stackrel{def}{=} (H_1 + \cdots + H_n)$ is passive.

Proof of (ii) Assume that $k$ out of the $n$ systems $H_i$ are strictly passive, $1 \leq k \leq n$. By relabeling the systems, if necessary, we can assume that these are the systems $H_1, H_2, \cdots, H_k$. It follows that

$$
\begin{aligned}
\langle x, Hx \rangle_T &= \langle x, H_1 x + \cdots + H_n x \rangle_T \\
&= \langle x, H_1 x \rangle_T + \cdots + \langle x, H_k x \rangle_T + \cdots + \langle x, H_n x \rangle_T \\
&\geq \delta_1 \langle x, x \rangle_T + \cdots + \delta_k \langle x, x \rangle_T + \beta_1 + \cdots + \beta_n \\
&= (\delta_1 + \cdots + \delta_k) \|x_T\|_{\mathcal{X}} + (\beta_1 + \cdots + \beta_n)
\end{aligned}
$$

and the result follows.

Proof of (iii): Consider the following inner product:

$$
\begin{aligned}
\langle u, y \rangle_T &= \langle e + H_2 y, y \rangle_T \\
&= \langle e, y \rangle_T + \langle H_2 y, y \rangle_T \\
&= \langle e, H_1 e \rangle_T + \langle y, H_2 y \rangle_T \geq (\beta_1 + \beta_2).
\end{aligned}
$$

This completes the proof.                                                                                □

**Remarks:** In general, the number of systems in parts (i) and (ii) of Theorem 8.2 cannot be assumed to be infinite. The validity of these results in case of an infinite sequence of systems depends on the properties of the inner product. It can be shown, however, that if the inner product is the standard inner product in $\mathcal{L}_2$, then this extension is indeed valid. The proof is omitted since it requires some relatively advanced results on Lebesgue integration.

### 8.3.1    Passivity and Small Gain

The purpose of this section is to show that, in an inner product space, the concept of passivity is closely related to the norm of a certain operator to be defined.

In the following theorem $\mathcal{X}_e$ is an inner product space, and the gain of a system $H : \mathcal{X}_e \to \mathcal{X}_e$ is the gain induced by the norm $\|x\|^2 = \langle x, x \rangle$.

**Theorem 8.3** *Let $H : \mathcal{X}_e \to \mathcal{X}_e$, and assume that $(I + H)$ is invertible in $\mathcal{X}_e$, that is, assume that $(I + H)^{-1} : \mathcal{X}_e \to \mathcal{X}_e$. Define the function $S : \mathcal{X}_e \to \mathcal{X}_e$:*

$$
S = (H - I)(I + H)^{-1}. \tag{8.16}
$$

*We have:*

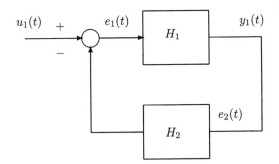

Figure 8.5: The Feedback System $S_1$.

*(a) H is passive if and only if the gain of S is at most 1, that is, S is such that*

$$\|(Sx)_T\|_{\mathcal{X}} \leq \|x_T\|_{\mathcal{X}} \qquad \forall x \in \mathcal{X}_e, \forall T \in \mathcal{X}_e. \tag{8.17}$$

*(b) H is strictly passive and has finite gain if and only if the gain of S is less than 1.*

**Proof**: See the Appendix.

## 8.4 Stability of Feedback Interconnections

In this section we exploit the concept of passivity in the stability analysis of feedback interconnections. To simplify our proofs, we assume without loss of generality that the systems are initially relaxed, and so the constant $\beta$ in Definitions 8.2 and 8.3 is identically zero.

Our first result consists of the simplest form of the passivity theorem. The simplicity of the theorem stems from considering a feedback system with one single input $u_1$, as shown in Figure 8.5 ($u_2 = 0$ in the feedback system used in Chapter 6).

**Theorem 8.4** : *Let $H_1, H_2 : \mathcal{X}_e \to \mathcal{X}_e$ and consider the feedback interconnection defined by the following equations:*

$$e_1 = u_1 - H_2 e_2 \tag{8.18}$$
$$y_1 = H_1 e_1. \tag{8.19}$$

*Under these conditions, if $H_1$ is passive and $H_2$ is strictly passive, then $y_1 \in \mathcal{X}$ for every $x \in \mathcal{X}$.*

**Proof**: We have

$$
\begin{aligned}
\langle u_1, y_1 \rangle_T &= \langle u_1, H_1 e_1 \rangle_T \\
&= \langle e_1 + H_2 e_2, H_1 e_1 \rangle_T \\
&= \langle e_1, H_1 e_1 \rangle_T + \langle H_2 e_2, H_1 e_1 \rangle_T \\
&= \langle e_1, H_1 e_1 \rangle_T + \langle H_2 y_1, y_1 \rangle_T
\end{aligned}
$$

but

$$
\begin{aligned}
\langle e_1, H_1 e_1 \rangle_T &\geq 0 \\
\langle H_2 y_1, y_1 \rangle_T &\geq \delta \| y_{1T} \|_{\mathcal{X}}^2
\end{aligned}
$$

since $H_1$ and $H_2$ are passive and strictly passive, respectively. Thus

$$
\langle u_1, y_1 \rangle_T \geq \delta \| y_{1T} \|_{\mathcal{X}}^2
$$

By the Schwarz inequality, $| \langle u_1, y_1 \rangle_T | \leq \| u_{1T} \|_{\mathcal{X}} \, \| y_{1T} \|_{\mathcal{X}}$. Hence

$$
\| u_{1T} \|_{\mathcal{X}} \| y_{1T} \|_{\mathcal{X}} \geq \delta \| y_{1T} \|_{\mathcal{X}}^2
$$

$$
\Rightarrow \| y_{1T} \|_{\mathcal{X}} \leq \delta^{-1} \| u_{1T} \|_{\mathcal{X}} \tag{8.20}
$$

Therefore, if $u_1 \in \mathcal{X}$, we can take limits as $T$ tends to infinity on both sides of inequality (8.20) to obtain

$$
\| y_1 \| \geq \delta^{-1} \| u_1 \|
$$

which shows that if $u_1$ is in $\mathcal{X}$, then $y_1$ is also in $\mathcal{X}$.  □

**Remarks**: Theorem 8.4 says exactly the following. Assuming that the feedback system of equations (8.18)–(8.19) admits a solution, then if $H_1$ is passive and $H_2$ is strictly passive, the output $y_1$ is bounded whenever the input $u_1$ is bounded. According to our definition of input–output stability, this implies that the *closed-loop system* seen as a mapping from $u_1$ to $y_1$ is input-output-stable. The theorem, however, does not guarantee that the error $e_1$ and the output $y_2$ are bounded. For these two signals to be bounded, we need a stronger assumption; namely, the strictly passive system must also have finite gain. We consider this case in our next theorem.

**Theorem 8.5** : *Let $H_1, H_2 : \mathcal{X}_e \to \mathcal{X}_e$ and consider again the feedback system of equations (8.18)–(8.19). Under these conditions, if both systems are passive and one of them is (i) strictly passive and (ii) has finite gain, then $e_1$, $e_2$, $y_1$, and $y_2$ are in $\mathcal{X}$ whenever $x \in \mathcal{X}$.*

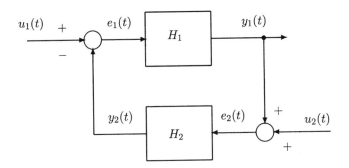

Figure 8.6: The Feedback System $S$.

**Proof:** We prove the theorem assuming that $H_1$ is passive and $H_2$ is strictly passive with finite gain. The opposite case (i.e., $H_2$ is passive and $H_1$ is strictly passive with finite gain) is entirely similar. Proceeding as in Theorem 8.4 (equation (8.20), we obtain

$$\Rightarrow \|y_{1T}\|_{\mathcal{X}} \leq \delta^{-1}\|u_{1T}\|_{\mathcal{X}} \tag{8.21}$$

so that $y_1 \in \mathcal{X}$ whenever $u_1 \in \mathcal{X}$. Also $y_2 = H_2 y_1$. Then $\|y_{2T}\|_{\mathcal{X}} = \|(H_2 y_1)_T\|_{\mathcal{X}}$, and since $H_2$ has finite gain, we obtain

$$\|y_{2T}\|_{\mathcal{X}} \leq \gamma(H_2)\|y_{1T}\|_{\mathcal{X}} \leq \gamma(H_2)\delta^{-1}\|u_{1T}\|_{\mathcal{X}}$$

Thus, $y_2 \in \mathcal{X}$ whenever $u_1 \in \mathcal{X}$. Finally, from equation (8.18) we have

$$\|e_{1T}\|_{\mathcal{X}} \leq \|u_{1T}\|_{\mathcal{X}} + \|(H_2 y_1)_T\|_{\mathcal{X}} \leq \|u_{1T}\|_{\mathcal{X}} + \gamma(H_2)\|y_{1T}\|_{\mathcal{X}}$$

which, taking account of (8.21), implies that $e_1 \in \mathcal{X}$ whenever $u_1 \in \mathcal{X}$. □

**Remarks:** Theorem 8.5 is general enough for most purposes. For completeness, we include the following theorem, which shows that the result of Theorem 8.5 is still valid if the feedback system is excited by two external inputs (Figure 8.6).

**Theorem 8.6** : *Let $H_1, H_2 : \mathcal{X}_e \to \mathcal{X}_e$ and consider the feedback system of equations*

$$e_1 = u_1 - H_2 e_2 \tag{8.22}$$
$$e_2 = u_2 + H_1 e_1 \tag{8.23}$$

*Under these conditions, if both systems are passive and one of them is (i) strictly passive and (ii) has finite gain, then $e_1$, $e_2$, $y_1$, and $y_2$ are in $\mathcal{X}$ whenever $x1, x_2 \in \mathcal{X}$.*

**Proof**: Omitted.

## 8.5   Passivity of Linear Time-Invariant Systems

In this section, we examine in some detail the implications of passivity and strict passivity in the context of linear time-invariant systems. Throughout this section, we will restrict attention to the space $\mathcal{L}_2$.

**Theorem 8.7** *Consider a linear time-invariant system $H : \mathcal{L}_{2e} \to \mathcal{L}_{2e}$ defined by $Hx = h * x$, where $h \in \mathcal{A}$ , $x \in \mathcal{L}_{2e}$. We have*

(i) $H$ is passive if and only if $\Re[\widehat{H}(\jmath\omega)] \geq 0$ $\ \forall \omega \in \mathbb{R}$.

(ii) $H$ is strictly passive if and only if $\exists \delta > 0$ such that $\Re[\widehat{H}(\jmath\omega)] \geq \delta$ $\ \forall \omega \in \mathbb{R}$.

**Proof**: The elements of $\mathcal{A}$ have a Laplace transform that is free of poles in the closed right half plane. Thus, points of the form $s = \jmath\omega$ belong to the region of convergence of $\widehat{H}(s)$, and $\widehat{H}(\jmath\omega)$ is the Fourier transform of $h(t)$. We also recall two properties of the Fourier transform:

(a) If $f$ is real-valued, then the real and imaginary parts of $\widehat{F}(\jmath\omega)$, denoted $\Re[\widehat{F}(\jmath\omega)]$ and $\Im[\widehat{F}(\jmath\omega)]$, respectively, are even and odd functions on $\omega$, respectively. In other words

$$\Re[\widehat{F}(\jmath\omega)] = \Re[\widehat{F}(-\jmath\omega)], \quad \Im[\widehat{F}(\jmath\omega)] = -\Im[\widehat{F}(-\jmath\omega)]$$

(b) (Parseval's relation)

$$\int_{-\infty}^{\infty} f(t)g(t) \, \mathrm{d}t \; = \; \frac{1}{2\pi} \int_{-\infty}^{\infty} \widehat{F}(\jmath\omega)\widehat{G}(\jmath\omega)^* \, \mathrm{d}\omega$$

where $\widehat{G}(\jmath\omega)^*$ represents the complex conjugate of $\widehat{G}(\jmath\omega)$.

Also, by assumption, $h \in \mathcal{A}$ , which implies that $H$ is causal and stable. Thus, according to Theorem 8.1, $H$ is passive (strictly passive) if and only if it is positive (strictly positive), and we can drop all truncations in the passivity definition. With this in mind, we have

$$
\begin{aligned}
\langle x, Hx \rangle &= \langle x, h * x \rangle \\
&= \int_{-\infty}^{\infty} x(t)[h(t) * x(t)] \, dt \\
&= \frac{1}{2\pi} \int_{-\infty}^{\infty} \widehat{X}(\jmath\omega)[\widehat{H}(\jmath\omega)\widehat{X}(\jmath\omega)]^* \, d\omega \\
&= \frac{1}{2\pi} \int_{-\infty}^{\infty} |\widehat{X}(\jmath\omega)|^2 \; \widehat{H}(\jmath\omega)^* \, d\omega \\
&= \frac{1}{2\pi} \int_{-\infty}^{\infty} \Re[\widehat{H}(\jmath\omega)] \, |\widehat{X}(\jmath\omega)|^2 \, d\omega + \frac{\jmath}{2\pi} \int_{-\infty}^{\infty} \Im[\widehat{H}(\jmath\omega)] \, |\widehat{X}(\jmath\omega)|^2 \, d\omega
\end{aligned}
$$

and since $\Im[\widehat{H}(\jmath\omega)]$ is an odd function of $\omega$, the second integral is zero. It follows that

$$
\langle x, Hx \rangle = \frac{1}{2\pi} \int_{-\infty}^{\infty} \Re[\widehat{H}(\jmath\omega)] \, |\widehat{X}(\jmath\omega)|^2 \, d\omega
$$

and noticing that

$$
\langle x, x \rangle = \frac{1}{2\pi} \int_{-\infty}^{\infty} |\widehat{X}(\jmath\omega)|^2 \, d\omega
$$

we have that

$$
\langle x, Hx \rangle \geq \inf_{\omega} \Re[\widehat{H}(\jmath\omega)]
$$

from where the sufficiency of conditions (i) and (ii) follows immediately. To prove necessity, assume that $\Re[\widehat{H}(\jmath\omega)] < 0$ at some frequency $\omega = \omega^*$. By the continuity of the Fourier transform as a function of $\omega$, it must be true that $\Re[\widehat{H}(\jmath\omega)] < 0 \quad \forall \omega \in |\omega - \omega^*| < \epsilon$, for some $\epsilon > 0$. We can now construct $\widehat{X}(\jmath\omega$ as follows:

$$
\begin{aligned}
\widehat{X}(\jmath\omega) &\geq M \quad \forall \omega \in |\omega - \omega^*| < \epsilon \\
\widehat{X}(\jmath\omega) &\leq m \quad \text{elsewhere}
\end{aligned}
$$

It follows that $\widehat{X}(\jmath\omega)$ has energy concentrated in the frequency interval where $\Re[\widehat{H}(\jmath\omega)]$ is negative and thus $\langle x, Hx \rangle < 0$ for appropriate choice of $M$ and $m$. This completes the proof. $\qquad \square$

Theorem 8.7 was stated and proved for single-input–single-output systems. For completeness, we state the extension of this result to multi-input–multi-output systems. The proof follows the same lines and is omitted.

**Theorem 8.8** *Consider a multi-input–multi-output linear time-invariant system $H : \mathcal{L}_{2e} \to \mathcal{L}_{2e}$ defined by $Hx = h * x$, where $h \in \mathcal{A}, x \in \mathcal{L}_{2e}$. We have,*

(i) $H$ is passive if and only if $[\widehat{H}(\jmath\omega)] + \widehat{H}(\jmath\omega)^*] \geq 0 \quad \forall\omega \in \mathbb{R}$.

(ii) $H$ is strictly passive if and only if $\exists\delta > 0$ such that

$$\lambda_{\min}[\widehat{H}(\jmath\omega)] + \widehat{H}(\jmath\omega)^*] \geq \delta \quad \forall\omega \in \mathbb{R}$$

It is important to notice that Theorem 8.7 was proved for systems whose impulse response is in the algebra $\mathcal{A}$. In particular, for finite-dimensional systems, this algebra consists of systems with all of their poles in the open left half of the complex plane. Thus, Theorem 8.7 says nothing about whether a system with transfer function

$$\widehat{H}(s) = \frac{\alpha s}{s^2 + \omega_0^2} \quad \alpha > 0,\ \omega_0 \geq 0 \tag{8.24}$$

is passive. This transfer function, in turn, is the building block of a very important class of system to be described later. Our next theorem shows that the class of systems with a transfer function of the form (8.24) is indeed passive, regardless of the particular value of $\alpha$ and $\omega$.

**Theorem 8.9** *Consider the system $H : \mathcal{L}_{2e} \to \mathcal{L}_{2e}$ defined by its transfer function*

$$\widehat{H}(s) = \frac{\alpha s}{s^2 + \omega_0^2} \quad \alpha > 0,\ \omega_0 \geq 0.$$

*Under these conditions, $H$ is passive.*

**Proof**: By Theorem 8.3, $H$ is passive if and only if

$$\|\widehat{S}\|_\infty = \left\|\frac{1 - H}{1 + H}\right\|_\infty \leq 1$$

but for the given $\widehat{H}(s)$

$$\widehat{S}(s) = \frac{s^2 - \alpha s + \omega_0^2}{s^2 + \alpha s + \omega_0^2}$$

$$\Rightarrow \quad \widehat{S}(\jmath\omega) = \frac{(\omega_0^2 - \omega^2) - \jmath\alpha\omega}{(\omega_0^2 - \omega^2) + \jmath\alpha\omega}$$

which has the form $(a - \jmath b)/(a + \jmath b)$. Thus $|\widehat{S}(\jmath\omega)| = 1 \quad \forall\omega$, which implies that $\|\widehat{S}\|_\infty = 1$, and the theorem is proved. $\qquad\square$

**Remarks**: Systems with a transfer function of the form (8.24) are oscillatory; i.e., if excited, the output oscillates without damping with a frequency $\omega = \omega_0$. A transfer function of this form is the building block of an interesting class of systems known as *flexible structures*. A linear time-invariant model of a flexible structure has the following form:

$$\widehat{H}(s) = \sum_{i=1}^{\infty} \frac{\alpha_i s}{s^2 + \omega_i^2}. \tag{8.25}$$

Examples of flexible structures include flexible manipulators and space structures. They are challenging to control given the infinite-dimensional nature of their model. It follows from Theorem 8.2 that, working in the space $\mathcal{L}_2$, a system with transfer function as in (8.25) is passive.

## 8.6   Strictly Positive Real Rational Functions

According to the results in the previous section, a (causal and stable) LTI system $H$ is strictly passive if and only if $\widehat{H}(\jmath\omega) \geq \delta > 0 \ \forall \omega \in \mathbb{R}$. This is a very severe restriction, rarely satisfied by physical systems. Indeed, no strictly proper system can satisfy this condition. This limitation, in turn, severely limits the applicability of the passivity theorem. Consider, for example, the case in which a passive plant (linear or not) is to be controlled via a linear time-invariant controller. If stability is to be enforced by the passivity theorem, then only controllers with relative degree zero can qualify as possible candidates. This is indeed a discouraging result. Fortunately, help in on the way: in this section we introduce the concept of strict positive realness (SPRness), which, roughly speaking, lies somewhere between passivity and strict passivity. It will be shown later that the feedback combination of a passive system with an SPR one, is also stable, thus relaxing the conditions of the passivity theorem.

In the sequel, $\mathcal{P}^n$ denotes the set of all polynomials on $n$th degree in the undetermined variable $s$.

**Definition 8.5** *Consider a rational function* $\widehat{H}(s) = p(s)/q(s)$*, where* $p(\cdot) \in \mathcal{P}^n$*, and* $q(\cdot) \in \mathcal{P}^m$*. Then* $\widehat{H}(s)$ *is said to be positive real (PR), if*

$$\Re[\widehat{H}(s)] \geq 0 \quad \text{for } \Re[s] \geq 0.$$

$\widehat{H}(s)$ *is said to be strictly positive real (SPR) if there exists* $\epsilon > 0$ *such that* $\widehat{H}(s - \epsilon)$ *is PR.*

**Remarks:** Definition 8.5 is rather difficult to use since it requires checking the real part of $\widehat{H}(s)$ for all possible values of $s$ in the closed right half plane. For this reason, frequency-domain conditions for SPRness are usually preferred. We now state these conditions as an alternative definition for SPR rational functions.

**Definition 8.6** *Consider a rational function* $\widehat{H}(s) = p(s)/q(s)$, *where* $p(\cdot) \in \mathcal{P}^n$, *and* $q(\cdot) \in \mathcal{P}^m$. *Then* $\widehat{H}(s)$ *is said to be in the class* $\mathcal{Q}$ *if*

*(i)* $q(\cdot)$ *is a Hurwitz polynomial (i.e. the roots of* $q(\cdot)$ *are in the open left half plane), and*

*(ii)* $\Re[\widehat{H}(\jmath\omega)] > 0$, $\quad \forall \omega \in [0, \infty)$.

**Definition 8.7** $\widehat{H}(s)$, *is said to be weak SPR if it is in the class* $\mathcal{Q}$ *and the degree of the numerator and denominator polynomials differ by at most 1.* $\widehat{H}(s)$ *is said to be SPR if it is weak SPR and in addition, one of the following conditions is satisfied:*

*(i)* $n = m$, *that is,* $p$ *and* $q$ *have the same degree.*

*(ii)* $n = m + 1$, *that is,* $\widehat{H}(s)$ *is strictly proper, and*

$$\lim_{\omega \to \infty} \omega^2 \Re[\widehat{H}(\jmath\omega)] > 0$$

It is important to notice the difference among the several concepts introduced above. Clearly, if $\widehat{H}(s)$ is SPR (or even weak SPR), then $\widehat{H}(s) \in \mathcal{Q}$. The converse is, however, not true. For example, the function $\widehat{H}_1(s) = (s + 1)^{-1} + s^3 \in \mathcal{Q}$, but it is not SPR according to Definition 8.5. In fact, it is not even positive real. The necessity of including condition (ii) in Definition 8.7 was pointed out by Taylor in Reference [79]. The importance of this condition will became more clear soon.

We now state an important result, known as the Kalman–Yakubovich lemma.

**Lemma 8.1** *Consider a system of the form*

$$\begin{aligned} \dot{x} &= Ax + Bu, & x \in \mathbb{R}^n, \ u \in \mathbb{R}^m \\ y &= Cx + Du, & u \in \mathbb{R}^m \end{aligned}$$

*and assume that (i) the eigenvalues of* $A$ *lie in the left half of the complex plane, (ii)* $(A, B)$ *is controllable, and (iii)* $(C, A)$ *is observable. Then* $\widehat{H} = C(sI - A)^{-1}B + D$ *is SPR if and only if there exist a symmetric positive definite matrix* $P \in \mathbb{R}^{n \times n}$, *and matrices* $Q \in \mathbb{R}^{m \times m}$, $W \in \mathbb{R}^{m \times m}$, *and* $\epsilon > 0$ *sufficiently small such that*

$$\begin{aligned} PA + A^T P &= -QQ^T - \epsilon P & (8.26) \\ PB + W^T Q &= C & (8.27) \\ W^T W &= D + D^T & (8.28) \end{aligned}$$

**Remarks**: In the special case of single-input–single-output systems of the form

$$\dot{x} = Ax + Bu$$
$$y = Cx$$

the conditions of lemma 8.1 can be simplified as follows: $\widehat{H} = C(sI - A)^{-1}B$ is SPR if and only if there exist symmetric positive definite matrices $P$ and $L$ a real matrix $Q$, and $\mu$ sufficiently small such that

$$PA + A^T P = -QQ^T - \mu L \qquad (8.29)$$
$$PB = C^T \qquad (8.30)$$

**Proof**: The proof is available in many references and is omitted. In Chapter 9, however, we will state and prove a result that can be considered as the nonlinear counterpart of this theorem. See Theorem 9.2.

**Theorem 8.10** *Consider the feedback interconnection of Figure 8.5, and assume that*

*(i) $H_1$ is linear time-invariant, strictly proper, and SPR.*

*(ii) $H_2$ is passive (and possibly nonlinear).*

*Under these assumptions, the feedback interconnection is input-output-stable.*

**Proof**: The proof consists of employing a type I loop transformation with $K = -\epsilon$ and showing that if $\epsilon > 0$ is small enough, then the two resulting subsystem satisfy the following conditions:

- $\widehat{H}_1'(s) = \widehat{H}_1(s)/[1 - \epsilon \widehat{H}_1(s)]$ is *passive*
- $\widehat{H}_2' = H_2 + \epsilon I$ is strictly passive.

Thus, stability follows by theorem 8.4. The details are in the Appendix. $\qquad \Box$

Theorem 8.10 is very useful. According to this result, the conditions of the passivity theorem can be relaxed somewhat. Indeed, when controlling a passive plant, a linear time-invariant SPR controller can be used instead of a strictly passive one. The significance of the result stems from the fact that SPR functions can be strictly proper, while strictly passive functions cannot. The following example shows that the loop transformation approach used in the proof of Theorem 8.10 (see the Appendix) will fail if the linear system is *weak* SPR, thus emphasizing the importance of condition (ii) in Definition 8.7.

**Example 8.4** *Consider the linear time-invariant system $\widehat{H}(s) = (s+c)/[(s+b)(s+b)]$, and let $\widehat{H}'(s) = H(s)/[1 - \epsilon H(s)]$. We first investigate the SPR condition on the system $\widehat{H}(s)$. We have*

$$\widehat{H}(\jmath\omega) = \frac{c + \jmath\omega}{(ab - \omega^2) + \jmath\omega(a+b)}$$

$$\Re e[\widehat{H}(\jmath\omega)] = \frac{c(ab - \omega^2) + \omega^2(a+b)}{(ab - \omega^2)^2 + \omega^2(a+b)^2}$$

$$\Rightarrow \quad \lim_{\omega\to\infty} \omega^2 \Re e[\widehat{H}(\jmath\omega)] = a + b - c$$

*from here we conclude that*

(i) *$\widehat{H}(s)$ is SPR if and only if $a + b > c$.*

(ii) *$\widehat{H}(s)$ is weak SPR if $a + b = c$.*

(iii) *$\widehat{H}(s)$ is not SPR if $a + b < c$.*

*We now consider the system $\widehat{H}'(s)$, after the loop transformation. We need to see whether $\widehat{H}'(s)$ is passive (i.e., $\Re e[\widehat{H}'(\jmath\omega)] \geq 0$). To analyze this condition, we proceed as follows*

$$\Re e[\widehat{H}'(\jmath\omega)] = \frac{1}{2}[\widehat{H}(\jmath\omega) + \widehat{H}(-\jmath\omega)] = \frac{(abc - \epsilon c^2) + \omega^2(a + b - c - \epsilon)}{(ab - \epsilon c - \omega^2)^2 + (a + b - \epsilon)^2} \geq 0$$

*if and only if*

$$abc - \epsilon c^2 \geq 0 \tag{8.31}$$
$$a + b - c - \epsilon \geq 0. \tag{8.32}$$

*If $a + b > c$, we can always find an $\epsilon > 0$ that satisfies (8.31)–(8.32). However, if $\widehat{H}(s)$ is weak SPR, then $a + b = c$, and no such $\epsilon > 0$ exists.* $\qquad\square$

## 8.7  Exercises

(8.1) Prove Theorem 8.4 in the more general case when $\beta \neq 0$ in Definitions 8.2 and 8.3.

(8.2) Prove Theorem 8.5 in the more general case when $\beta \neq 0$ in Definitions 8.2 and 8.3.

## Notes and References

This chapter introduced the concept of passivity in its purest (input–output) form. Our presentation closely follows Reference [21], which is an excellent source on the input–output theory of systems in general, and passivity results in particular. Early results on stability of passive systems can be found in Zames [98]. Strictly positive real transfer functions and the Kalman–Yakubovich lemma 8.1, play a very important role in several areas of system theory, including stability of feedback systems, adaptive control, and even the synthesis of passive networks. The proof of the Kalman–Yakubovich lemma can be found in several works. See for example Narendra and Anaswami [56], Vidyasagar [88] or Anderson [1]. Reference [1], in particular, contains a very thorough coverage of the Kalman-Yakubovich lemma and related topics. See References [35] and [78] for a detailed treatment of frequency domain properties of SPR functions. Example 8.4 is based on unpublished work by the author in collaboration with Dr. C. Damaren (University of Toronto).

# Chapter 9

# Dissipativity

In Chapter 8 we introduced the concept of a *passive system* (FIgure 9.1). Specifically, given an inner product space $\mathcal{X}$, a system $H : \mathcal{X}_e \to \mathcal{X}_e$ is said to be passive if $\langle u, Hu \rangle_T \geq \beta$. This concept was motivated by circuit theory, where $u$ and $y$ are, respectively, the voltage $v(t)$ and current $i(t)$ across a network or vice versa. Thus, in that case (assuming $\mathcal{X} = \mathcal{L}_2$)

$$\langle u, Hu \rangle_T = \langle v(t), i(t) \rangle_T = \int_{-\infty}^{T} v(t)i(t)\, \mathrm{d}t$$

which represents the energy *supplied* to the network at time $T$ or, equivalently, the energy *absorbed* by the network during the same time interval. For more general classes of dynamical systems, "passivity" is a somewhat restrictive property. Many systems fail to be passive simply because $\langle u, y \rangle$ may not constitute a suitable candidate for an energy function. In this chapter we pursue these ideas a bit further and postulate the existence of an *input energy* function and introduce the concept of *dissipative dynamical system* in terms of a non-negativity condition on this function. We will also depart from the classical input–output

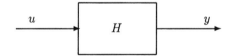

Figure 9.1: A Passive system.

theory of systems and consider state space realizations. The use of an internal description will bring more freedom in dealing with initial conditions on differential equations and will also allow us to study connections between input–output stability, and stability in the sense of Lyapunov.

## 9.1   Dissipative Systems

Throughout most of this chapter we will assume that the dynamical systems to be studied are given by a state space realization of the form

$$\psi : \begin{cases} \dot{x} = f(x, u), & u \in \mathcal{U}, \ x \in X \\ y = h(x, u), & y \in \mathcal{Y} \end{cases} \tag{9.1}$$

where

$\mathcal{U}$ is the input space of functions, $u \in \mathcal{U} : \Omega \subset \mathbb{R} \to \mathbb{R}^m$. In other words, the functions in $\mathcal{U}$ map a subset of the real numbers into $\mathbb{R}^m$.

$\mathcal{Y}$ the output space, which consists of functions, $y \in \mathcal{U} : \Omega \subset \mathbb{R} \to \mathbb{R}^p$.

$X$ The set $X \subset \mathbb{R}^n$ represents the state space.

Associated with this system we have defined a function $w(t) = w(u(t), y(t)) : \mathcal{U} \times \mathcal{Y} \to \mathbb{R}$, called the *supply rate*, that satisfies

$$\int_{t_0}^{t_1} |w(t)| \, dt \ < \ \infty,$$

that is, $w(\cdot)$ is a locally integrable function of the input and output $u$ and $y$ of the system $\psi$.

**Definition 9.1** *A dynamical system $\psi$ is said to be dissipative with respect to the supply rate $w(t)$ if there exists a function $\phi : X \to \mathbb{R}^+$, called the* storage function, *such that for all $x_0 \in X$ and for all inputs $u \in \mathcal{U}$ we have*

$$\phi(x_1) \leq \phi(x_0) + \int_{t_0}^{t_1} w(t) \ dt. \tag{9.2}$$

Inequality (9.2) is called the *dissipation inequality*, and the several terms in (9.2) represent the following:

- $\phi(\cdot)$: the storage function; $\phi(x(t^*))$ represents the "energy" stored by the system $\psi$ at time $t^*$.

- $\int_{t_0}^{t_1} w(t) \ dt$: represents the energy externally supplied to the system $\psi$ during the interval $[t_0, t_1]$.

Thus, according to (9.2), the stored energy $\phi(x_1)$ at time $t_1 \geq t_0$ is, at most, equal to the sum of the energy $\phi(x_0)$ initially stored at time $t_0$, plus the total energy externally supplied during the interval $[t_0, t_1]$. In this way there is no internal "creation" of energy. It is important to notice that if a motion is such that it takes the system $\psi$ from a particular state to the same terminal state along a certain trajectory in the state space, then we have (since $x_1 = x_0$)

$$\phi(x_0) \ \leq \ \phi(x_0) + \oint w(t) \ dt$$

$$\Rightarrow \quad \oint w(t) dt \ \geq \ 0 \tag{9.3}$$

where $\oint$ indicates a closed trajectory with identical initial and final states. Inequality (9.3) states that in order to complete a closed trajectory, a dissipative system requires external energy.

## 9.2 Differentiable Storage Functions

In general, the storage function $\phi$ of a dissipative system, defined in Definition 9.1 need not be differentiable. Throughout the rest of this chapter, however, we will see that many important results can be obtained by strengthening the conditions imposed on $\phi$. First we notice that if $\phi$ is continuously differentiable, then dividing (9.2) by $(t_1 - t_0)$, and denoting $\phi(x_i)$ the value of $\phi(x)$ when $t = t_i$, we can write

$$\frac{\phi(x_1) - \phi(x_0)}{t_1 - t_0} \ \leq \ \frac{1}{t_1 - t_0} \int_{t_0}^{t_1} w(t) \ dt \tag{9.4}$$

but

$$\lim_{t_1 \to t_0} \frac{\phi(x_1) - \phi(x_0)}{t_1 - t_0} \ = \ \frac{d\phi(x)}{dt} \ = \ \frac{\partial\phi(x)}{\partial x} f(x, u)$$

and thus (9.4) is satisfied if and only if

$$\frac{\partial\phi(x)}{\partial x} f(x, u) \leq w(t) = w(u, y) = w(u, h(x, u)) \quad \forall x, u. \tag{9.5}$$

Inequality (9.5) is called the *differential dissipation inequality*, and constitutes perhaps the most widely used form of the dissipation inequality. Assuming that $\phi$ is differentiable, we can restate definition 9.1 as follows.

**Definition 9.2** *(Dissipativity re-stated) A dynamical system $\psi$ is said to be dissipative with respect to the supply rate $\omega(t) = \omega(u, y)$ if there exist a continuously differentiable function $\phi : X \to \mathbb{R}^+$, called the storage function, that satisfies the following properties:*

(i) *There exist class $\mathcal{K}_\infty$ functions $\alpha_1$ and $\alpha_2$ such that*

$$\alpha_1(\|x\|) \le \phi(x) \ \le \ \alpha_2(\|x\|) \qquad \forall x \in \mathbb{R}^n$$

(ii)

$$\frac{\partial \phi}{\partial x} f(x, u) \ \le \ \omega(u, y) \qquad \forall x \in \mathbb{R}^n, \ u \in \mathbb{R}^m, \ \text{and} \ y = h(x, u).$$

In other words; defining property (i) simply states that $\phi(\cdot)$ is positive definite, while property (ii) is the differential dissipation inequality.

### 9.2.1   Back to Input-to-State Stability

In Chapter 7 we studied the important notion of input-to-state stability. We can now review this concept as a special case of a dissipative system. In the following lemma we assume that the storage function corresponding to the supply rate $\omega$ is differentiable.

**Lemma 9.1** *A system $\psi$ is input-to-state stable if and only if it is dissipative with respect to the supply rate*

$$\omega(t) = -\alpha_3(\|x\|) + \sigma(\|u\|)$$

*where $\alpha_3$ and $\sigma$ are class $\mathcal{K}_\infty$ functions.*

**Proof:** The proof is an immediate consequence of Definition 9.2 and Lemma 7.6.          □

## 9.3   $QSR$ Dissipativity

So far we have paid little attention to the supply rate $w(t)$. There are, however, several interesting candidates for this function. In this section we study a particularly important function $w(\cdot)$ and some of its implications. We will see that concepts such as passivity and small gain are special cases of this supply rate.

**Definition 9.3** *Given constant matrices $Q \in \mathbb{R}^{p \times p}, S \in \mathbb{R}^{p \times m}$, and $R \in \mathbb{R}^{m \times m}$ with $Q$ and $R$ symmetric, we define the supply rate $w(t) = w(u, y)$ as follows:*

$$
\begin{aligned}
w(t) &= y^T Q y + 2 y^T S u + u^T R u \\
&= [y^T, u^T] \begin{bmatrix} Q & S \\ S^T & R \end{bmatrix} \begin{bmatrix} y \\ u \end{bmatrix}.
\end{aligned}
\tag{9.6}
$$

It is immediately obvious that

$$
\int_0^\infty w(t) \, dt = \langle y, Qy \rangle + 2\langle y, Su \rangle + \langle u, Ru \rangle
\tag{9.7}
$$

and

$$
\int_0^T w(t) \, dt = \langle y, Qy \rangle_T + 2\langle y, Su \rangle_T + \langle u, Ru \rangle_T
\tag{9.8}
$$

moreover, using time invariance, $w(t)$ defined in (9.6) is such that

$$
\int_{t_0}^{t_0+T} w(t) \, dt = \int_0^T w(t) \, dt.
\tag{9.9}
$$

Thus, we can now state the following definition

**Definition 9.4** *The system $\psi$ is said to be QSR-dissipative if there exist a storage function $\phi : X \to \mathbb{R}^+$ such that $\forall x(0) = x_0 \in X$ and for all $u \in \mathcal{U}$ we have that*

$$
\int_0^T w(t) \, dt = \langle y, Qy \rangle_T + 2\langle y, Su \rangle_T + \langle u, Ru \rangle_T \geq \phi(x_1) - \phi(x_0).
\tag{9.10}
$$

Definition 9.4 is clearly a special case of Definition (9.1) with one interesting twist. Notice that with this supply rate, the state space realization of the system $\psi$ is no longer essential or necessary. Indeed, $QSR$ dissipativity can be interpreted as an input–output property. Instead of pursuing this idea, we will continue to assume that the input output relationship is obtained from the state space realization $\psi$ as defined in (9.1). As we will see, doing so will allow us to study connections between certain input–output properties and stability in the sense of Lyapunov.

We now single out several special cases of interest, which appear for specific choices of the parameters $QS$ and $R$.

1- **Passive systems:** The system $\psi$ is passive if and only if it is dissipative with respect to $Q = 0, R = 0$, and $S = \frac{1}{2}I$. Equivalently, $\psi$ is passive if and only if it is $(0, \frac{I}{2}, 0)$-dissipative. The equivalence is immediate since in this case (9.10) implies that

$$
\langle y, u \rangle_T \geq \phi(x_1) - \phi(x_0) \geq -\phi(x_0)
\tag{9.11}
$$
$$
\text{(since } \phi(x) > 0 \ \forall x, \text{ by assumption)}
$$

Thus, defining $\beta \overset{def}{=} - \phi(x_0)$, (9.11) is identical to Definition (8.2). This formulation is also important in that it gives $\beta$ a precise interpretation: $\beta$ is the stored energy at time $t = 0$, given by the initial conditions $x_0$.

2- **Strictly passive systems:** The system $\psi$ is strictly passive if and only if it is dissipative with respect to $Q = 0, R = -\delta$, and $S = \frac{1}{2}I$. To see this, we substitute these values in (9.10) and obtain

$$\langle y, u \rangle_T + \langle u, -\delta u \rangle_T \geq \phi(x_1) - \phi(x_0) \geq -\phi(x_0) \overset{def}{=} \beta$$

or

$$\langle u, y \rangle_T \geq \delta \langle u, u \rangle_T + \beta = \delta \|u\|_T^2 + \beta.$$

3- **Finite-gain-stable:** The system $\psi$ is finite-gain-stable if and only if it is dissipative with respect to $Q = -\frac{1}{2}I, R = \frac{\gamma^2}{2}I$, and $S = 0$. To see this, we substitute these values in (9.10) and obtain

$$-\frac{1}{2}\langle y, y \rangle_T + \frac{\gamma^2}{2}\langle u, u \rangle_T \geq \phi(x_1) - \phi(x_0) \geq -\phi(x_0)$$
$$\Rightarrow \quad -\langle y, y \rangle_T \geq -\gamma^2 \langle u, u \rangle_T - 2\phi(x_0)$$
$$\text{or} \quad \langle y, y \rangle_T \leq \gamma^2 \langle u, u \rangle_T + 2\phi(x_0)$$
$$\Rightarrow \quad \|y_T\|_{\mathcal{L}_2}^2 \leq \gamma^2 \|u_T\|_{\mathcal{L}_2}^2 + 2\phi(x_0)$$
$$\Longleftrightarrow \quad \|y_T\|_{\mathcal{L}_2} \leq \sqrt{\gamma^2 \|u_T\|_{\mathcal{L}_2}^2 + 2\phi(x_0)}$$

and since for $a, b > 0$, $\sqrt{a^2 + b_2} \leq (a + b)$, then defining $\beta = \sqrt{2\phi(x_0)}$, we conclude that

$$\|y_T\|_{\mathcal{L}_2} \leq \gamma \|u_T\|_{\mathcal{L}_2} + \beta.$$

We have already encountered passive, strictly passive, and finite-gain-stable systems in previous chapters. Other cases of interest, which appear frequently in the literature are described in paragraphs 4 and 5.

4- **Strictly output-passive systems:** The system $\psi$ is said to be strictly output passive if it is dissipative with respect to $Q = -\epsilon I, R = 0$, and $S = \frac{1}{2}I$. In this case, substituting these values in (9.10), we obtain

$$-\epsilon \langle y, y \rangle_T + \langle y, u \rangle_T \geq \phi(x_1) - \phi(x_0) \geq -\phi(x_0)$$

or

$$\int_0^T u^T y \, dt = \langle u, y \rangle_T \geq \epsilon \langle y, y \rangle_T + \beta.$$

5- **Very strictly-passive Systems**: The system $\psi$ is said to be very strictly passive if it is dissipative with respect to $Q = -\epsilon I, R = -\delta I$, and $S = \frac{1}{2}I$. In this case, substituting these values in (9.10), we obtain

$$-\epsilon\langle y, y\rangle_T - \delta\langle u, u\rangle_T + \langle y, u\rangle_T \geq \phi(x_1) - \phi(x_0) \geq -\phi(x_0) \stackrel{def}{=} \beta$$

or

$$\int_0^T u^T y \, dt = \langle u, y\rangle_T \geq \delta\langle u, u\rangle_T + \epsilon\langle y, y\rangle_T + \beta.$$

The following lemma states a useful results, which is a direct consequence of these definitions.

**Lemma 9.2** *If $\psi$ is strictly output passive, then it has a finite $\mathcal{L}_2$ gain.*

**Proof**: The proof is left as an exercise (Exercise 9.2).

## 9.4 Examples

### 9.4.1 Mass–Spring System with Friction

Consider the mass–spring system moving on a horizontal surface, shown in Figure 9.2. Assuming for simplicity that the friction between the mass and the surphase is negligible, we obtain the following equation of the motion:

$$m\ddot{x} + \beta\dot{x} + kx = f$$

where $m$ represents the mass, $k$ is the spring constant, $\beta$, the viscous friction force associated with the spring, and $f$ is an external force. Defining state variables $x_1 = x$, and $\dot{x}_1 = x_2$, and assuming that the desired output variable is the velocity vector, $x_2$, we obtain the following state space realization:

$$\psi : \begin{cases} \dot{x}_1 &= x_2 \\ \dot{x}_2 &= -\frac{k}{m}x_1 - \frac{\beta}{m}x_2 + \frac{f}{m} \\ y &= x_2 \end{cases}$$

To study the dissipative properties of this system, we proceed to find the total energy stored in the system at any given time. We have

$$E = \frac{1}{2}kx_1^2 + \frac{1}{2}mx_2^2$$

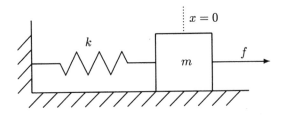

Figure 9.2: Mass-spring system.

where $\frac{1}{2}mx_2^2$ represents the kinetic energy of the mass and $\frac{1}{2}kx_1^2$ is the energy stored by the spring. Since the energy is a positive quantity, we propose $E$ as a "possible" storage function; thus, we define:

$$\phi \overset{def}{=} = E = \frac{1}{2}kx_1^2 + \frac{1}{2}mx_2^2.$$

Since $\phi$ is continuously differentiable with respect to $x_1$ and $x_2$, we can compute the time derivative of $\phi$ along the trajectories of $\psi$:

$$
\begin{aligned}
\dot{\phi} &= \frac{\partial \phi}{\partial x}\dot{x} \\
&= \begin{bmatrix} kx_1 & mx_2 \end{bmatrix} \begin{bmatrix} x_2 \\ -\frac{k}{m}x_1 - \frac{\beta}{m}x_2 + \frac{f}{m} \end{bmatrix} \\
&= -\beta x_2^2 + xf \\
&= -\beta y^2 + yf.
\end{aligned}
$$

Thus,

$$\int_0^t \dot{\phi}\, dt = E(t) \geq 0$$

and it follows that the mass–spring system with output $\dot{x} = x_2$ is dissipative with respect to the supply rate

$$\omega(t) = yf - \beta y^2.$$

It is immediately evident that this supply rate corresponds to $Q = -\beta$, $S = \frac{1}{2}$, and $R = 0$, from where we conclude that the mass-spring system $\psi$ is *strictly output-passive*.

## 9.4.2 Mass–Spring System without Friction

Consider again the mass–spring system of the previous example, but assume that $\beta = 0$. In this case, the state space realization reduces to

$$\psi : \begin{cases} \dot{x}_1 &= x_2 \\ \dot{x}_2 &= -\frac{k}{m}x_1 + \frac{f}{m} \\ y &= x_2 \end{cases}$$

Proceeding as in the previous example, we define

$$\phi \overset{def}{=} = E = \frac{1}{2}kx_1^2 + \frac{1}{2}mx_2^2.$$

Differentiating $\phi$ along the trajectories of $\psi$, we obtain:

$$\dot{\phi} = x_2 f = yf$$

since once again,

$$\int_0^t \dot{\phi}\, dt = E(t) \geq 0.$$

We conclude that the mass–spring system with output $\dot{x} = x_2$ is dissipative with respect to the supply rate

$$w(t) = yf$$

which corresponds to $Q = 0$, $S = \frac{1}{2}$, and $R = 0$. This implies that the mass–spring system $\psi$ is *passive*.

## 9.5 Available Storage

Having defined dissipative systems, along with supply rates and storage functions, we now turn our attention to a perhaps more abstract question. Given a dissipative dynamical system, we ask: What is the maximum amount of energy that can be extracted from it, at any given time? Willems [91] termed this quantity the "available storage." This quantity plays an important conceptual role in the theory of dissipative systems and appears in the proofs of certain theorems. For completeness, we now introduce this concept. This section is not essential and can be skipped in a first reading of this chapter.

**Definition 9.5** *The* available storage, *$\phi_a$ of a dynamical system $\psi$ with supply rate $w$ is defined by*

$$\phi_a(x) = \sup_{\substack{u(\cdot) \\ T \geq 0}} - \int_0^T w(u(t), y(t))\, dt, \qquad x(0) = x. \qquad (9.12)$$

As defined, $\phi_a(x)$ denotes the energy that can be extracted from $\psi$, starting from the initial state $x$ at $t = 0$.

The following theorem is important in that it provides a (theoretical) way of checking whether or not a system is dissipative, in terms of the available storage.

**Theorem 9.1** *A dynamical system $\psi$ is dissipative if and only if for all $x \in X$ the available storage $\phi_a(x)$ is finite. Moreover, for a dissipative system we have that $0 \le \phi_a \le S$, and thus $\phi_a$ itself is a possible storage function.*

**Proof:**

**Sufficiency**: Assume first that $\phi_a$ is finite. Since the right-hand side of (9.12) contains the zero element (obtained setting $T = 0$) we have that $\phi_a \ge 0$. To show that $\psi$ is dissipative, we now consider an arbitrary input $u^* : [0,T] \to \mathbb{R}^n$ that takes the dynamical system $\psi$ from the initial state $x_0$ at $t = 0$ to a final state $x_1$ at $t = T$, and show that the $\phi_a$ satisfies

$$\phi_a(x_1) \le \phi_a(x_0) + \int_0^T \omega(t)\ \mathrm{d}t.$$

Thus $\phi_a$ is itself a storage function. To see this, we compare $\phi_a(x_0)$ and $\phi_a(x_1)$, the "energies" that can be extracted from $\psi$ starting at the states $x_0$ and $x_1$, respectively. We have

$$
\begin{aligned}
\phi(x_0) &= \sup_{\substack{u(\cdot) \\ T \ge 0}} -\int_0^T \omega(u(t), y(t))\, \mathrm{d}t, \\
&\ge -\int_0^{t_1} \omega(u^*, y)\ \mathrm{d}t \; - \sup_{\substack{u(\cdot) \\ T \ge 0}} \int_0^T \omega(u(t), y(t))\, \mathrm{d}t, \\
&= -\int_0^{t_1} \omega(u^*, y)\ \mathrm{d}t \; + \; \phi_a(x_1).
\end{aligned}
$$

This means that when extracting energy from the system $\psi$ at time $x_0$, we can follow an "optimal" trajectory that maximizes the energy extracted following an arbitrary trajectory, or we can force the system to go from $x_0$ to $x_1$ and then extract whatever energy is left in $\psi$ with initial state $x_1$. This second process is clearly nonoptimal, thus leading to the result.

**Necessity**: Assume now that $\psi$ is dissipative. This means that $\exists \phi \ge 0$ such that for <u>all</u> $u(\cdot)$ we have that

$$\phi(x_0) + \int_0^T \omega(u, y)\ \mathrm{d}t \; \ge \; \phi(x(T)) \; \ge \; 0.$$

Thus,

$$\phi(x_0) \geq \sup_{\substack{u(\cdot) \\ T \geq 0}} - \int_0^T w(u(t), y(t)) \, dt, \; = \; \phi_a(x_0).$$

## 9.6 Algebraic Condition for Dissipativity

We now turn our attention to the issue of checking the dissipativity condition for a given system. The notion of available storage gave us a theoretical answer to this riddle in Theorem 9.1. That result, however, is not practical in applications. Our next theorem provides a result that is in the same spirit as the Kalman–Yakuvovich lemma studied in Chapter 8; specifically, it shows that under certain assumptions, dissipativeness can be characterized in terms of the coefficients of the state space realization of the system $\psi$. Moreover, it will be shown that in the special case of linear passive systems, this characterization of dissipativity, leads in fact, to the Kalman–Yakuvovich lemma.

Throughout the rest of this section we will make the following assumptions:

1- We assume that both $f(\cdot, \cdot)$ and $h(\cdot, \cdot)$ in the state space realization $\psi$ are *affine functions* of the input $u$, that is, we assume that $\psi$ is of the form

$$\psi : \begin{cases} \dot{x} = f(x) + g(x)u \\ y = h(x) + j(x)u \end{cases} \tag{9.13}$$

where $x \in \mathbb{R}^n, u \in \mathbb{R}^m$, and $y \in \mathbb{R}^p$, respectively.

2- The state space of the system (9.13) is *reachable* from the origin. This means that given any $x_1 \in \mathbb{R}^n$ and $t = t_1 \in \mathbb{R}^+$ there exists a $t_0 \leq t_1$ and an input $u \in \mathcal{U}$ such that the state can be driven from the origin at $t = 0$ to $x = x_1$ at $t = t_1$.

3- Whenever the system $\psi$ is dissipative with respect to a supply rate of the form (9.6), the available storage $\phi_a(x)$ is a differentiable function of $x$.

These assumptions, particularly 1 and 3, bring, of course, some restrictions on the class of systems considered. The benefit is a much more explicit characterization of the dissipative condition.

**Theorem 9.2** *The nonlinear system $\psi$ given by (9.13) is QSR-dissipative (i.e. dissipative with supply rate given by (9.6)) if there exists a differentiable function $\phi : \mathbb{R}^n \to \mathbb{R}$ and function $L : \mathbb{R}^n \to \mathbb{R}^q$ and $W : \mathbb{R}^n \to \mathbb{R}^{q \times m}$ satisfying*

$$\phi(x) > 0 \; \forall x \neq 0, \quad \phi(0) = 0 \tag{9.14}$$

$$\frac{\partial \phi}{\partial x} f(x) = h^T(x)Qh(x) - L^T(x)L(x) \tag{9.15}$$

$$\frac{1}{2}g^T(\frac{\partial \phi}{\partial x})^T = \widehat{S}^T h(x) - W^T L(x) \tag{9.16}$$

$$\widehat{R} = W^T W \tag{9.17}$$

*for all x, where*

$$\widehat{R} = R + j^T(x)S + S^T j(x) + j^T(x)Qj(x) \tag{9.18}$$

*and*

$$\widehat{S}(x) = Qj(x) + S. \tag{9.19}$$

**Proof**: To simplify our proof we assume that $j(x) = 0$ in the state space realization (9.13) (see exercise 9.3). In the case of linear systems this assumption is equivalent to assuming that $D = 0$ in the state space realization, something that is true in most practical cases. With this assumption we have that

$$\widehat{R} = R, \quad \text{and} \quad \widehat{S} = S,$$

We prove sufficiency. The necessity part of the proof can be found in the Appendix. Assuming that $S$, $L$ and $W$ satisfy the assumptions of the theorem, we have that

$$
\begin{aligned}
w(u, y) &= y^T QY + u^T Ru + 2y^T Su \\
&= h^T Qh + u^T Ru + 2h^T Su \quad \text{substituting (9.13)} \\
&= [\frac{\partial \phi}{\partial x} f(x) + L^T L] + u^T Ru + 2h^T Su \quad \text{substituting (9.15)} \\
&= \frac{\partial \phi}{\partial x} f(x) + L^T L + u^T Ru + 2u^T S^T h \\
&= \frac{\partial \phi}{\partial x} f(x) + L^T L + u^T W^T Wu + 2u^T [\frac{1}{2}g^T(\frac{\partial \phi}{\partial x})^T + W^T L] \\
&\qquad\qquad \text{substituting (9.17) and (9.16)} \\
&= \frac{\partial \phi}{\partial x}[f(x) + gu] + L^T L + u^T W^T Wu + 2u^T W^T L \\
&= \frac{\partial \phi}{\partial x}\dot{x} + (L + Wu)^T(L + Wu) \\
&= \dot{\phi} + (L + Wu)^T(L + Wu) \tag{9.20}
\end{aligned}
$$

$$\Rightarrow \quad \int_0^t w(t)\, dt \quad = \quad \phi(x(t)) - \phi(x_0) + \int_0^t (L + Wu)^T (L + Wu)\, dt$$

$$\geq \quad \phi(x(t)) - \phi(x_0)$$

and setting $x_0 = 0$ implies that

$$\int_0^t w(t)\, dt \geq \phi(x(t)) \geq 0.$$

$\square$

**Corollary 9.1** *If the system $\psi$ is dissipative with respect to the supply rate (9.6), then there exist a real function $\phi$ satisfying $\phi(x) > 0 \; \forall x \neq 0$, $\phi(0) = 0$, such that*

$$\frac{d\phi}{dt} = -(L + Wu)^T (L + Wu) + w(u, y) \qquad (9.21)$$

**Proof**: A direct consequence of Theorem 9.2. Notice that (9.21) is identical to (9.20) in the sufficiency part of the proof. $\square$

### 9.6.1 Special Cases

We now consider several cases of special interest.

**Passive systems**: Now consider the passivity supply rate $w(u, y) = u^T y$ (*i.e.* $Q = R = 0$, and $S = \frac{1}{2}$ in (9.6)), and assume as in the proof of Theorem (9.2) that $j(x) = 0$. In this case, Theorem 9.2 states that the nonlinear system $\psi$ is passive if and only if

$$\frac{\partial \phi}{\partial x} f(x) = -L^T(x) L(x)$$

$$g^T \left( \frac{\partial \phi}{\partial x} \right)^T = h(x)$$

or, equivalently

$$\frac{\partial \phi}{\partial x} f(x) \leq 0 \qquad (9.22)$$

$$\frac{\partial \phi}{\partial x} g = h^T(x). \qquad (9.23)$$

Now assume that, in addition, the system $\psi$ is linear then $f(x) = Ax$, $g(x) = B$, and $h(x) = Cx$. Guided by our knowledge of Lyapunov stability of linear systems, we define the storage function $\phi(x) = x^T P x$, $P = P^T \geq 0$. Setting $u = 0$, we have that

$$\frac{\partial \phi}{\partial x} f(x) = x^T [A^T P + PA] x$$

which implies that (9.22)–(9.23) are satisfied if and only if

$$A^T P + P A \leq 0$$
$$B^T P = C^T.$$

Therefore, for passive systems, Theorem 9.2 can be considered as a nonlinear version of the Kalman-Yakubovich lemma discussed in Chapter 8.

**Strictly output passive systems**: Now consider the strictly output passivity supply rate $\omega(u, y) = u^T y - \epsilon y^T y$ (i.e., $Q = -\epsilon I$, $R = 0$, and $S = \frac{1}{2} I$ in (9.6)), and assume once again that $j(x) = 0$. In this case Theorem 9.2 states that the nonlinear system $\psi$ is strictly output-passive if and only if

$$\frac{\partial \phi}{\partial x} f(x) = -\epsilon h^T(x) h(x) - L^T(x) L(x)$$
$$g^T \left( \frac{\partial \phi}{\partial x} \right)^T = h(x)$$

or, equivalently,

$$\frac{\partial \phi}{\partial x} f(x) \leq -\epsilon h^T(x) h(x) \qquad (9.24)$$
$$\frac{\partial \phi}{\partial x} g = h^T(x). \qquad (9.25)$$

**Strictly Passive systems**: Finally consider the strict passivity supply rate $\omega(u, y) = u^T y - \delta u^T u$ (i.e., $Q = 0$, $R = -\delta I$, and $S = \frac{1}{2} I$ in (9.6)), and assume once again that $j(x) = 0$. In this case Theorem 9.2 states that the nonlinear system $\psi$ is strictly passive if and only if

$$\frac{\partial \phi}{\partial x} f(x) = -L^T(x) L(x)$$
$$g^T \left( \frac{\partial \phi}{\partial x} \right)^T = h(x) - 2 W^T L$$

with

$$\widehat{R} = R = W^T W = -\delta I$$

which can never be satisfied since $W^T W \geq 0$ and $\delta > 0$. It then follows that no system of the form (9.13), with $j = 0$, can be strictly passive.

## 9.7 Stability of Dissipative Systems

Throughout this section we analyze the possible implications of stability (in the sense of Lyapunov) for dissipative dynamical systems. Throughout this section, we assume that the storage function $\phi : X \to \mathbb{R}^+$ is differentiable and satisfies the differential dissipation inequality (9.5).

In the following theorem we consider a dissipative system $\psi$ with storage function $\phi$ and assume that $x_e$ is an equilibrium point for the *unforced* systems $\psi$, that is, $f(x_e) = f(x_e, 0) = 0$.

**Theorem 9.3** *Let $\psi$ be a dissipative dynamical system with respect to the (continuously differentiable) storage function $\phi : X \to \mathbb{R}^+$, which satisfies (9.5), and assume that the following conditions are satisfied:*

*(i) $x_e$ is a strictly local minimum for $\phi$:*

$$\phi(x_e) < \phi(x) \quad \forall x \text{ in a neigborhood of } x_e$$

*(ii) The supply rate $w = w(u, y)$ is such that*

$$w(0, y) \leq 0 \quad \forall y.$$

*Under these conditions $x_e$ is a stable equilibrium point for the unforced systems $\dot{x} = f(x, 0)$.*

**Proof:** Define the function $V(x) \overset{def}{=} \phi(x) - \phi(x_e)$. This function is continuously differentiable, and by condition (i) is positive definite $\forall x$ in a neighborhood of $x_e$. Also, the time derivative of $V$ along the trajectories of $\psi$ is given by

$$\dot{V}(x) = \frac{\partial \phi(x)}{\partial x} f(x, u)$$

thus, by (9.5) and condition (ii) we have that $\dot{V}(x) \leq 0$ and stability follows by the Lyapunov stability theorem. □

Theorem 9.3 is important not only in that implies the stability of dissipative systems (with an equilibrium point satisfying the conditions of the theorem) but also in that it suggests the use of the storage function $\phi$ as a means of constructing Lyapunov functions. It is also important in that it gives a clear connection between the concept of dissipativity and stability in the sense of Lyapunov. Notice also that the general class of dissipative systems discussed in theorem 9.3 includes $QSR$ dissipative systems as a special case. The property of $QSR$ dissipativity, however, was introduced as an input–output property. Thus, dissipativity provides a very important link between the input–output theory of system and stability in the sense of Lyapunov.

**Corollary 9.2** *Under the conditions of Theorem 9.3, if in addition no solution of $\dot{x} = f(x)$ other than $x(t) = x_e$ satisfies $w(0, y) = w(0, h(x, 0)) = 0$, then $x_e$ is asymptotically stable.*

**Proof:** Under the present conditions, $\dot{V}(x)$ is strictly negative and all trajectories of $\psi = f(x_e, 0)$, and satisfies $\dot{V}(x) = 0$ if and only if $x(t) = x_e$, and asymptotic stability follows from LaSalle's theorem.                                                          $\square$

Much more explicit stability theorems can be derived for $QSR$ dissipative systems. We now present one such a theorem (theorem ), due to Hill and Moylan [30].

**Definition 9.6** *([30]) A state space realization of the form $\psi$ is said to be zero-state detectable if for any trajectory such that $u \equiv 0, y \equiv 0$ we have that $x(t) \equiv 0$.*

**Theorem 9.4** *Let the system*

$$\psi : \begin{cases} \dot{x} = f(x) + g(x)u \\ y = h(x) \end{cases}$$

*be dissipative with respect to the supply rate $w(t) = y^T Q y + 2y^T S u + u^T R u$ and zero state detectable. Then, the free system $\dot{x} = f(x)$ is*

(i) *Lyapunov-stable if $Q \leq 0$.*

(ii) *Asymptotically stable if $Q < 0$.*

**Proof:** From Theorem 9.2 and Corollary 9.1 we have that if $\psi$ is $QSR$ dissipative, then there exists $\phi > 0$ such that

$$\frac{d\phi}{dt} = -(L + Wu)^T (L + Wu) + \omega(u, y)$$

and setting $u = 0$,

$$\frac{d\phi}{dt} = -L^T L + h^T(x) Q h(x)$$

along the trajectories of $\dot{x} = f(x)$. Thus, stability follows from the Lyapunov stability theorem. Assume now that $Q < 0$. In this case we have that

$$\frac{d\phi}{dt} = 0 \qquad \Rightarrow \qquad h^T(x) Q h(x)$$

$$\Longleftrightarrow \quad h(x) = 0$$

$$\Longleftrightarrow \quad x = 0$$

by the zero-state detectability condition.                                       □

The following corollary is then an immediate consequence of Theorem 9.6 and the special cases discussed in Section 9.2.

**Corollary 9.3** *Given a zero state detectable state space realization*

$$\psi : \begin{cases} \dot{x} = f(x) + g(x)u \\ y = h(x) + j(x)u \end{cases}$$

*we have*

(i) *If $\phi$ is passive, then the unforced system $\dot{x} = f(x)$ is Lyapunov-stable.*

(ii) *If $\phi$ is strictly passive, then $\dot{x} = f(x)$ is Lyapunov-stable.*

(iii) *If $\phi$ is finite-gain-stable, then $\dot{x} = f(x)$ is asymptotically stable.*

(iv) *If $\phi$ is strictly output-passive, then $\dot{x} = f(x)$ is asymptotically stable.*

(v) *If $\phi$ is very strictly passive, then $\dot{x} = f(x)$ is asymptotically stable.*

## 9.8   Feedback Interconnections

In this section we consider the feedback interconnections and study the implication of different forms of $QSR$ dissipativity on closed-loop stability in the sense of Lyapunov. We will see that several important results can be derived rather easily, thanks to the machinery developed in previous sections.

Throughout this section we assume that state space realizations $\psi_1$ and $\psi_2$ each have the form

$$\psi_i : \begin{cases} \dot{x}_i = f_i(x_i) + g(x_i)u_i, & u \in \mathbb{R}^m, \ x \in \mathbb{R}^n \\ y_i = h_i(x_i) + j(x_i)u_i, & y \in \mathbb{R}^m \end{cases}$$

for $i = 1, 2$. Notice that, for compatibility reasons, we have assumed that both systems are "squared"; that is, they have the same number of inputs and outputs. We will also make the following assumptions about each state space realization:

- $\psi_1$ and $\psi_2$ are zero-state-detectable, that is, $u \equiv 0$ and $y \equiv 0$ implies that $x(t) \equiv 0$.

- $\psi_1$ and $\psi_2$ are completely reachable, that is, for any given $x_1$ and $t_1$ there exists a $t_0 \leq t_1$ and an input function $u(\cdot) \in \mathcal{U} \to \mathbb{R}^m$ such that the state can be driven from $x(t_0) = 0$ to $x(t_1) = x_1$.

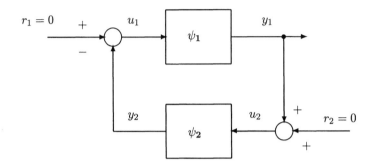

Figure 9.3: Feedback interconnection.

Having laid out the ground rules, we can now state the main theorem of this section.

**Theorem 9.5** *Consider the feedback interconnection (Figure 9.3) of the systems $\psi_1$ and $\psi_2$, and assume that both systems are dissipative with respect to the supply rate*

$$\omega_i(u_i, y_i) = y_i^T Q_i y_i + 2 y_i^T S_i u_i + u_i^T R_i u_i. \tag{9.26}$$

*Then the feedback interconnection of $\psi_1$ and $\psi_2$ is Lyapunov stable (asymptotically stable) if the matrix $\widehat{Q}$ defined by*

$$\widehat{Q} = \begin{bmatrix} Q_1 + \alpha R_2 & -S_1 + \alpha S_2^T \\ -S_1^T + \alpha S_2 & R_1 + \alpha Q_2 \end{bmatrix} \tag{9.27}$$

*is negative semidefinite (negative definite) for some $\alpha > 0$.*

**Proof of Theorem 9.5**: Consider the Lyapunov function candidate

$$\phi(x_1, x_2) = \phi(x_1) + \alpha \phi(x_2) \qquad \alpha > 0$$

where $\phi(x_1)$ and $\phi(x_2)$ are the storage function of the systems $\psi_1$ and $\psi_2$, respectively. Thus, $\phi(x_1, x_2)$ is positive definite by construction. The derivative of $\phi(\cdot, \cdot)$ along the trajectories of the composite state $[x_1, x_2]^T$ is given by

$$\begin{aligned}
\dot{\phi} &= \dot{\phi}_1(x_1) + \alpha \dot{\phi}_2(x_2) \\
&\leq \omega_1(u_1, y_1) + \alpha \omega_2(u_2, y_2) \\
&= (y_1^T Q_1 y_1 + 2 y_1^T S_1 u_1 + u_1^T R_1 u_1) + \alpha(y_2^T Q_2 y_2 + 2 y_2^T S_2 u_2 + u_2^T R_2 u_2) \quad (9.28)
\end{aligned}$$

Substituting $u_1 = -y_2$ and $u_2 = y_1$ into (9.28), we obtain

$$\dot{\phi} = [\; y_1^T \quad y_2^T \;] \left[ \begin{array}{cc} Q_1 + \alpha R_2 & -S_1 + \alpha S_2^T \\ -S_1^T + \alpha S_2 & R_1 + \alpha Q_2 \end{array} \right] \left[ \begin{array}{c} y_1^T \\ y_2^T \end{array} \right]$$

and the result follows. $\square$

This theorem is very powerful and includes several cases of special interest, which we now spell out in Corollaries 9.4 and 9.5.

**Corollary 9.4** *Under the conditions of Theorem 9.5, we have that*

*(a) If both $\phi_1$ and $\phi_2$ are passive, then the feedback system is Lyapunov stable.*

*(b) Asymptotic stability follows if, in addition, one of the following conditions are satisfied:*

    *(1) One of $\phi_1$ and $\phi_2$ is very strictly passive.*

    *(2) Both $\phi_1$ and $\phi_2$ are strictly passive.*

    *(3) Both $\phi_1$ and $\phi_2$ are strictly output passive.*

**Proof:** Setting $\alpha = 1$ in (9.27), we obtain the following:

(a) If both $\phi_1$ and $\phi_2$ are passive, then $Q_i = 0, R_i = 0$, and $S_i = \frac{1}{2}I$, for $i = 1, 2$. With these values, $\widehat{Q} = 0$, and the result follows.

(b1) Assuming that $\phi_1$ is passive and that $\phi_2$ is very strictly passive, we have

- $\phi_1$ passive: $Q_1 = 0, R_1 = 0$, and $S_1 = \frac{1}{2}I$.
- $\phi_2$ very strictly passive: $Q_2 = -\epsilon_2 I, R_2 = -\delta_2 I$, and $S_2 = \frac{1}{2}I$.

Thus

$$\widehat{Q} = \left[ \begin{array}{cc} -\delta_2 I & 0 \\ 0 & -\epsilon_2 I \end{array} \right]$$

which is negative definite, and the result follows by Theorem 9.5. The case $\phi_2$ passive and $\phi_1$ very strictly passive is entirely analogous.

(b2) If both $\phi_1$ and $\phi_2$ are strictly passive, we have

- $Q_i = 0, R_2 = -\delta_i I$, and $S_i = \frac{1}{2}I$.

Thus

$$\widehat{Q} = \begin{bmatrix} -\delta_2 I & 0 \\ 0 & -\delta_1 I \end{bmatrix}$$

which is once again negative definite, and the result follows by Theorem 9.5.

(b3) If both $\phi_1$ and $\phi_2$ are strictly output-passive, we have:

- $Q_i = \epsilon_i I, R_2 = 0$, and $S_i = \frac{1}{2}I$.

Thus

$$\widehat{Q} = \begin{bmatrix} -\epsilon_1 I & 0 \\ 0 & -\epsilon_2 I \end{bmatrix}$$

which is negative definite, and the result follows.                                □

Corollary 9.4 is significant in that it identifies closed-loop stability-asymptotic stability in the sense of Lyapunov for combinations of passivity conditions.

The following corollary is a Lyapunov version of the small gain theorem.

**Corollary 9.5** *Under the conditions of Theorem 9.5, if both $\phi_1$ and $\phi_2$ are finite-gain-stable with gains $\gamma_1$ and $\gamma_2$, the feedback system is Lyapunov-stable (asymptotically stable) if $\gamma_1\gamma_2 \leq 1$ ($\gamma_1\gamma_2 < 1$).*

**Proof:** Under the assumptions of the corollary we have that

- $Q_i = I, R_1 = \frac{\gamma_i}{2}$, and $S_i = 0$.

Thus $\widehat{Q}$ becomes

$$\widehat{Q} = \begin{bmatrix} (\alpha\gamma_2^2 - 1)I & 0 \\ 0 & (\gamma_1^2 - \alpha)I \end{bmatrix}$$

Thus, $\widehat{Q}$ is negative semidefinite, provided that

$$\gamma_1^2 \leq \alpha \tag{9.29}$$
$$\alpha\gamma_2^2 \leq 1 \tag{9.30}$$

and substituting (9.29) into (9.30) leads to the stability result. The case of asymptotic stability is, of course, identical.

□

## 9.9 Nonlinear $\mathcal{L}_2$ Gain

Consider a system $\psi$ of the form

$$\psi : \begin{cases} \dot{x} = f(x) + g(x)u \\ y = h(x) \end{cases} \tag{9.31}$$

As discussed in Section 9.2, this system is finite-gain-stable with gain $\gamma$, if it is dissipative with supply rate $\omega = \frac{1}{2}(\gamma^2 \|u\|^2 - \|y\|^2)$. Assuming that the storage function corresponding to this supply rate is differentiable, the differential dissipation inequality (9.5) implies that

$$\dot{\phi}(t) = \frac{\partial \phi(x)}{\partial x} \dot{x} \leq \omega(t) \leq \frac{1}{2}\gamma^2 \|u\|^2 - \frac{1}{2}\|y\|^2 \leq 0. \tag{9.32}$$

Thus, substituting (9.31) into (9.32) we have that

$$\frac{\partial \phi(x)}{\partial x} f(x) + \frac{\partial \phi(x)}{\partial x} g(x)u \leq \omega(t) \leq \frac{1}{2}\gamma^2 \|u\|^2 - \frac{1}{2}\|y\|^2. \tag{9.33}$$

Adding and subtracting $\frac{1}{2}\gamma^2 u^T u$ and $\frac{1}{2\gamma^2}\frac{\partial \phi}{\partial x}gg^T(\frac{\partial \phi}{\partial x})^T$ to the the left hand side of (9.33) we obtain

$$
\begin{aligned}
\frac{\partial \phi(x)}{\partial x} f(x) + \frac{\partial \phi(x)}{\partial x} g(x)u &= \frac{\partial \phi(x)}{\partial x} f(x) + \frac{\partial \phi(x)}{\partial x} g(x)u + \frac{1}{2}\gamma^2 u^T u - \frac{1}{2}\gamma^2 u^T u \\
&\quad + \frac{1}{2\gamma^2}\frac{\partial \phi}{\partial x}gg^T\left(\frac{\partial \phi}{\partial x}\right)^T - \frac{1}{2\gamma^2}\frac{\partial \phi}{\partial x}gg^T\left(\frac{\partial \phi}{\partial x}\right)^T \\
&= -\frac{1}{2}\gamma^2 \left\| u - \frac{1}{2\gamma^2}g^T(\frac{\partial \phi}{\partial x})^T \right\|^2 + \frac{\partial \phi}{\partial x}f(x) \\
&\quad + \frac{1}{2\gamma^2}\frac{\partial \phi}{\partial x}gg^T\left(\frac{\partial \phi}{\partial x}\right)^T + \frac{1}{2}\gamma^2 \|u\|^2
\end{aligned} \tag{9.34}
$$

Therefore, substituting (9.34) into (9.33) results in

$$-\frac{1}{2}\gamma^2 \left\| u - \frac{1}{2\gamma^2}g^T\left(\frac{\partial \phi}{\partial x}\right)^T \right\|^2 + \frac{\partial \phi}{\partial x}f(x) + \frac{1}{2\gamma^2}\frac{\partial \phi}{\partial x}gg^T\left(\frac{\partial \phi}{\partial x}\right)^T \leq -\frac{1}{2}\|y\|^2$$

or, equivalently

$$\frac{\partial \phi}{\partial x}f(x) + \frac{1}{2\gamma^2}\frac{\partial \phi}{\partial x}gg^T\left(\frac{\partial \phi}{\partial x}\right)^T + \frac{1}{2}\|y\|^2 \leq \frac{1}{2}\gamma^2 \left\| u - \frac{1}{2\gamma^2}g^T\left(\frac{\partial \phi}{\partial x}\right)^T \right\|^2 \leq 0. \tag{9.35}$$

This result is important. According to (9.35), if the system $\psi$ given by (9.31) is finite-gain-stable with gain $\gamma$, then it must satisfy the so-called Hamilton–Jacobi inequality:

$$\mathcal{H} \stackrel{def}{=} \frac{\partial \phi}{\partial x} f(x) + \frac{1}{2\gamma^2} \frac{\partial \phi}{\partial x} gg^T \left( \frac{\partial \phi}{\partial x} \right)^T + \frac{1}{2} \|y\|^2 \leq 0 \qquad (9.36)$$

Finding a function $\phi$ that maximizes $\mathcal{H}$ and satisfies inequality (9.36) is, at best, very difficult. Often, we will be content with "estimating" an upper bound for $\gamma$. This can be done by "guessing" a function $\phi$ and then finding an approximate value for $\gamma$; a process that resembles that of finding a suitable Lyapunov function. The true $\mathcal{L}_2$ gain of $\psi$, denoted $\gamma^*$, is bounded above by $\gamma$, i.e.,

$$0 \leq \gamma^* \leq \gamma$$

**Example 9.1** *Consider the following system:*

$$\begin{cases} \dot{x}_1 &= x_1^2 x_2 - 2x_1 x_2^2 - x_1^3 + \beta x_1 u\,, \quad \beta > 0 \\ \dot{x}_2 &= --x_1^3 - x_2^3 + \beta x_2 u \\ y &= x_1^2 + x_2^2 \end{cases}$$

*i.e.*

$$f(x) = \begin{bmatrix} x_1^2 x_2 - 2x_1 x_2^2 - x_1^3 \\ -x_1^3 - x_2^3 \end{bmatrix}, \quad g(x) = \begin{bmatrix} \beta x_1 \\ \beta x_2 \end{bmatrix}, \quad h(x) = (x_1^2 + x_2^2)$$

*To estimate the $\mathcal{L}_2$ norm, we consider the storage function "candidate" $\phi(x) = \frac{1}{2}(x_1^2 + x_2^2)$. With this function, the three terms in the Hamilton–Jacobi inequality (9.36) can be obtained as follows:*

(i) $\frac{\partial \phi}{\partial x} f(x)$

$$\begin{aligned} \frac{\partial \phi}{\partial x} f(x) &= [x_1, x_2] \begin{bmatrix} x_1^2 x_2 - 2x_1 x_2^2 - x_1^3 \\ -x_1^3 - x_2^3 \end{bmatrix} \\ &= -(x_1^2 + x_2^2)^2 \\ &= -\|x\|^4 \end{aligned}$$

(ii) $\frac{1}{2\gamma^2} \frac{\partial \phi}{\partial x} gg^T \left( \frac{\partial \phi}{\partial x} \right)^T$

$$\frac{\partial \phi}{\partial x} g(x) = [x_1, x_2] \begin{bmatrix} \beta x_1 \\ \beta x_2 \end{bmatrix} = \beta x_1^2 + \beta x_2^2 = \beta \|x\|^2$$

$$\Rightarrow \quad \frac{1}{2\gamma^2} \frac{\partial \phi}{\partial x} gg^T \left( \frac{\partial \phi}{\partial x} \right)^T = \frac{\beta^2}{2\gamma^2} \|x\|^4$$

(iii) $\frac{1}{2} \|y\|^2$

$$\frac{1}{2}\|y\|^2 = \frac{1}{2}(x_1^2 + x_2^2)^2 = \frac{1}{2}\|x\|^4$$

*Combining these results, we have that*

$$\mathcal{H} = -\|x\|^4 + \frac{\beta}{2\gamma^2}\|x\|^4 + \frac{1}{2}\|x\|^4 \;\leq\; 0$$

*or*

$$\frac{\beta}{2\gamma^2} \leq \frac{1}{2}$$

$$\Rightarrow \quad \gamma \geq \beta$$

$\square$

## 9.9.1   Linear Time-Invariant Systems

It is enlightening to consider the case of linear time-invariant systems as a special case. Assuming that this is the case, we have that

$$\psi : \begin{cases} \dot{x} = Ax + Bu \\ y = Cx \end{cases}$$

that is, $f(x) = Ax$, $g(x) = B$, and $h(x) = Cx$. Taking $\phi(x) = \frac{1}{2}x^T P x$, $(P > P, \; P = P^T)$ and substituting in (9.36), we obtain

$$\mathcal{H} = x^T \left[ PA + \frac{1}{2\gamma^2}PBB^T P + \frac{1}{2}C^T C \right] x \;\leq\; 0. \tag{9.37}$$

Taking the transpose of (9.37), we obtain

$$\mathcal{H}^T = x^T \left[ A^T P + \frac{1}{2\gamma^2}PBB^T P + \frac{1}{2}C^T C \right] x \;\leq\; 0. \tag{9.38}$$

Thus, adding (9.38) and (9.38), we obtain

$$\mathcal{H} + \mathcal{H}^T = x^T \left[ \underbrace{PA + A^T P + \frac{1}{2\gamma^2}PBB^T P + \frac{1}{2}C^T C}_{\mathcal{R}} \right] x \;\leq\; 0. \tag{9.39}$$

Thus, the system $\psi$ has a finite $\mathcal{L}_2$ gain less than or equal to $\gamma$ provided that for some $\gamma > 0$

$$PA + A^T P + \frac{1}{2\gamma^2} PBB^T P + \frac{1}{2} C^T C \leq 0. \tag{9.40}$$

The left-hand side of inequality (9.40) is well known and has played and will continue to play a very significant role in linear control theory. Indeed, the equality

$$\mathcal{R} \overset{def}{=} PA + A^T P + \frac{1}{2\gamma^2} PBB^T P + \frac{1}{2} C^T C = 0 \tag{9.41}$$

is known as the *Riccati equation*. In the linear case, further analysis leads to a stronger result. It can be shown that the linear time-invariant system $\psi$ has a finite gain less than or equal to $\gamma$ if and only if the Riccati equation (9.41) has a positive definite solution. See Reference [23] or [100].

## 9.9.2   Strictly Output Passive Systems

It was pointed out in Section 9.3 (lemma 9.2) that strictly output passive systems have a finite $\mathcal{L}_2$ gain. We now discuss how to compute the $\mathcal{L}_2$ gain $\gamma$. Consider a system of the form

$$\psi : \begin{cases} \dot{x} = f(x) + g(x)u \\ y = h(x) \end{cases}$$

and assume that there exists a differentiable function $\phi_1 > 0$ satisfying (9.24)–(9.25), i.e.,

$$\frac{\partial \phi}{\partial x} f(x) \leq -\epsilon h^T(x) h(x) \tag{9.42}$$

$$\frac{\partial \phi}{\partial x} g(x) \leq h^T(x) \tag{9.43}$$

which implies that $\psi$ is strictly output-passive. To estimate the $\mathcal{L}_2$ gain of $\psi$, we consider the Hamilton–Jabobi inequality (9.36)

$$\mathcal{H} \overset{def}{=} \frac{\partial \phi}{\partial x} f(x) + \frac{1}{2\gamma^2} \frac{\partial \phi}{\partial x} gg^T \left( \frac{\partial \phi}{\partial x} \right)^T + \frac{1}{2} \|y\|^2 \leq 0 \tag{9.44}$$

Letting $\phi = k\phi_1$ for some $k > 0$ and substituting (9.42)-(9.43) into (9.44), we have that $\psi$ has $\mathcal{L}_2$ gain less than or equal to $\gamma$ if and only if

$$-k\epsilon h^T(x) h(x) + \frac{1}{2\gamma^2} k^2 h^T(x) h(x) + \frac{1}{2} h^T(x) h(x) \leq 0$$

or, extracting the common factor $h^T(x)h(x)$

$$-k\epsilon + \frac{1}{2\gamma^2}k^2 + \frac{1}{2} \leq 0$$

$$\Longleftrightarrow \quad \gamma \geq \frac{k}{\sqrt{2k\epsilon - 1}}$$

and choosing $k = 1/\epsilon$ we conclude that

$$\gamma \geq \epsilon^{-1}.$$

## 9.10   Some Remarks about Control Design

Control systems design is a difficult subject in which designers are exposed to a multiplicity of specifications and design constraints. One of the main reasons for the use of feedback, however, is its ability to reduce the effect of undesirable exogenous signals over the system's output. Therefore, over the years a lot of research has been focused on developing design techniques that employ this principle as the main design criteria. More explicitly, control design can be viewed as solving the following problem: find a control law that

1- Stabilizes the closed-loop system, in some prescribed sense.

2- Reduces the effect of the exogenous input signals over the desired output, also in some specific sense.

Many problems can be cast in this form. To discuss control design in some generality, it is customary to replace the actual plant in the feedback configuration with a more general *system* known as the generalized plant. The standard setup is shown in Figure 9.4, where

$C=$ controller to be designed,

$G=$ generalized plant,

$y=$ measured output,

$z=$ output to be regulated (such as tracking errors and actuator signals),

$u=$ control signal, and

$d=$ "exogenous" signals (such as disturbances and sensor noise).

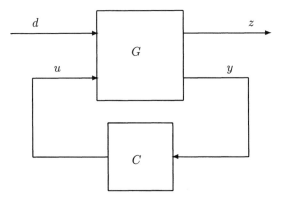

Figure 9.4: Standard setup.

To fix ideas and see that this standard setup includes the more "classical " feedback configuration, consider the following example:

**Example 9.2** *Consider the feedback system shown in Figure 9.5.*

*Suppose that our interest is to design a controller C to reduce the effect of the ex-ogenous disturbance d over the signals $y_2$ and $e_2$ (representing the input and output of the plant). This problem can indeed be studied using the standard setup of Figure 9.4. To see this, we must identify the inputs and outputs in the generalized plant G. In our case, the several variables in the standard setup of Figure 9.4 correspond to the following variables in Figure 9.5:*

- $y = e_1$ *(the controller input)*

- $u = y_1$ *(the controller output)*

- *d: disturbance (same signal in both Figure 9.5 and the generalized setup of Figure 9.4).*

- $z = \begin{bmatrix} e_2 \\ y_2 \end{bmatrix}$, *i.e., z is a vector whose components are input and output of the plant.*

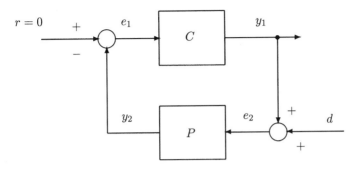

Figure 9.5: Feedback Interconnection used in example 9.2.

A bit of work shows that the feedback system of Figure 9.5 can be redrawn as shown in Figure 9.6, which has the form of the standard setup of Figure 9.4. The control design problem can now be restated as follows. Find $C$ such that

(i) stabilizes the feedback system of Figure 9.6, in some specific sense, and

(ii) reduces the effect of the exogenous signal $d$ on the desired output $z$.                    □

A lot of research has been conducted since the early 1980s on the synthesis of controllers that provide an "optimal" solution to problems such as the one just described. The properties of the resulting controllers depend in an essential manner on the spaces of functions used as inputs and outputs. Two very important cases are the following:

- **$\mathcal{L}_2$ signals:**

  In this case the problem is to find a stabilizing controller that minimizes the $\mathcal{L}_2$ gain of the (closed-loop) system mapping $d \to z$. For linear time-invariant systems, we saw that the $\mathcal{L}_2$ gain is the $H_\infty$ norm of the transfer function mapping $d \to z$, and the theory behind the synthesis of these controllers is known as $H_\infty$ optimization. The $H_\infty$ optimization theory was initiated in 1981 by G. Zames [99], and the synthesis of $H_\infty$ controllers was solved during the 1980s, with several important contributions by several authors. See References [25] and [100] for a comprehensive treatment of the $H_\infty$ theory for linear time-invariant systems.

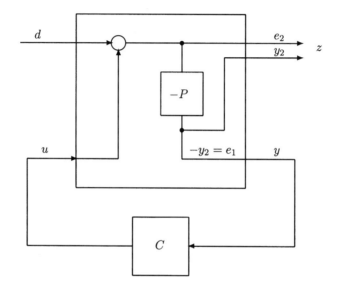

Figure 9.6: Standard setup.

- $\mathcal{L}_\infty$ **signals:**

  Similarly, in this case the problem is to find a stabilizing controller that minimizes the $\mathcal{L}_\infty$ gain of the (closed-loop) system mapping $d \to z$. For linear time-invariant systems, we saw that the $\mathcal{L}_\infty$ gain is the $\mathcal{L}_1$ norm of the transfer function mapping $d \to z$, and the theory behind the synthesis of these controllers is known as $\mathcal{L}_1$ optimization. The $\mathcal{L}_1$ optimization theory was proposed in 1986 by M. Vidyasagar [89]. The first solution of the $\mathcal{L}_1$-synthesis problem was obtained in a series of papers by M. A. Dahleh and B. Pearson [16]–[19]. See also the survey [20] for a comprehensive treatment of the $\mathcal{L}_1$ theory for linear time-invariant systems.

All of these references deal exclusively with linear time-invariant systems. In the remainder of this chapter we present an introductory treatment of the $\mathcal{L}_2$ control problem for nonlinear systems.

We will consider the standard configuration of Figure 6.4 and consider a nonlinear system $\psi$ of the form

$$\psi : \begin{cases} \dot{x} &=& f(x, u, d) \\ y &=& g(x, u, d) \\ z &=& h(x, u, d) \end{cases} \tag{9.45}$$

where $u$ and $d$ represent the control and exogenous input, respectively, and $d$ and $y$ represent the measured and regulated output, respectively. We look for a controller $C$ that stabilizes the closed-loop system and minimizes the $\mathcal{L}_2$ gain of the mapping from $d$ to $z$. This problem can be seen as an extension of the $H_\infty$ optimization theory to the case of nonlinear plants, and therefore it is often referred to as the nonlinear $H_\infty$-optimization problem.

## 9.11  Nonlinear $\mathcal{L}_2$-Gain Control

Solving the nonlinear $\mathcal{L}_2$-gain design problem as described above is very difficult. Instead, we shall content ourselves with solving the following suboptimal control problem. Given a "desirable" exogenous signal attenuation level, denoted by $\gamma_1$, find a control $C_1$ such that the mapping from $d \to w$ has an $\mathcal{L}_2$ gain less than or equal to $\gamma_1$. If such a controller $C_1$ exists, then we can choose $\gamma_2 < \gamma_1$ and find a new controller $C_2$ such that the mapping from $d \to w$ has an $\mathcal{L}_2$ gain less than or equal to $\gamma_2$. Iterating this procedure can lead to a controller $C$ that approaches the "optimal" (i.e., minimum) $\gamma$.

We will consider the *state feedback* suboptimal $\mathcal{L}_2$-gain optimization problem. This problem is sometimes referred to as the *full information case*, because the full state is assumed to be available for feedback. We will consider a state space realization of the form

$$\psi : \begin{cases} \dot{x} = a(x) + b(x)u + g(x)d, & u \in \mathbb{R}^m, d \in \mathbb{R}^r \\ z = \begin{bmatrix} h(x) \\ u \end{bmatrix} \end{cases} \qquad (9.46)$$

where $a$, $b$, $g$ and $h$ are assumed to be $C^k$, with $k \geq 2$.

**Theorem 9.6** *The closed-loop system of equation (9.46) has a finite $\mathcal{L}_2$ gain $\leq \gamma$ if and only if the Hamilton-Jacobi inequality $\widehat{\mathcal{H}}$ given by*

$$\widehat{\mathcal{H}} = \frac{\partial \phi}{\partial x} a(x) + \frac{1}{2} \frac{\partial \phi}{\partial x} \left[ \frac{1}{\gamma^2} g(x) g^T(x) - b(x) b^T(x) \right] \left( \frac{\partial \phi}{\partial x} \right)^T + \frac{1}{2} h^T(x) h(x) \leq 0 \qquad (9.47)$$

*has a solution $\phi \geq 0$. The control law is given by*

$$u = -b^T(x) \left( \frac{\partial \phi}{\partial x} \right)^T (x). \qquad (9.48)$$

**Proof**: We only prove sufficiency. See Reference [85] for the necessity part of the proof.

Assume that $\phi \geq 0$ is a solution of $\widehat{\mathcal{H}}$. Substituting $u$ into the system equations (9.46), we obtain

$$\dot{x} = a(x) - b(x) b^T(x) \left( \frac{\partial \phi}{\partial x} \right)^T + g(x) d \qquad (9.49)$$

$$z = \begin{bmatrix} h(x) \\ -b^T(x) \left( \frac{\partial \phi}{\partial x} \right)^T \end{bmatrix}. \qquad (9.50)$$

Substituting (9.49) and (9.50) into the Hamilton–Jacobi inequality (9.36) with

$$f(x) = a(x) - b(x) b^T(x) \left( \frac{\partial \phi}{\partial x} \right)^T (x)$$

$$h(x) = \begin{bmatrix} h(x) \\ -b^T(x) \left( \frac{\partial \phi}{\partial x} \right)^T \end{bmatrix}$$

results in

$$\mathcal{H} = \frac{\partial \phi}{\partial x} \left( a(x) - b(x) b^T(x) \left( \frac{\partial \phi}{\partial x} \right)^T \right) + \frac{1}{2\gamma^2} \frac{\partial \phi}{\partial x} g(x) g^T(x) \left( \frac{\partial \phi}{\partial x} \right)^T$$

$$+ \frac{1}{2} \left[ h(x)^T - \frac{\partial \phi}{\partial x} b(x) \right] \begin{bmatrix} h(x) \\ -b^T(x) \left( \frac{\partial \phi}{\partial x} \right)^T \end{bmatrix}$$

which implies (9.47), and so the closed-loop system has a finite $\mathcal{L}_2$ gain $\gamma$.  $\square$

## 9.12 Exercises

(9.1) Prove Lemma 9.1.

(9.2) Prove Lemma 9.2.

(9.3) Consider the pendulum-on-a-cart system discussed in Chapter 1. Assuming for simplicity that the moment of inertia of the pendulum is negligible, and assuming that the output equation is

$$y = \dot{x} = x_2$$

  (a) Find the kinetic energy $K_1$ of the cart.

  (b) Find the kinetic energy $K_2$ of the pendulum.

  (c) Find the potential energy $P$ in the cart-pendulum system.

  (d) Using the previous results, find the total energy $E = K_1 + K_2 + P = K + P$ stored in the cart-pendulum system.

  (e) Defining variables

$$q = \begin{bmatrix} x \\ \theta \end{bmatrix}, \quad M = \begin{bmatrix} M + m & ml\cos\theta \\ ml\cos & ml^2\theta \end{bmatrix}, \quad u = \begin{bmatrix} f \\ 0 \end{bmatrix}$$

  show that the energy stored in the pendulum can be expressed as

$$E = \frac{1}{2}\dot{q}^T M(q)\dot{q} + mgl(\cos\theta - 1)$$

  (f) Computing the derivative of $E$ along the trajectories of the system, show that the pendulum-on-a-cart system with input $u = f$ and output $\dot{x}$ is a passive system.

(9.4) Consider the system

$$\psi : \begin{cases} \dot{x}_1 &= -\alpha x_1 - x_2 + u, \quad \alpha > 0 \\ \dot{x}_2 &= x_1 - x_1^2 x_2 \\ y &= x_1 \end{cases}$$

  (a) Determine whether $\psi$ is

  (i) passive.

  (ii) strictly passive.

  (iii) strictly output-passive.

  (iv) has a finite $\mathcal{L}_2$ gain.

  (b) If the answer to part (a)(iv) is affirmative, find an upper bound on the $\mathcal{L}_2$ gain.

(9.5) Consider the system

$$\psi : \begin{cases} \dot{x}_1 &= -x_1 + x_2^2 + u, \quad \alpha > 0 \\ \dot{x}_2 &= -x_1 x_2 - x_2 \\ y &= \beta x_1 \qquad\qquad \beta > 0. \end{cases}$$

(a) Determine whether $\psi$ is

(i) passive.

(ii) strictly passive.

(iii) strictly output-passive.

(iv) has a finite $\mathcal{L}_2$ gain.

(b) If the answer to part (a)(iv) is affirmative, find an upper bound on the $\mathcal{L}_2$ gain.

# Notes and References

The theory of dissipativity of systems was initiated by Willems is his seminal paper [91], on which Section 9.1 is based. The beauty of the Willems formulation is its generality. Reference [91] considers a very general class of nonlinear systems and defines *dissipativity* as an extension of the concept of passivity as well as supply rates and storage functions as generalizations of input power and stored energy, respectively. Unlike the classical notion of passivity, which was introduced as an input–output concept, state space realizations are central to the notion of dissipativity in Reference [91].

The definition of $QSR$ dissipativity is due to Hill and Moylan [30], [29]. Employing the $QSR$ supply rate, dissipativity can be interpreted as an input–output property. Section 9.3 is based on Hill and Moylan's original work [30]. Theorem 9.1 is from Willems [91], Theorem 1. The connections between dissipativity and stability in the sense of Lyapunov were extensively explored by Willems, [93], [94], as well as [91]. Theorem 9.3 follows Theorem 6 in [91]. See also Section 3.2 in van der Schaft [85]. Section 9.6 follows Reference [30]. Stability of feedback interconnections was studies in detail by Hill and Moylan. Section 9.8 follows Reference [31] very closely. Nonlinear $\mathcal{L}_2$ gain control (or nonlinear $H_\infty$ control) is currently a very active area of research. Sections 9.9 and 9.11 follow van der Schaft [86], and [87]. See also References [6] and [38] as well as the excellent monograph [85] for a comprehensive treatment of this beautiful subject.

# Chapter 10

# Feedback Linearization

In this chapter we look at a class of control design problems broadly described as *feedback linearization*. The main problem to be studied is: Given a nonlinear dynamical system, find a transformation that renders a new dynamical system that is linear time-invariant. Here, by *transformation* we mean a control law plus possibly a change of variables. Once a linear system is obtained, a secondary control law can be designed to ensure that the overall closed-loop system performs according to the specifications. This time, however, the design is carried out using the new linear model and any of the well-established linear control design techniques.

Feedback linearization was a topic of much research during the 1970s and 1980s. Although many successful applications have been reported over the years, feedback linearization has a number of limitations that hinder its use, as we shall see. Even with these shortcomings, feedback linearization is a concept of paramount importance in nonlinear control theory. The intention of this chapter is to provide a brief introduction to the subject. For simplicity, we will limit our attention to single-input–single-output systems and consider only *local results*. See the references listed at the end of this chapter for a more complete coverage.

## 10.1 Mathematical Tools

Before proceeding, we need to review a few basic concepts from differential geometry. Throughout this chapter, whenever we write $D \subset \mathbb{R}^n$, we assume that $D$ is an open and connected subset of $\mathbb{R}^n$.

We have already encountered the notion of *vector field* in Chapter 1. A vector field

255

is a function $f : D \subset \mathbb{R}^n \to \mathbb{R}^n$, namely, a mapping that assigns an $n$-dimensional vector to every point in the $n$-dimensional space. Defined in this way, a vector field is an $n$-dimensional "column." It is customary to label <u>covector field</u> to the transpose of a vector field. Throughout this chapter we will assume that all functions encountered are sufficiently smooth, that is, they have continuous partial derivatives of any required order.

### 10.1.1 Lie Derivative

When dealing with stability in the sense of Lyapunov we made frequent use of the notion of "time derivative of a scalar function $V$ along the trajectories of a system $\dot{x} = f(x)$." As we well know, given $V : D \to \mathbb{R}$ and $\dot{x} = f(x)$, we have

$$\dot{V} = \frac{\partial V}{\partial x} f(x) = \nabla V f(x) = L_f V(x).$$

A slightly more abstract definition leads to the concept of *Lie derivative*.

**Definition 10.1** *Consider a scalar function $h : D \subset \mathbb{R}^n \to \mathbb{R}$ and a vector field $f : D \subset \mathbb{R}^n \to \mathbb{R}^n$. The Lie derivative of $h$ with respect to $f$, denoted $L_f h$, is given by*

$$L_f h(x) = \frac{\partial h}{\partial x} f(x). \tag{10.1}$$

Thus, going back to Lyapunov functions, $\dot{V}$ is merely the Lie derivative of $V$ with respect to $f(x)$. The Lie derivative notation is usually preferred whenever higher order derivatives need to be calculated. Notice that, given two vector fields $f, g : D \subset \mathbb{R}^n \to \mathbb{R}^n$ we have that

$$L_f h(x) = \frac{\partial h}{\partial x} f(x), \quad L_g h(x) = \frac{\partial h}{\partial x} g(x)$$

and

$$L_g L_f h(x) = L_g[L_f h(x)] = \frac{\partial (L_f h)}{\partial x} g(x)$$

and in the special case $f = g$,

$$L_f L_f h(x) = L_f^2 h(x) = \frac{\partial (L_f h)}{\partial x} f(x).$$

**Example 10.1** *Let*

$$h(x) = \frac{1}{2}(x_1^2 + x_2^2)$$

$$f(x) = \begin{bmatrix} -x_2 \\ -x_1 - \mu(1 - x_1^2)x_2 \end{bmatrix}, \quad g(x) = \begin{bmatrix} -x_1 - x_1 x_2^2 \\ -x_2 + x_1^2 x_2 \end{bmatrix}.$$

*Then, we have*

- $L_f h(x)$:

$$L_f h(x) \;=\; \frac{\partial h}{\partial x} f(x)$$

$$= \; [x_1 \;\; x_2] \begin{bmatrix} -x_2 \\ -x_1 - \mu(1 - x_1^2)x_2 \end{bmatrix}$$

$$= \; -\mu(1 - x_1^2)x_2^2.$$

- $L_g h(x)$:

$$L_f h(x) \;=\; \frac{\partial h}{\partial x} g(x)$$

$$= \; [x_1 \;\; x_2] \begin{bmatrix} -x_1 - x_1 x_2^2 \\ -x_2 + x_1^2 x_2 \end{bmatrix}$$

$$= \; -(x_1^2 + x_2^2).$$

- $L_f L_g h(x)$:

$$L_f L_g h(x) \;=\; \frac{\partial (L_g h)}{\partial x} f(x)$$

$$= \; -2 \,[x_1 \;\; x_2] \begin{bmatrix} -x_2 \\ -x_1 - \mu(1 - x_1^2)x_2 \end{bmatrix}$$

$$= \; 2\mu(1 - x_1^2)x_2^2.$$

- $L_g^2 h(x)$:

$$L_g^2 h(x) \;=\; \frac{\partial (L_g h)}{\partial x} g(x)$$

$$= \; -2 \,[x_1 \;\; x_2] \begin{bmatrix} -x_1 - x_1 x_2^2 \\ -x_2 + x_1^2 x_2 \end{bmatrix}$$

$$= \; 2(x_1^2 + x_2^2).$$

$\square$

### 10.1.2  Lie Bracket

**Definition 10.2** *Consider the vector fields $f, g : D \subset \mathbb{R}^n \to \mathbb{R}^n$. The Lie bracket of $f$ and $g$, denoted by $[f, g]$, is the vector field defined by*

$$[f, g](x) = \frac{\partial g}{\partial x} f(x) - \frac{\partial f}{\partial x} g(x). \tag{10.2}$$

**Example 10.2** *Letting*

$$f(x) = \begin{bmatrix} -x_2 \\ -x_1 - \mu(1 - x_1^2)x_2 \end{bmatrix}, \quad g(x) = \begin{bmatrix} x_1 \\ x_2 \end{bmatrix}$$

*we have*

$$
\begin{aligned}
[f, g](x) &= \frac{\partial g}{\partial x} f(x) - \frac{\partial f}{\partial x} g(x) \\
&= \begin{bmatrix} 1 & 0 \\ 0 & 1 \end{bmatrix} \begin{bmatrix} -x_2 \\ -x_1 - \mu(1 - x_1^2)x_2 \end{bmatrix} - \begin{bmatrix} 0 & -1 \\ -1 + 2\mu x_1 x_2 & -\mu(1 - x_1^2) \end{bmatrix} \begin{bmatrix} x_1 \\ x_2 \end{bmatrix} \\
&= \begin{bmatrix} 0 \\ -2\mu x_1^2 x_2 \end{bmatrix}.
\end{aligned}
$$

$\square$

The following notation, frequently used in the literature, is useful when computing repeated bracketing:

$$[f, g](x) \overset{def}{=} ad_f g(x)$$

and

$$
\begin{aligned}
ad_f^0 g &= g \\
ad_f^i g &= [f, ad_f^{i-1} g].
\end{aligned}
$$

Thus,

$$
\begin{aligned}
ad_f^2 g &= [f, ad_f g] = [f, [f, g]] \\
ad_f^3 g &= [f, ad_f^2 g] = [f, [f, [f, g]]].
\end{aligned}
$$

The following lemma outlines several useful properties of Lie brackets.

**Lemma 10.1** : *Given* $f_{1,2} : D \subset \mathbb{R}^n \to \mathbb{R}^n$, *we have*

(i) *Bilinearity:*

    (a)       $[\alpha_1 f_1 + \alpha_2 f_2, g] = \alpha_1[f_1, g] + \alpha_2[f_2, g].$
    (b)       $[f, \alpha_1 g_1 + \alpha_2 g_2] = \alpha_1[f, g_1] + \alpha_2[f, g_2].$

(ii) *Skew commutativity:*

$$[f, g] = -[g, f].$$

(iii) *Jacobi identity: Given vector fields* $f$ *and* $g$ *and a real-valued function* $h$, *we obtain*

$$L_{[f,g]}h = L_f L_g h(x) - L_g L_f h.$$

*where* $L_{[f,g]}h$ *represents the Lie derivative of* $h$ *with respect to the vector* $[f, g]$.

### 10.1.3 Diffeomorphism

**Definition 10.3** : *(Diffeomorphism) A function $f : D \subset \mathbb{R}^n \to \mathbb{R}^n$ is said to be a diffeomorphism on $D$, or a local diffeomorphism, if*

(i) *it is continuously differentiable on $D$, and*

(ii) *its inverse $f^{-1}$, defined by*

$$f^{-1}\left(f(x)\right) = x \quad \forall x \in D$$

*exists and is continuously differentiable.*

The function $f$ is said to be a global diffeomorphism if in addition

(i) $D = \mathbb{R}^n$, and

(ii) $\lim_{x \to \infty} \|f(x)\| = \infty$.

The following lemma is very useful when checking whether a function $f(x)$ is a local diffeomorphism.

**Lemma 10.2** *Let $f(x) : D \in \mathbb{R}^n \to \mathbb{R}^n$ be continuously differentiable on $D$. If the Jacobian matrix $Df = \nabla f$ is nonsingular at a point $x_0 \in D$, then $f(x)$ is a diffeomorphism in a subset $\omega \subset D$.*

**Proof**: An immediate consequence of the inverse function theorem. □

### 10.1.4 Coordinate Transformations

For a variety of reasons it is often important to perform a coordinate change to transform a state space realization into another. For example, for a linear time-invariant system of the form

$$\dot{x} = Ax + Bu$$

we can define a coordinate change $z = Tx$, where $T$ is a nonsingular constant matrix. Thus, in the new variable $z$ the state space realization takes the form

$$\dot{z} = TAT^{-1}z + TBu \;=\; \bar{A}z + \bar{B}u.$$

This transformation was made possible by the nonsingularity of $T$. Also because of the existence of $T^{-1}$, we can always recover the original state space realization. Given a nonlinear state space realization of an affine system of the form

$$\dot{x} = f(x) + g(x)u$$

a diffeomorphism $T(x)$ can be used to perform a coordinate transformation. Indeed, assuming that $T(x)$ is a diffeomorphism and defining $z = T(x)$, we have that

$$\dot{z} = \frac{\partial T}{\partial x}\dot{x} = \frac{\partial T}{\partial x}\left[f(x) + g(x)u\right].$$

Given that $T$ is a diffeomorphism, we have that $\exists T^{-1}$ from which we can recover the original state space realization, that is, knowing $z$

$$x = T^{-1}(z).$$

**Example 10.3** *Consider the following state space realization*

$$\dot{x} = \begin{bmatrix} 0 \\ x_1 \\ x_2 - 2x_1 x_2 + x_1^2 \end{bmatrix} + \begin{bmatrix} 1 \\ -2x_1 \\ 4x_1 x_2 \end{bmatrix} u$$

*and consider the coordinate transformation:*

$$x = T(x) = \begin{bmatrix} x_1 \\ x_1^2 + x_2 \\ x_2^2 + x_3 \end{bmatrix}$$

$$\Rightarrow \frac{\partial T}{\partial x}(x) = \begin{bmatrix} 1 & 0 & 0 \\ 2x_1 & 1 & 0 \\ 0 & 2x_2 & 1 \end{bmatrix}.$$

*Therefore we have,*

$$\dot{z} = \begin{bmatrix} 1 & 0 & 0 \\ 2x_1 & 1 & 0 \\ 0 & 2x_2 & 1 \end{bmatrix} \begin{bmatrix} 0 \\ x_1 \\ x_2 - 2x_1 x_2 + x_1^2 \end{bmatrix} + \begin{bmatrix} 1 & 0 & 0 \\ 2x_1 & 1 & 0 \\ 0 & 2x_2 & 1 \end{bmatrix} \begin{bmatrix} 1 \\ -2x_1 \\ 4x_1 x_2 \end{bmatrix} u$$

*and*

$$\dot{z} = \begin{bmatrix} 0 \\ x_1 \\ x_1^2 + x_2 \end{bmatrix} + \begin{bmatrix} 1 \\ 0 \\ 0 \end{bmatrix} u$$

*and substituting $x_1 = z_1$, $x_2 = z_2 - z_1^2$, we obtain*

$$\dot{z} = \begin{bmatrix} 0 \\ z_1 \\ z_2 \end{bmatrix} + \begin{bmatrix} 1 \\ 0 \\ 0 \end{bmatrix} u.$$

□

## 10.1.5 Distributions

Throughout the book we have made constant use of the concept of vector space. The backbone of linear algebra is the notion of linear independence in a vector space. We recall, from Chapter 2, that a finite set of vectors $S = \{x_1, x_2, \cdots, x_p\}$ in $\mathbb{R}^n$ is said to be linearly *dependent* if there exist a corresponding set of real numbers $\{\lambda_i\}$, not all zero, such that

$$\sum_i \lambda_i x_i = \lambda_1 x_1 + \lambda_2 x_2 + \cdots + \lambda_p x_p = 0.$$

On the other hand, if $\sum_i \lambda_i x_i = 0$ implies that $\lambda_i = 0$ for each $i$, then the set $\{x_i\}$ is said to be linearly *independent*.

Given a set of vectors $S = \{x_1, x_2, \cdots, x_p\}$ in $\mathbb{R}^n$, a linear combination of those vector defines a new vector $x \in \mathbb{R}^n$, that is, given real numbers $\lambda_1, \lambda_2, \cdots \lambda_p$,

$$x = \lambda_1 x_1 + \lambda_2 x_2 + \cdots + \lambda_p x_p \qquad \in \mathbb{R}^n$$

the set of <u>all</u> linear combinations of vectors in $S$ generates a subspace $\mathcal{M}$ of $\mathbb{R}^n$ known as the *span* of $S$ and denoted by span$\{S\} = $ span$\{x_1, x_2, \cdots, x_p\}$, i.e.,

$$\text{span}\{x_1, x_2, \cdots, x_p\} = \{x \in \mathbb{R}^n : x = \lambda_1 x_1 + \lambda_2 x_2 + \cdots + \lambda_p x_p, \ \lambda_i \in \mathbb{R}\}.$$

The concept of distribution is somewhat related to this concept.

Now consider a differentiable function $f : D \subset \mathbb{R}^n \to \mathbb{R}^n$. As we well know, this function can be interpreted as a vector field that assigns the $n$-dimensional vector $f(x)$ to each point $x \in D$. Now consider "$p$" vector fields $f_1, f_2, \cdots, f_p$ on $D \subset \mathbb{R}^n$. At any fixed point $x \in D$ the functions $f_i$ generate vectors $f_1(x), f_2(x), \cdots, f_p(x) \in \mathbb{R}^n$ and thus

$$\Delta(x) = \text{span}\{f_1(x), \cdots, f_p(x)\}$$

is a subspace of $\mathbb{R}^n$. We can now state the following definition.

**Definition 10.4** *(Distribution) Given an open set $D \subset \mathbb{R}^n$ and smooth functions $f_1, f_2, \cdots, f_p$, $D \to \mathbb{R}^n$, we will refer to as a smooth distribution $\Delta$ to the process of assigning the subspace*

$$\Delta = span\{f_1, f_2, \cdots, f_p\}$$

*spanned by the values of $x \in D$.*

We will denote by $\Delta(x)$ the "values" of $\Delta$ at the point $x$. The dimension of the distribution $\Delta(x)$ at a point $x \in D$ is the dimension of the subspace $\Delta(x)$. It then follows that

$$\dim(\Delta(x)) = \text{rank}([f_1(x), f_2(x), \cdots, f_p(x)])$$

i.e., the dimension of $\Delta(x)$ is the rank of the matrix $[f_1(x), f_2(x), \cdots, f_p(x)]$.

**Definition 10.5** *A distribution $\Delta$ defined on $D \subset \mathbb{R}^n$ is said to be nonsingular if there exists an integer $d$ such that*

$$dim(\Delta(x)) = d \qquad \forall x \in D.$$

*If this condition is not satisfied, then $\Delta$ is said to be of variable dimension.*

**Definition 10.6** *A point $x_0$ of $D$ is said to be a regular point of the distribution $\Delta$ if there exist a neighborhood $D_0$ of $x_0$ with the property that $\Delta$ is nonsingular on $D_0$. Each point of $D$ that is not a regular point is said to be a singularity point.*

**Example 10.4** *Let $D = \{x \in \mathbb{R}^2 : x_1 + x_2 \neq 0\}$ and consider the distribution $\Delta = span\{f_1, f_2\}$, where*

$$f_1 = \begin{bmatrix} 1 \\ 0 \end{bmatrix}, \quad f_2 = \begin{bmatrix} 1 \\ x_1 + x_2 \end{bmatrix}.$$

*We have*

$$dim(\Delta(x)) = rank\left(\begin{bmatrix} 1 & 1 \\ 0 & x_1 + x_2 \end{bmatrix}\right).$$

*Then $\Delta$ has dimension 2 everywhere in $\mathbb{R}^2$, except along the line $x_1 + x_2 = 0$. It follows that $\Delta$ is nonsingular on $D$ and that every point of $D$ is a regular point.* □

**Example 10.5** *Consider the same distribution used in the previous example, but this time with $D = \mathbb{R}^2$. From our analysis in the previous example, we have that $\Delta$ is not regular since $dim(\Delta(x))$ is not constant over $D$. Every point on the line $x_1 + x_2 = 0$ is a singular point.* □

**Definition 10.7** *(Involutive Distribution) A distribution $\Delta$ is said to be involutive if $g_1 \in \Delta$ and $g_2 \in \Delta \Rightarrow [g_1, g_2] \in \Delta$.*

It then follows that $\Delta = \mathrm{span}\{f_1, f_2, \cdots, f_p\}$ is involutive if and only if

$$\mathrm{rank}\left([f_1(x), \cdots, f_p(x)]\right) \equiv \mathrm{rank}\left([f_1(x), \cdots, f_p(x), [f_i, f_j]]\right), \quad \forall x \text{ and all } i, j$$

**Example 10.6** *Let $D = \mathbb{R}^3$ and $\Delta = \mathrm{span}\{f_1, f_2\}$ where*

$$f_1 = \begin{bmatrix} 1 \\ 0 \\ x_1^2 \end{bmatrix} \quad f_2 = \begin{bmatrix} 0 \\ x_1 \\ 1 \end{bmatrix}.$$

*Then it can be verified that $\dim(\Delta(x)) = 2 \; \forall x \in D$. We also have that*

$$[f_1, f_2] = \frac{\partial f_2}{\partial x} f_1 - \frac{\partial f_1}{\partial x} f_2 = \begin{bmatrix} 0 \\ 1 \\ 0 \end{bmatrix}.$$

*Therefore $\Delta$ is involutive if and only if*

$$\mathrm{rank}\left(\begin{bmatrix} 1 & 0 \\ 0 & x_1 \\ x_1^2 & 1 \end{bmatrix}\right) = \mathrm{rank}\left(\begin{bmatrix} 1 & 0 & 0 \\ 0 & x_1 & 1 \\ x_1^2 & 1 & 0 \end{bmatrix}\right).$$

*This, however is not the case, since*

$$\mathrm{rank}\left(\begin{bmatrix} 1 & 0 \\ 0 & x_1 \\ x_1^2 & 1 \end{bmatrix}\right) = 2 \text{ and } \mathrm{rank}\left(\begin{bmatrix} 1 & 0 & 0 \\ 0 & x_1 & 1 \\ x_1^2 & 1 & 0 \end{bmatrix}\right) = 3$$

*and hence $\Delta$ is not involutive.* □

**Definition 10.8** *(Complete Integrability) A linearly independent set of vector fields $f_1, \cdots, f_p$ on $D \subset \mathbb{R}^n$ is said to be completely integrable if for each $x_0 \in D$ there exists a neighborhood $N$ of $x_0$ and $n - p$ real-valued smooth functions $h_1(x), h_2(x), \cdots, h_{n-p}(x)$ satisfying the partial differentiable equation*

$$\frac{\partial h_j}{\partial x} f_i(x) = 0 \qquad 1 \le i \le p, \; 1 \le j \le n - p$$

*and the gradients $\nabla h_i$ are linearly independent.*

The following result, known as the *Frobenius theorem*, will be very important in later sections.

**Theorem 10.1** *(Frobenius Theorem) Let $f_1, f_2, \cdots, f_p$ be a set of linearly independent vector fields. The set is completely integrable if and only if it is involutive.*

**Proof:** the proof is omitted. See Reference [36].

**Example 10.7** *[36] Consider the set of partial differential equations*

$$2x_3 \frac{\partial h}{\partial x_1} - \frac{\partial h}{\partial x_2} = 0$$

$$-x_1 \frac{\partial h}{\partial x_1} - 2x_2 \frac{\partial h}{\partial x_2} + x_3 \frac{\partial h}{\partial x_3} = 0$$

*which can be written as*

$$\begin{bmatrix} \frac{\partial h}{\partial x_1} & \frac{\partial h}{\partial x_2} & \frac{\partial h}{\partial x_3} \end{bmatrix} \begin{bmatrix} 2x_3 & -x_1 \\ -1 & -2x_2 \\ 0 & x_3 \end{bmatrix} = 0$$

*or*

$$\nabla h \begin{bmatrix} f_1 & f_2 \end{bmatrix}$$

*with*

$$f_1 = \begin{bmatrix} 2x_3 \\ -1 \\ 0 \end{bmatrix}, \quad f_2 = \begin{bmatrix} -x_1 \\ -2x_2 \\ x_3 \end{bmatrix}.$$

*To determine whether the set of partial differential equations is solvable or, equivalently, whether $[f_1, f_2]$ is completely integrable, we consider the distribution $\Delta$ defined as follows:*

$$\Delta = span \left( \begin{bmatrix} 2x_3 \\ -1 \\ 0 \end{bmatrix}, \begin{bmatrix} -x_1 \\ -2x_2 \\ x_3 \end{bmatrix} \right).$$

*It can be checked that $\Delta$ has dimension 2 everywhere on the set $D$ defined by $D = \{x \in \mathbb{R}^3 : x_1^2 + x_2^2 \neq 0\}$. Computing the Lie bracket $[f_1, f_2]$, we obtain*

$$[f_1, f_2] = \begin{bmatrix} -4x_3 \\ 2 \\ 0 \end{bmatrix}$$

*and thus*

$$\begin{bmatrix} f_1 & f_2 & [f_1, f_2] \end{bmatrix} = \begin{bmatrix} 2x_3 & -x_1 & -4x_3 \\ -1 & -2x_2 & 2 \\ 0 & x_3 & 0 \end{bmatrix}$$

*which has rank 2 for all $x \in \mathbb{R}^3$. It follows that the distribution is involutive, and thus it is completely integrable on D, by the Frobenius theorem.* $\square$

## 10.2   Input–State Linearization

Throughout this section we consider a dynamical systems of the form

$$\dot{x} = f(x) + g(x)u$$

and investigate the possibility of using a state feedback control law plus a coordinate transformation to transform this system into one that is linear time-invariant. We will see that not every system can be transformed by this technique. To grasp the idea, we start our presentation with a very special class of systems for which an input–state linearization law is straightforward to find. For simplicity, we will restrict our presentation to single-input systems.

### 10.2.1   Systems of the Form $\dot{x} = Ax + B\omega(x)[u - \phi(x)]$

First consider a nonlinear system of the form

$$\dot{x} = Ax + B\omega(x)[u - \phi(x)] \tag{10.3}$$

where $A \in \mathbb{R}^{n \times n}, B \in \mathbb{R}^{n \times 1}, \phi : D \subset \mathbb{R}^n \to \mathbb{R}^1, \omega : D \subset \mathbb{R}^n \to \mathbb{R}$. We also assume that $\omega \neq 0 \quad \forall x \in D$, and that the pair $(A, B)$ is controllable. Under these conditions, it is straightforward to see that the control law

$$u = \phi(x) + \omega^{-1}(x)v \tag{10.4}$$

renders the system

$$\dot{x} = Ax + Bv$$

which is linear time-invariant and controllable. $\square$

The beauty of this approach is that it splits the feedback effort into two components that have very different purpose.

1- The feedback law (10.4) was obtained with the sole purpose of linearizing the original state equation. The resulting linear time-invariant system may or may not have "desirable" properties. Indeed, the resulting system may or may not be stable, and may or may not behave as required by the particular design.

2- Once a linear system is obtained, a secondary control law can be applied to stabilize the resulting system, or to impose any desirable performance. This secondary law, however, is designed using the resulting linear time-invariant system, thus taking advantage of the very powerful and much simpler techniques available for control design of linear systems.

**Example 10.8** *First consider the nonlinear mass–spring system of example 1.2*

$$\begin{cases} \dot{x}_1 = x_2 \\ \dot{x}_2 = -\frac{k}{m}x_1 - \frac{k}{m}a^2 x_1^3 - \frac{\omega}{m}x_2 + \frac{f(t)}{m} \end{cases}$$

*which can be written in the form*

$$\begin{bmatrix} \dot{x}_1 \\ \dot{x}_2 \end{bmatrix} = \begin{bmatrix} 0 & x_2 \\ -\frac{k}{m} & -\frac{\omega}{m} \end{bmatrix} \begin{bmatrix} x_1 \\ x_2 \end{bmatrix} + \begin{bmatrix} 0 \\ \frac{1}{m} \end{bmatrix} [f - ka^2 x_1^3].$$

*Clearly, this system is of the form (10.3) with $\omega = 1$ and $\phi(x) = ka^2 x_1^3$. It then follows that the linearizing control law is*

$$u = ka^2 x_1^3 + v$$

□

**Example 10.9** *Now consider the system*

$$\begin{cases} \dot{x}_1 = x_2 \\ \dot{x}_2 = -ax_1 + bx_2 + \cos x_1(u - x_2^2) \end{cases}$$

*Once again, this system is of the form (10.3) with $\omega = \cos x_1$ and $\phi(x) = x_2^2$. Substituting into (10.4), we obtain the linearizing control law:*

$$u = x_2^2 + \cos^{-1} x_1\, v$$

*which is well defined for $-\frac{\pi}{2} < x_1 < \frac{\pi}{2}$.*

□

## 10.2.2   Systems of the Form $\dot{x} = f(x) + g(x)u$

Now consider the more general case of affine systems of the form

$$\dot{x} = f(x) + g(x)u. \tag{10.5}$$

Because the system (10.5) does not have the simple form (10.4), there is no obvious way to construct the input-state linearization law. Moreover, it is not clear in this case whether such a linearizing control law actually exists. We will pursue the input–state linearization of these systems as an extension of the previous case. Before proceeding to do so, we formally introduce the notion of input-state linearization.

**Definition 10.9** *A nonlinear system of the form (10.5) is said to be input-state linearizable if there exist a diffeomorphism* $T : D \subset \mathbb{R}^n \to \mathbb{R}^n$ *defining the coordinate transformation*

$$z = T(x) \tag{10.6}$$

*and a control law of the form*

$$u = \phi(x) + \omega^{-1}(x)v \tag{10.7}$$

*that transform (10.5) into a state space realization of the form*

$$\dot{z} = Az + Bv.$$

We now look at this idea in more detail. Assuming that, after the coordinate transformation (10.6), the system (10.5) takes the form

$$
\begin{aligned}
\dot{z} &= Az + B\bar{\omega}(z)\left[u - \bar{\phi}(z)\right] \\
&= Az + B\omega(x)\left[u - \phi(x)\right]
\end{aligned}
\tag{10.8}
$$

where $\bar{\omega}(z) = \omega(T^{-1}(z))$ and $\bar{\phi}(z) = \phi(T^{-1}(z))$. We have:

$$\dot{z} = \frac{\partial T}{\partial x}\dot{x} = \frac{\partial T}{\partial x}[f(x) + g(x)u]. \tag{10.9}$$

Substituting (10.6) and (10.9) into (10.8), we have that

$$\frac{\partial T}{\partial x}[f(x) + g(x)u] = AT(x) + B\omega(x)[u - \phi(x)] \tag{10.10}$$

must hold $\forall x$ and $u$ of interest. Equation (10.10) is satisfied if and only if

$$
\begin{aligned}
\frac{\partial T}{\partial x}f(x) &= AT(x) - B\omega(x)\phi(x) \tag{10.11} \\
\frac{\partial T}{\partial x}g(x) &= B\omega(x). \tag{10.12}
\end{aligned}
$$

From here we conclude that any coordinate transformation $T(x)$ that satisfies the system of differential equations (10.11)(10.12) for some $\phi$, $\omega$, $A$ and $B$ transforms, via the coordinate transformation $z = T(x)$, the system

$$\dot{x} = f(x) + g(x)u \tag{10.13}$$

into one of the form

$$\dot{z} = Az + B\bar{\omega}(z)[u - \bar{\phi}(z)] \tag{10.14}$$

Moreover, any coordinate transformation $z = T(x)$ that transforms (10.13) into (10.14) must satisfy the system of equations (10.11)-(10.12).

**Remarks**: The procedure just described allows for a considerable amount of freedom when selecting the coordinate transformation. Consider the case of single-input systems, and recall that our objective is to obtain a system of the form

$$\dot{z} = Az + B\omega(x)[u - \phi(x)]$$

The $A$ and $B$ matrices in this state space realization are, however, non-unique and therefore so is the diffeomorphism $T$. Assuming that the matrices $(A, B)$ form a controllable pair, we can assume, without loss of generality, that $(A, B)$ are in the following so-called controllable form:

$$A_c = \begin{bmatrix} 0 & 1 & 0 & \cdots & 0 \\ 0 & 0 & 1 & \cdots & 0 \\ \vdots & \vdots & & & \vdots \\ \vdots & \vdots & & 0 & 1 \\ 0 & 0 & \cdots & 0 & 0 \end{bmatrix}_{n \times n}, \quad B_c = \begin{bmatrix} 0 \\ 0 \\ \vdots \\ \vdots \\ 1 \end{bmatrix}_{n \times 1}$$

Letting

$$T(x) = \begin{bmatrix} T_1(x) \\ T_2(x) \\ \vdots \\ T_n(x) \end{bmatrix}_{n \times 1}$$

with $A = A_c$, $B = B_c$ and $z = T(x)$, the right-hand side of equations (10.11)-(10.12) becomes

$$A_c T(x) - B_c \omega(x)\phi(x) = \begin{bmatrix} T_2(x) \\ T_3(x) \\ \vdots \\ T_n(x) \\ -\phi(x)\omega(x) \end{bmatrix} \tag{10.15}$$

and

$$B_c \omega(x) = \begin{bmatrix} 0 \\ 0 \\ \vdots \\ 0 \\ \omega(x) \end{bmatrix}. \tag{10.16}$$

Substituting (10.15) and (10.16) into (10.11) and (10.12), respectively, we have that

$$
\begin{aligned}
\frac{\partial T_1}{\partial x} f(x) &= T_2(x) \\
\frac{\partial T_2}{\partial x} f(x) &= T_3(x) \\
&\;\;\vdots \\
\frac{\partial T_{n-1}}{\partial x} f(x) &= T_n(x) \\
\frac{\partial T_n}{\partial x} f(x) &= -\phi(x)\omega(x)
\end{aligned}
\tag{10.17}
$$

and

$$
\begin{aligned}
\frac{\partial T_1}{\partial x} g(x) &= 0 \\
\frac{\partial T_2}{\partial x} g(x) &= 0 \\
&\;\;\vdots \\
\frac{\partial T_{n-1}}{\partial x} g(x) &= 0 \\
\frac{\partial T_n}{\partial x} g(x) &= \omega(x) \neq 0.
\end{aligned}
\tag{10.18}
$$

Thus the components $T_1, T_2, \cdots, T_n$ of the coordinate transformation $T$ must be such that

(i)

$$
\begin{aligned}
\frac{\partial T_i}{\partial x} g(x) &= 0 \quad \forall i = 1, 2, \cdots, n-1. \\
\frac{\partial T_n}{\partial x} g(x) &\neq 0.
\end{aligned}
$$

(ii)

$$
\frac{\partial T_i}{\partial x} f(x) = T_{i+1} \quad i = 1, 2, \cdots, n-1.
$$

(iii) The functions $\phi$ and $\omega$ are given by

$$\omega(x) = \frac{\partial T_n}{\partial x} g(x), \quad \phi(x) = -\frac{(\partial T_n/\partial x)f(x)}{(\partial T_n/\partial x)g(x)}.$$

## 10.3   Examples

**Example 10.10** *Consider the system*

$$\dot{x} = \begin{bmatrix} e^{x_2} - 1 \\ ax_1^2 \end{bmatrix} + \begin{bmatrix} 0 \\ 1 \end{bmatrix} u = f(x) + g(x)u.$$

*To find a feedback linearization law, we seek a transformation* $T = [T_1, T_2]^T$ *such that*

$$\frac{\partial T_1}{\partial x} g = 0 \tag{10.19}$$

$$\frac{\partial T_2}{\partial x} g \neq 0 \tag{10.20}$$

*with*

$$\frac{\partial T_1}{\partial x} f(x) = T_2. \tag{10.21}$$

*In our case, (10.19) implies that*

$$\frac{\partial T_1}{\partial x} g = \begin{bmatrix} \frac{\partial T_1}{\partial x_1} & \frac{\partial T_1}{\partial x_2} \end{bmatrix} \begin{bmatrix} 0 \\ 1 \end{bmatrix} = \frac{\partial T_1}{\partial x_2} = 0$$

*so that* $T_1 = T_1(x_1)$ *(independent of* $x_2$*). Taking account of (10.21), we have that*

$$\frac{\partial T_1}{\partial x} f(x) = T_2$$

$$\Rightarrow \begin{bmatrix} \frac{\partial T_1}{\partial x_1} & \frac{\partial T_1}{\partial x_2} \end{bmatrix} f(x) = \begin{bmatrix} \frac{\partial T_1}{\partial x_1} & 0 \end{bmatrix} \begin{bmatrix} e^{x_2} - 1 \\ ax_1^2 \end{bmatrix} = T_2$$

$$\Rightarrow T_2 = \frac{\partial T_1}{\partial x_1}(e^{x_2} - 1).$$

*To check that (10.20) is satisfied we notice that*

$$\frac{\partial T_2}{\partial x} g(x) = \begin{bmatrix} \frac{\partial T_2}{\partial x_1} & \frac{\partial T_2}{\partial x_2} \end{bmatrix} g(x) = \frac{\partial T_2}{\partial x_2} = \frac{\partial}{\partial x_2}\left(\frac{\partial T_1}{\partial x_1}(e^{x_2} - 1)\right) \neq 0$$

*provided that* $\frac{\partial T_1}{\partial x_1} \neq 0$. *Thus we can choose*

$$T_1(x) = x_1$$

*which results in*

$$T = \begin{bmatrix} x_1 \\ e^{x_2} - 1 \end{bmatrix}.$$

*Notice that this coordinate transformation maps the equilibrium point at the origin in the x plane into the origin in the z plane.*

*The functions $\phi$ and $\omega$ can be obtained as follows:*

$$\omega = \frac{\partial T_2}{\partial x} g(x) = e^{x_2}$$

$$\phi = -\frac{(\partial T_2/\partial x) f(x)}{(\partial T_2/\partial x) g(x)} = -ax_1^2.$$

*It is easy to verify that, in the z-coordinates*

$$\begin{cases} z_1 = x_1 \\ z_2 = e^{x_2} - 1 \end{cases}$$

$$\begin{cases} \dot{z}_1 = z_2 \\ \dot{z}_2 = az_1^2 z_2 + az_1^2 + (z_2 + 1)u \end{cases}$$

*which is of the form*

$$\dot{z} = Az + B\omega(z)[u - \phi(z)]$$

*with*

$$A = A_c = \begin{bmatrix} 0 & 1 \\ 0 & 0 \end{bmatrix}, \quad B = B_c = \begin{bmatrix} 0 \\ 1 \end{bmatrix}$$

$$\omega(z) = z_2 + 1, \quad \phi(z) = -az_1^2.$$

□

**Example 10.11** *(Magnetic Suspension System)*

*Consider the magnetic suspension system of Chapter 1. The dynamic equations are*

$$\begin{bmatrix} \dot{x}_1 \\ \dot{x}_2 \\ \dot{x}_3 \end{bmatrix} = \begin{bmatrix} x_2 \\ g - \frac{k}{m}x_2 - \frac{\lambda\mu x_3^2}{2m(1+\mu x_1)^2} \\ \frac{1+\mu x_1}{\lambda}\left(-Rx_3 + \frac{\lambda\mu}{(1+\mu x_1)^2}x_2 x_3\right) \end{bmatrix} + \begin{bmatrix} 0 \\ 0 \\ \left(\frac{1+\mu x_1}{\lambda}\right) \end{bmatrix}$$

*which is of the form* $\dot{x} = \bar{f}(x) + \bar{g}(x)u$. *To find a feedback linearizing law, we seek a transformation* $T = [T_1 \ T_2 \ T_3]^T$ *such that*

$$\frac{\partial T_1}{\partial x}\bar{g}(x) = 0 \tag{10.22}$$

$$\frac{\partial T_2}{\partial x}\bar{g}(x) = 0 \tag{10.23}$$

$$\frac{\partial T_2}{\partial x}\bar{g}(x) \neq 0 \tag{10.24}$$

*with*

$$\frac{\partial T_1}{\partial x}\bar{f}(x) = T_2 \tag{10.25}$$

$$\frac{\partial T_2}{\partial x}\bar{f}(x) = T_3 \tag{10.26}$$

$$\frac{\partial T_2}{\partial x}\bar{f}(x) = -\phi(x)\omega(x). \tag{10.27}$$

*Equation (10.22) implies that*

$$\begin{bmatrix} \frac{\partial T_1}{\partial x_1} & \frac{\partial T_1}{\partial x_2} & \frac{\partial T_1}{\partial x_3} \end{bmatrix} \begin{bmatrix} 0 \\ 0 \\ \left(\frac{1+\mu x_1}{\lambda}\right) \end{bmatrix}$$

$$\Rightarrow \quad \frac{\partial T_1}{\partial x_3}\left(\frac{1+\mu x_1}{\lambda}\right) = 0$$

*so* $T_1$ *is not a function of* $x_3$. *To proceed we arbitrarily choose* $T_1 = x_1$. *We will need to verify that this choice satisfies the remaining linearizing conditions (10.22)–(10.27).*

*Equation (10.25) implies that*

$$\frac{\partial T_1}{\partial x}\bar{f}(x) = T_2 \quad \Rightarrow \quad [1 \ 0 \ 0]\bar{f}(x) = T_2$$

*and thus* $T_2 = x_2$. *We now turn to equation (10.26). We have that*

$$\frac{\partial T_2}{\partial x}\bar{f}(x) = T_3$$

*and substituting values, we have that*

$$T_3 = g - \frac{k}{m}x_2 - \frac{\lambda \mu x_3^2}{2m(1+\mu x_1)^2}.$$

*To verify that (10.23)–(10.24) we proceed as follows:*

$$\frac{\partial T_2}{\partial x} \bar{g}(x) = [0 \ 1 \ 0] \begin{bmatrix} 0 \\ 0 \\ \left(\frac{1+\mu x_1}{\lambda}\right) u \end{bmatrix} = 0$$

*so that (10.23) is satisfied. Similarly*

$$\frac{\partial T_3}{\partial x} \bar{g}(x) = \frac{\mu x_3}{m(1 + \mu x_1^2)} \neq 0$$

*provided that $x_3 \neq 0$. Therefore all conditions are satisfied in $D = \{x \in \mathbb{R}^3 : x_3 \neq 0\}$. The coordinate transformation is*

$$T = \begin{bmatrix} x_1 \\ x_2 \\ g - \frac{k}{m}x_2 - \frac{\lambda \mu x_3^2}{2m(1+\mu x_1)^2} \end{bmatrix}.$$

*The function $\phi$ and $\omega$ are given by*

$$\omega(x) = \frac{\partial T_3}{\partial x} g(x) \quad \phi(x) = -\frac{(\partial T_3/\partial x)f(x)}{(\partial T_3/\partial x)g(x)}.$$

□

## 10.4 Conditions for Input–State Linearization

In this section we consider a system of the form

$$\dot{x} = f(x) + g(x)u \tag{10.28}$$

where $f, g : D \to \mathbb{R}^n$ and discuss under what conditions on $f$ and $g$ this system is input–state linearizable.

**Theorem 10.2** *The system (10.28) is input–state linearizable on $D_0 \subset D$ if and only if the following conditions are satisfied:*

*(i) The vector fields $\{g(x), ad_f g(x), \cdots, ad_f^{n-1}g(x)\}$ are linearly independent in $D_0$. Equivalently, the matrix*

$$C = [g(x), ad_f g(x), \cdots, ad_f^{n-1}g(x)]_{n \times n}$$

*has rank $n$ for all $x \in D_0$.*

(ii) The distribution $\Delta = span\{g, ad_fg, \cdots, ad_f^{n-2}g\}$ is involutive in $D_0$.

**Proof:** See the Appendix.

**Example 10.12** *Consider again the system of Example 10.10*

$$\dot{x} = \begin{bmatrix} e^{x_2} - 1 \\ ax_1^2 \end{bmatrix} + \begin{bmatrix} 0 \\ 1 \end{bmatrix} u = f(x) + g(x)u.$$

*We have,*

$$ad_fg = [f, g] = \frac{\partial g}{\partial x}f(x) - \frac{\partial f}{\partial x}g(x)$$

$$ad_fg = \begin{bmatrix} e^{x_2} \\ 0 \end{bmatrix}.$$

*Thus, we have that*

$$\{g, ad_fg\} = \left\{ \begin{bmatrix} 0 \\ 1 \end{bmatrix}, \begin{bmatrix} e^{x_2} \\ 0 \end{bmatrix} \right\}$$

*and*

$$rank(C) = rank\left( \begin{bmatrix} 0 & e^{x_2} \\ 1 & 0 \end{bmatrix} \right) = 2, \quad \forall x \in \mathbb{R}^2.$$

*Also the distribution $\Delta$ is given by*

$$\Delta = span\{g\} = span \left( \begin{bmatrix} 0 \\ 1 \end{bmatrix} \right)$$

*which is clearly involutive in $\mathbb{R}^2$. Thus conditions (i) and (ii) of Theorem 10.2 are satisfied $\forall x \in \mathbb{R}^2$.* □

**Example 10.13** *Consider the linear time-invariant system*

$$\dot{x} = Ax + Bu$$

*i.e., $f(x) = Ax$, and $g(x) = B$. Straightforward computations show that*

$$ad_fg = [f, g] = \frac{\partial g}{\partial x}f(x) - \frac{\partial f}{\partial x}g(x) = -AB$$

$$ad_f^2g = [f, [f, g]] = A^2B$$

$$\vdots$$

$$ad_f^ig = (-1)^i A^i B.$$

*Thus,*

$$\{g(x),\ ad_f g(x),\cdots,\ ad_f^{n-1} g(x)\} = \{B, -AB, A^2 B, \cdots, (-1)^{n-i} A^{n-i} B\}$$

*and therefore, the vectors* $\{g(x),\ ad_f g(x),\cdots,\ ad_f^{n-1} g(x)\}$ *are linearly independent if and only if the matrix*

$$C = [B,\ AB,\ A^2 B,\ \cdots, A^{n-1} B]_{n \times n}$$

*has rank $n$. Therefore, for linear systems condition (i) of theorem 10.2 is equivalent to the controllability of the pair $(A, B)$. Notice also that conditions (ii) is trivially satisfied for any linear time-invariant system, since the vector fields are constant and so $\delta$ is always involutive.* □

## 10.5 Input–Output Linearization

So far we have considered the input-state linearization problem, where the interest is in linearizing the mapping from input to state. Often, such as in a tracking control problem, our interest is in a certain output variable rather than the state. Consider the system

$$\begin{cases} \dot{x} = f(x) + g(x)u \\ y = h(x) \end{cases} \qquad \begin{array}{l} f, g : D \subset \mathbb{R}^n \to \mathbb{R}^n \\ h : D \subset \mathbb{R}^n \to \mathbb{R} \end{array} \qquad (10.29)$$

Linearizing the state equation, as in the input-state linearization problem, does not necessarily imply that the resulting map from input $u$ to output $y$ is linear. The reason, of course, is that when deriving the coordinate transformation used to linearize the state equation we did not take into account the nonlinearity in the output equation.

In this section we consider the problem of finding a control law that renders a linear differential equation relating the input $u$ to the output $y$. To get a better grasp of this principle, we consider the following simple example.

**Example 10.14** *Consider the system of the form*

$$\begin{cases} \dot{x}_1 = x_2 \\ \dot{x}_2 = -x_1 - ax_1^2 x_2 + (x_2 + 1)u \\ y = x_1. \end{cases}$$

*We are interested in the input–output relationship, so we start by considering the output equation $y = x_1$. Differentiating, we obtain*

$$\dot{y} = \dot{x}_1 = x_2$$

*which does not contain $u$. Differentiating once again, we obtain*

$$\ddot{y} = \dot{x}_2 = -x_1 - ax_1^2 x_2 + (x_2 + 1)u.$$

*Thus, letting*

$$u \overset{def}{=} \frac{1}{x_2 + 1}\left[v + ax_1^2 x_2\right] \quad (x_2 \neq 1)$$

*we obtain*

$$\ddot{y} = -x_1 + v$$

*or*

$$\ddot{y} + y = v$$

*which is a linear differential equation relating $y$ and the new input $v$. Once this linear system is obtained, linear control techniques can be employed to complete the design.* □

This idea can be easily generalized. Given a system of the form (10.29) where $f, g : D \subset \mathbb{R}^n \to \mathbb{R}^n$ and $h : D \subset \mathbb{R}^n \to \mathbb{R}$ are sufficiently smooth, the approach to obtain a linear input–output relationship can be summarized as follows.

- Differentiate the output equation to obtain

$$
\begin{aligned}
\dot{y} &= \frac{\partial h}{\partial x}\dot{x} \\
&= \frac{\partial h}{\partial x}f(x) + \frac{\partial h}{\partial x}g(x)u \\
&= L_f h(x) + L_g h(x)u
\end{aligned}
$$

There are two cases of interest:

  - **CASE (1):** $\frac{\partial h}{\partial x} \neq 0 \in D$. In this case we can define the control law

$$u = \frac{1}{L_g h(x)}\left[-L_f h + v\right]$$

   that renders the linear differential equation

$$\dot{y} = v.$$

  - **CASE (2):** $\frac{\partial h}{\partial x} = 0 \in D$. In this case we continue to differentiate $y$ until $u$ appears explicitly:

$$\ddot{y} \stackrel{def}{=} y^{(2)} = \frac{d}{dt}[\frac{\partial h}{\partial x} f(x)] = L_f^2 h(x) + L_g L_f h(x) u.$$

If $L_g L_f h(x) = 0$, we continue to differentiate until, for some integer $r \leq n$

$$y^{(r)} = L_f^r h(x) + L_g L_f^{(r-1)} h(x) u$$

with $L_g L_f^{(r-1)} h(x) \neq 0$. Letting

$$u = \frac{1}{L_g L_f^{(r-1)} h(x)} \left[ -L_f^r h + v \right]$$

we obtain the linear differential equation

$$y^{(r)} = v. \tag{10.30}$$

The number of differentiations of $y$ required to obtain (10.30) is important and is called the *relative degree* of the system. We now define this concept more precisely.

**Definition 10.10** *A system of the form*

$$\begin{cases} \dot{x} = f(x) + g(x)u & f, g : D \subset \mathbb{R}^n \to \mathbb{R}^n \\ y = h(x) & h : D \subset \mathbb{R}^n \to \mathbb{R} \end{cases}$$

*is said to have a relative degree $r$ in a region $D_0 \subset D$ if*

$$\begin{aligned} L_g L_f^i h(x) &= 0 & \forall i, \ 0 \leq i \leq r-1, \quad \forall x \in D_0 \\ L_g L_f^{r-1} h(x) &\neq 0 & \forall x \in D_0 \end{aligned}$$

**Remarks:**

(a) Notice that if $r = n$, that is, if the relative degree is equal to the number of states, then denoting $h(x) = T_1(x)$, we have that

$$\dot{y} = \frac{\partial T_1}{\partial x} [f(x) + g(x)u].$$

The assumption $r = n$ implies that $\frac{\partial T_1}{\partial x} g(x) = 0$. Thus, we can define

$$\dot{y} = \frac{\partial T_1}{\partial x} f(x) \stackrel{def}{=} T_2$$

$$y^{(2)} = \frac{\partial T_2}{\partial x} [f(x) + g(x)u], \qquad \frac{\partial T_2}{\partial x} g(x) = 0$$

$$\Rightarrow y^{(2)} = \frac{\partial T_2}{\partial x} f(x) \stackrel{def}{=} T_3$$

and iterating this process

$$y^{(n)} = \frac{\partial T_n}{\partial x} f(x) + \frac{\partial T_n}{\partial x} g(x) u.$$

Therefore

$$\frac{\partial T_1}{\partial x} g(x) = 0$$

$$\frac{\partial T_2}{\partial x} g(x) = 0$$

$$\vdots$$

$$\frac{\partial T_{n-1}}{\partial x} g(x) = 0$$

$$\frac{\partial T_n}{\partial x} g(x) \neq 0$$

which are the input–state linearization conditions (10.17). Thus, if the relative degree $r$ equals the number of states $n$, then input–output linearization leads to input–state linearization.

(b) When applied to single-input–single-output systems, the definition of relative degree coincides with the usual definition of relative degree for linear time-invariant systems, as we shall see.

**Example 10.15** *Consider again the system of Example 10.14*

$$\begin{cases} \dot{x}_1 = x_2 \\ \dot{x}_2 = -x_1 - ax_1^2 x_2 + (x_2 + 1)u \\ y = x_1 \end{cases}$$

*we saw that*

$$\dot{y} = x_2$$
$$\ddot{y} = -x_1 - ax_1^2 x_2 + (x_2 + 1)u$$

*hence the system has relative degree 2 in* $D_0 = \{x \in \mathbb{R}^2 : x_2 \neq 1\}$.  □

**Example 10.16** *Consider the linear time-invariant system defined by*

$$\begin{cases} \dot{x} = Ax + Bu \\ y = Cx \end{cases}$$

where

$$A = \begin{bmatrix} 0 & 1 & 0 & \cdots & & 0 \\ 0 & 0 & 1 & & & 0 \\ \vdots & & & & & \vdots \\ 0 & 0 & \cdots & & 0 & 1 \\ -q_0 & -q_1 & \cdots & & & -q_{n-1} \end{bmatrix}_{n \times n} , \quad B = \begin{bmatrix} 0 \\ 0 \\ \vdots \\ 0 \\ 1 \end{bmatrix}_{n \times 1}$$

$$C = \begin{bmatrix} p_0 & p_1 & \cdots & p_m & 0 & \cdots \end{bmatrix}_{1 \times n}, \quad p_m \neq 0.$$

The transfer function associated with this state space realization can be easily seen to be

$$\hat{H}(s) = C(sI - A)^{-1}B = \frac{p_m s^m + p_{m-1} s^{m-1} + \cdots + p_0}{s^n + q_{n-1} s^{s-1} + \cdots + q_0}, \quad m < nc$$

The relative degree of this system is the excess of poles over zeros, that is, $r = n - m > 0$. We now calculate the relative degree using definition 10.10. We have

$$\dot{y} = C\dot{x} = CAx + CBu$$

$$CB = \begin{bmatrix} p_0 & p_1 & \cdots & p_m & 0 & \cdots \end{bmatrix} \begin{bmatrix} 0 \\ 0 \\ \vdots \\ 0 \\ 1 \end{bmatrix} = \begin{cases} p_m, & \text{if } m = n - 1 \\ 0, & \text{otherwise} \end{cases}$$

Thus, if $CB = p_m$, we conclude that $r = n - m = 1$. Assuming that this is not the case, we have that $\dot{y} = CAx$, and we continue to differentiate $y$:

$$y^{(2)} = CA\dot{x} = CA^2x + CABu$$

$$CAB = \begin{bmatrix} p_0 & p_1 & \cdots & p_m & 0 & \cdots \end{bmatrix} \begin{bmatrix} 0 \\ 0 \\ \vdots \\ 1 \\ -a_{n-1} \end{bmatrix} = \begin{cases} p_m, & \text{if } m = n - 2 \\ 0, & \text{otherwise} \end{cases}$$

If $CAB = p_m$, then we conclude that $r = n - m = 2$. Assuming that this is not the case, we continue to differentiate. With every differentiation, the "1" entry in the column matrix $A^i B$ moves up one row. Thus, given the form of the $C$ matrix, we have that

$$CA^{i-1}B = \begin{cases} 0 & \text{for } i = 1, 2, \cdots, n - m - 1 \\ p_m & \text{for } i = n - m \end{cases}$$

*It then follows that*

$$y^{(r)} = CA^r x + CA^{r-1} Bu, \quad CA^{r-1}B \neq 0, \ r = n - m$$

*and we conclude that $r = n - m$, that is, the relative degree of the system is the excess number of poles over zeros.*  □

**Example 10.17** *Consider the linear time-invariant system defined by*

$$\begin{cases} \dot{x} = Ax + Bu \\ y = Cx \end{cases}$$

*where*

$$A = \begin{bmatrix} 0 & 1 & 0 & 0 & 0 \\ 0 & 0 & 1 & 0 & 0 \\ 0 & 0 & 0 & 1 & 0 \\ 0 & 0 & 0 & 0 & 1 \\ -2 & -3 & -5 & -1 & -4 \end{bmatrix}_{5 \times 5}, \quad B = \begin{bmatrix} 0 \\ 0 \\ 0 \\ 0 \\ 1 \end{bmatrix}_{5 \times 1}$$

$$C = \begin{bmatrix} 7 & 2 & 6 & 0 & 0 \end{bmatrix}_{1 \times 5}$$

*To compute the relative degree, we compute successive derivatives of the output equation:*

$$\dot{y} = CAx + CBu, \quad CB = 0$$
$$y^{(2)} = CA^2 x + CABu, \quad CAB = 0$$
$$y^{(3)} = CA^3 x + CA^2 Bu, \quad CA^2 B = 6$$

*Thus, we have that $r = 3$. It is easy to verify that the transfer function associated with this state space realization is*

$$\widehat{H}(s) = C(sI - A)^{-1}B = \frac{6s^2 + 2s + 7}{s^5 + 4s^4 + s^3 + 5s^2 + 3s + 2}$$

*which shows that the relative degree equals the excess number of poles over zeros.*  □

## 10.6  The Zero Dynamics

On the basis of our exposition so far, it would appear that input–output linearization is rather simple to obtain, at least for single-input–single-output (SISO) systems, and that all SISO systems can be linearized in this form. In this section we discuss in more detail the internal dynamics of systems controlled via input–output linearization. To understand the main idea, consider first the SISO linear time-invariant system

$$\begin{cases} \dot{x} = Ax + Bu \\ y = Cx. \end{cases} \tag{10.31}$$

To simplify matters, we can assume without loss of generality that the system is of third order and has a relative degree $r = 2$. In companion form, equation (10.31) can be expressed as follows:

$$\begin{bmatrix} \dot{x}_1 \\ \dot{x}_2 \\ \dot{x}_3 \end{bmatrix} = \begin{bmatrix} 0 & 1 & 0 \\ 0 & 0 & 1 \\ -q_0 & -q_1 & -q_2 \end{bmatrix} \begin{bmatrix} x_1 \\ x_2 \\ x_3 \end{bmatrix} + \begin{bmatrix} 0 \\ 0 \\ 1 \end{bmatrix} u$$

$$y = \begin{bmatrix} p_0 & p_1 & 0 \end{bmatrix} \begin{bmatrix} x_1 \\ x_2 \\ x_3 \end{bmatrix}.$$

The transfer function associated with this system is

$$\widehat{H}(s) = \frac{p_0 + p_1 s}{q_0 + q_1 s + q_2 s^2 + q_3 s^3} \, u.$$

Suppose that our problem is to design $u$ so that $y$ tracks a desired output $y_d$. Ignoring the fact that the system is linear time-invariant, we proceed with our design using the input–output linearization technique. We have:

$$\begin{aligned} y &= p_0 x_1 + p_1 x_2 \\ \Rightarrow \dot{y} &= p_0 \dot{x}_1 + p_1 \dot{x}_2 \\ &= p_0 x_2 + p_1 x_3 \\ \ddot{y} &= p_0 \dot{x}_2 + p_1 \dot{x}_3 \\ &= p_0 x_3 + p_1(-q_0 x_1 - q_1 x_2 - q_2 x_3) + p_1 u. \end{aligned}$$

Thus, the control law

$$u = \left[ q_0 x_1 + q_1 x_2 + q_2 x_3 - \frac{p_0}{p_1} x_3 \right] + \frac{1}{p_1} v$$

produces the simple double integrator

$$\ddot{y} = v.$$

Since we are interested in a tracking control problem, we can define the tracking error $e = y - y_d$ and choose $v = -k_1 e - k_2 \dot{e} + \ddot{y}_d$. With this input $v$ we have that

$$u = \left[ q_0 x_1 + q_1 x_2 + q_2 x_3 - \frac{p_0}{p_1} x_3 \right] + \frac{1}{p_1}[-k_1 e - k_2 \dot{e} + \ddot{y}_d]$$

which renders the exponentially stable tracking error closed-loop system

$$\ddot{e} + k_2\dot{e} + k_1 e = 0. \tag{10.32}$$

A glance at the result shows that the order of the closed-loop tacking error (10.32) is the same as the relative order of the system. In our case, $r = 2$, whereas the original state space realization has order $n = 3$. Therefore, part of the dynamics of the original system is now unobservable after the input–output linearization. To see why, we reason as follows. During the design stage, based on the input–output linearization principle, the state equation (10.31) was manipulated using the input $u$. A look at this control law shows that $u$ consists of a state feedback law, and thus the design stage can be seen as a reallocation of the eigenvalues of the $A$-matrix via state feedback. At the end of this process observability of the three-dimensional state space realization (10.31) was lost, resulting in the (external) two-dimensional closed loop differential equation (10.32). Using elementary concepts of linear systems, we know that this is possible if during the design process one of the eigenvalues of the closed-loop state space realization coincides with one of the transmission zeros of the system, thus producing the (external) second-order differential equation (10.32). This is indeed the case in our example. To complete the three-dimensional state, we can consider the output equation

$$\begin{aligned} y &= p_0 x_1 + p_1 x_2 \\ &= p_0 x_1 + p_1 \dot{x}_1. \end{aligned}$$

Thus,

$$\dot{x}_1 = -\frac{p_0}{p_1} x_1 + \frac{1}{p_1} y = A_{id}\, x_1 + B_{id}\, y. \tag{10.33}$$

Equation (10.33) can be used to "complete" the three-dimensional state. Indeed, using $x_1$ along with $e$ and $\dot{e}$ as the "new" state variables, we obtain full information about the original state variables $x_1 - x_3$ through a one-to-one transformation.

The unobservable part of the dynamics is called the *internal dynamics*, and plays a very important role in the context of the input–output linearization technique. Notice that the output $y$ in equation (10.33) is given by $y = e + y_d$, and so $y$ is bounded since $e$ was stabilized by the input–output linearization law, and $y_d$ is the desired trajectory. Thus, $x_1$ will be bounded, provided that the first-order system (10.33) has an exponentially stable equilibrium point at the origin. As we well know, the internal dynamics is thus exponentially stable if the eigenvalues of the matrix $A_{id}$ in the state space realization (10.33) are in the left half plane. Stability of the internal dynamics is, of course, as important as the stability of the *external* tracking error (10.32), and therefore the effectiveness of the input-output linearization technique depends upon the stability of the internal dynamics. Notice also that the associated transfer function of this internal dynamics is

$$\widehat{H}_{id} = \frac{1}{p_0 + p_1 s}.$$

Comparing $\widehat{H}_{id}$ and the transfer function $\widehat{H}$ of the original third-order system, we conclude that the internal dynamics contains a pole whose location in the $s$ plane coincides with that of the zero of $\widehat{H}$, thus leading to the loss of observability. This also implies that the internal dynamics of the original system is exponentially stable, provided that the zeros of the transfer function $\widehat{H}_{id}$ are in the left-half plane. System that have all of their transfer function zeros are called *minimum phase*.

Our discussion above was based on a three dimensional example; however, our conclusions can be shown to be general, specifically, for linear time-invariant systems the stability of the internal dynamics is determined by the location of the system zeros.

Rather than pursuing the proof of this result, we study a nonlinear version of this problem. The extension to the nonlinear case is nontrivial since poles and zeros go hand in hand with the concept of transfer function, which pertains to linear time-invariant system only. We proceed as follows. Consider an $n$th-order system of the form

$$\begin{cases} \dot{x} = f(x) + g(x)u \\ y = h(x) \end{cases} \tag{10.34}$$

and assume that (10.34) has relative degree $r$. Now define the following nonlinear transformation

$$z = T(x) = \begin{bmatrix} \mu_1(x) \\ \vdots \\ \mu_{n-r}(x) \\ --- \\ \psi_1 \\ \vdots \\ \psi_r \end{bmatrix} \stackrel{def}{=} \begin{bmatrix} \mu(x) \\ --- \\ \psi(x) \end{bmatrix} = \begin{bmatrix} \eta \\ --- \\ \xi \end{bmatrix}, \quad \eta \in \mathbb{R}^{n-r}, \; \xi \in \mathbb{R}^r \tag{10.35}$$

where

$$\begin{aligned} \psi_1 &= h(x) \\ \psi_2 &= \frac{\partial \psi_1}{\partial x} f(x) \\ &\vdots \\ \psi_{i+1} &= \frac{\partial \psi_i}{\partial x} f(x) \qquad i = 1, 2, \cdots, r-1 \end{aligned}$$

and $\mu_1 - \mu_{n-r}$ are chosen so that $T(x)$ is a diffeomorphism on a domain $D_0 \subset D$ and

$$\frac{\partial \mu_i}{\partial x} g(x) = 0, \quad \text{for } 1 \leq i \leq n - r, \ \forall x \in D_0. \tag{10.36}$$

It will be shown later that this change of coordinates transfers the original system (10.34) into the following so-called normal form:

$$\begin{aligned}
\dot{\eta} &= f_0(\eta, \xi) \tag{10.37}\\
\dot{\xi} &= A_c \xi + B_c \omega [u - \phi(x)] \tag{10.38}\\
y &= C_c \xi \tag{10.39}
\end{aligned}$$

where

$$f_0(\eta, \xi) = \frac{\partial \mu}{\partial x} \Big|_{x=T^{-1}(z)}$$

and

$$\omega = \frac{\partial \psi_r}{\partial x} g(x), \quad \phi = -\frac{(\partial \psi_r/\partial x) \, f(x)}{(\partial \psi_r/\partial x) \, g(x)}.$$

The normal form is conceptually important in that it splits the original state $x$ into two parts, namely, $\eta$ and $\xi$, with the following properties:

- $\xi$ represents the *external dynamics*, which can be linearized by the input

$$u^* = \phi(x) + \omega^{-1} v \tag{10.40}$$

- $\eta$ represents the *internal dynamics*. When the input $u^*$ is applied, equations (10.37)–(10.39) take the form

$$\begin{aligned}
\dot{\eta} &= f_0(\eta, \xi)\\
\dot{\xi} &= A_c \xi + B_c v\\
y &= C_c \xi
\end{aligned}$$

and thus, $\eta$ is unobservable from the output. From here we conclude that the stability of the internal dynamics is determined by the autonomous equation

$$\dot{\eta} = f_0(\eta, 0).$$

This dynamical equation is referred to as the *zero dynamics*, which we now formally introduce.

**Definition 10.11** *Given a dynamical system of the form (10.37)–(10.39) (i.e., represented in normal form), the autonomous equation*

$$\dot{\eta} = f_0(\eta, 0)$$

*is called the* zero dynamics.

Because of the analogy with the linear system case, systems whose zero dynamics is stable are said to be *minimum phase*.

**Summary**: The stability properties of the zero dynamics plays a very important role whenever input–output linearization is applied. Input–output linearization is achieved via partial cancelation of nonlinear terms. Two cases should be distinguished:

- $r = n$: If the relative degree of the nonlinear system is the same as the order of the system, then the nonlinear system can be fully linearized and, input–output linearization can be successfully applied. This analysis, of course, ignores robustness issues that always arise as a result of imperfect modeling.

- $r < n$: If the relative degree of the nonlinear system is lower than the order of the system, then only the external dynamics of order $r$ is linearized. The remaining $n - r$ states are unobservable from the output. The stability properties of the internal dynamics is determined by the zero dynamics. Thus, whether input–output linearization can be applied successfully depends on the stability of the zero dynamics. If the zero dynamics is not asymptotically stable, then input–output linearization does not produce a control law of any practical use.

Before discussing some examples, we notice that setting $y \equiv 0$ in equations (10.37)–(10.39) we have that

$$
\begin{aligned}
y \equiv 0 \quad &\Longleftrightarrow \quad \xi \equiv 0 \\
&\Longleftrightarrow \quad u(t) = \phi(x) \\
&\Rightarrow \quad \dot{\eta} = f_0(\eta, 0).
\end{aligned}
$$

Thus the zero dynamics can be defined as the internal dynamics of the system when the output is kept identically zero by a suitable input function. This means that the zero dynamics can be determined without transforming the system into normal form.

**Example 10.18** *Consider the system*

$$
\begin{cases}
\dot{x}_1 = -kx_1 - 2x_2 u \\
\dot{x}_2 = -x_2 + x_1 u \\
y = x_2
\end{cases}
$$

First we find the relative order. Differentiating $y$, we obtain

$$\dot{y} = \dot{x}_2 = -x_2 + x_1 u.$$

Therefore, $r = 1$. To determine the zero dynamics, we proceed as follows:

$$y = 0 \iff x_2 = 0 \iff u = 0.$$

Then, the zero dynamics is given by

$$\dot{x}_1 = -kx_1$$

which is exponentially stable (globally) if $k > 0$, and unstable if $k < 0$. □

**Example 10.19** Consider the system

$$\begin{cases} \dot{x}_1 = x_2 + x_1^2 \\ \dot{x}_2 = x_2^3 + u \\ \dot{x}_3 = x_1 + x_2^3 + \alpha x_3 y = x_1 \end{cases}$$

Differentiating $y$, we obtain

$$\begin{aligned} \dot{y} &= \dot{x}_1 = x_2 + x_1^2 \\ \ddot{y} &= 2x_1\dot{x}_1 + \dot{x}_2 = 2x_1(x_2 + x_1^2) + x_2^3 + u \end{aligned}$$

Therefore

$$r = 2$$

To find the zero dynamics, we proceed as follows:

$$y = 0 \iff x_1 = 0 \Rightarrow \dot{x}_1 = 0 = x_2 + x_1^2 \Rightarrow x_2 = 0$$

$$\dot{x}_2 = x_2^3 + u = 0 \iff u = -x_2^3.$$

Therefore the zero dynamics is given by

$$\dot{x}_3 = \alpha x_3.$$

Moreover, the zero dynamics is asymptotically stable if $\alpha < 0$, and unstable if $\alpha > 0$. □

## 10.7 Conditions for Input–Output Linearization

According to our discussion in Section 10.6, the input-output linearization procedure depends on the existence of a transformation $T(\cdot)$ that converts the original system of the form

$$\begin{cases} \dot{x} = f(x) + g(x)u, & f,g : D \subset \mathbb{R}^n \to \mathbb{R}^n \\ y = h(x), & h : D \subset \mathbb{R}^n \to \mathbb{R} \end{cases} \tag{10.41}$$

into the normal form (10.37)–(10.39). The following theorem states that such a transformation exists for any SISO system of relative degree $r < n$.

**Theorem 10.3** *Consider the system (10.41) and assume that it has relative degree $r < n$ $\forall x \in D_0 \subset D$. Then, for every $x_0 \in D_0$, there exist a neighborhood $\Omega$ of $x_0$ and smooth function $\mu_1, \cdots, \mu_{n-r}$ such that*

*(i)* $L_g \mu_i(x) = 0, \quad$ for $1 \leq i \leq n - r, \quad \forall x \in \Omega$

*(ii)*

$$T(x) = \begin{bmatrix} \mu_1(x) \\ \vdots \\ \mu_{n-r}(x) \\ -\,-\,- \\ \psi_1 \\ \vdots \\ \psi_r \end{bmatrix}$$

$$\begin{aligned} \psi_1 &= h(x) \\ \psi_2 &= L_f h(x) \\ &\vdots \\ \psi_r &= L_f^{r-1} h(x) \end{aligned}$$

*is a diffeomorphism on $\Omega$.* $\qquad\square$

**Proof:** See the Appendix.

## 10.8 Exercises

(10.1) Prove Lemma 10.2.

(10.2) Consider the magnetic suspension system of Example 10.11. Verify that this system is input–state linearizable.

(10.3) Consider again the magnetic suspension system of Example 10.11. To complete the design, proceed as follows:

(a) Compute the function $\omega$ and $\phi$ and express the system in the form $\dot{z} = Az + Bv$.

(b) Design a state feedback control law to stabilize the ball at a desired position.

(10.4) Determine whether the system

$$\begin{cases} \dot{x}_1 = x_1^2 x_2^3 \\ \dot{x}_2 = u \end{cases}$$

is input–state linearizable.

(10.5) Consider the following system:

$$\begin{cases} \dot{x}_1 = x_1^2 + x_2 \\ \dot{x}_2 = x_3^2 + u \\ \dot{x}_3 = x_1 + x_2^3 + x_3^3 \end{cases}$$

(a) Determine whether the system is input–state linearizable.

(b) If the answer in part (a) is affirmative, find the linearizing law.

(10.6) Consider the following system

$$\begin{cases} \dot{x}_1 = x_1^2 + x_3 \\ \dot{x}_2 = x_1^2 x_2 \\ \dot{x}_3 = x_1 \sin x_2 + u \end{cases}$$

(a) Determine whether the system is input–state linearizable.

(b) If the answer in part (a) is affirmative, find the linearizing law.

(10.7) ([36]) Consider the following system:

$$\begin{cases} \dot{x}_1 = x_3 + x_2 x_3 \\ \dot{x}_2 = x_1 + (1 + x_2)u \\ \dot{x}_3 = x_2(1 + x_1) - x_3 + u \end{cases}$$

(a) Determine whether the system is input–state linearizable.

(b) If the answer in part (a) is affirmative, find the linearizing law.

(10.8) Consider the following system:

$$\begin{cases} \dot{x}_1 = x_1^2 + x_2 \\ \dot{x}_2 = x_3^2 + u \\ \dot{x}_3 = x_2 - \alpha x_3 \\ y = x_1 \end{cases}$$

(a) Find the relative order.

(b) Determine whether it is minimum phase.

(c) Using feedback linearization, design a control law to track a desired signal $y = y_{\text{ref}}$.

# Notes and References

The exact input–state linearization problem was solved by Brockett [12] for single-input systems. The multi-input case was developed by Jakubczyk and Respondek, [39], Su, [77] and Hunt et al. [34]. The notion of zero dynamics was introduced by Byrnes and Isidori [14].

For a complete coverage of the material in this chapter, see the outstanding books of Isidori [36], Nijmeijer and van der Schaft [57], and Marino and Tomei, [52]. Our presentation follows Isidori [36] with help from References [68], [88] and [41]. Section 10.1 follows closely References [36] and [11]. The linear time-invariant example of Section 6, used to introduce the notion of zero dynamics follows Slotine and Li [68].

# Chapter 11

# Nonlinear Observers

So far, whenever discussing state space realizations it was implicitly assumed that the state vector $x$ was available. For example, state feedback was used in Chapter 5, along with the backstepping procedure, assuming that the vector $x$ was measured. Similarly, the results of Chapter 10 on feedback linearization assume that the state $x$ is available for manipulation. Aside from state feedback, knowledge of the state is sometimes important in problems of fault detection as well as system monitoring. Unfortunately, the state $x$ is seldom available since rarely can one have a sensor on every state variable, and some form of state *reconstruction* from the available measured output is required. This reconstruction is possible provided certain "observability conditions" are satisfied. In this case, an *observer* can be used to obtain an estimate $\hat{x}$ of the true state $x$. While for linear time-invariant systems observer design is a well-understood problem and enjoys a well-established solution, the nonlinear case is much more challenging, and no universal solution exists. The purpose of this chapter is to provide an introduction to the subject and to collect some very basic results.

## 11.1 Observers for Linear Time-Invariant Systems

We start with a brief summary of essential results for linear time-invariant systems. Throughout this section we consider a state space realization of the form

$$\psi_l \begin{cases} \dot{x} = Ax + Bu \\ y = Cx \end{cases} \qquad \begin{aligned} & A \in \mathbb{R}^{n \times n}, \ B \in \mathbb{R}^{n \times 1} \\ & C \in \mathbb{R}^{1 \times n} \end{aligned} \qquad (11.1)$$

where, for simplicity, we assume that the system is single-input–single-output, and that $D = 0$ in the output equation.

### 11.1.1   Observability

**Definition 11.1** *The state space realization (11.1) is said to be observable if for any initial state $x_0$ and fixed time $t_1 > 0$, the knowledge of the input $u$ and output $y$ over $[0, t_1]$ suffices to uniquely determine the initial state $x_0$.*

Once $x_0$ is determined, the state $x(t_1)$ can be reconstructed using the well-known solution of the state equation:

$$x(t_1) = e^{At_1} x_0 + \int_0^{t_1} e^{A(t-\tau)} Bu(\tau) \, d\tau. \tag{11.2}$$

Also

$$y(t_1) = Cx(t_1) = Ce^{At_1} x_0 + C \int_0^{t_1} e^{A(t-\tau)} Bu(\tau) \, d\tau. \tag{11.3}$$

Notice that, for fixed $u = u^*$, equation (11.3) defines a linear transformation $L : \mathbb{R}^n \to \mathbb{R}$ that maps $x_0$ to $y$; that is, we can write

$$y(t) = (Lx_0)(t).$$

Thus, we argue that the state space realization (11.1) is observable if and only if the mapping $Lx_0$ is one-to-one. Indeed, if this is the case, then the inversion map $x_0 = L^{-1}y$ uniquely determines $x_0$. Accordingly,

$$\psi_l \text{ "observable"} \quad \text{if and only if} \quad Lx_1 = Lx_2 \Rightarrow x_1 = x_2.$$

Now consider two initial conditions $x_1$ and $x_2$. We have,

$$y(t_1) = (Lx_1)(t) = Ce^{At_1} x_1 + C \int_0^{t_1} e^{A(t-\tau)} Bu^*(\tau) \, d\tau \tag{11.4}$$

$$y(t_2) = (Lx_2)(t) = Ce^{At_1} x_2 + C \int_0^{t_1} e^{A(t-\tau)} Bu^*(\tau) \, d\tau \tag{11.5}$$

and

$$y_1 - y_2 = Lx_1 - Lx_2 = Ce^{At_1}(x_1 - x_2).$$

By definition, the mapping $L$ is one-to-one if

$$y_1 = y_2 \quad \Rightarrow \quad x_1 = x_2.$$

For this to be the case, we must have:

$$Ce^{At}x = 0 \quad \Longleftrightarrow \quad x = 0$$

or, equivalently

$$\mathcal{N}(Ce^{At}) = 0.$$

Thus, $L$ is one-to-one [and so (11.1) is observable] if and only if the null space of $Ce^{At}$ is empty (see Definition 2.9).

Notice that, according to this discussion, the observability properties of the state space realization (11.1) are independent of the input $u$ and/or the matrix $B$. Therefore we can assume, without loss of generality, that $u \equiv 0$ in (11.1). Assuming for simplicity that this is the case, (11.1) reduces to

$$\psi_l \begin{cases} \dot{x} = Ax \\ y = Cx. \end{cases} \tag{11.6}$$

Observability conditions can now be easily derived as follows. From the discussion above, setting $u = 0$, we have that the state space realization (11.6) [or (11.1)] is observable if and only if

$$y(t) = Ce^{At}x_0 \equiv 0 \quad \Rightarrow \quad x \equiv 0.$$

We now show that this is the case if and only if

$$\text{rank}(\mathcal{O}) \stackrel{def}{=} \text{rank} \left( \begin{bmatrix} C \\ CA \\ \vdots \\ CA^{n-1} \end{bmatrix} \right) = n \tag{11.7}$$

To see this, note that

$$\begin{aligned} y(t) &= Ce^{At}x_0 \\ &= C(I + tA + \frac{t^2}{2}A^2 + \cdots)x_0. \end{aligned}$$

By the Sylvester theorem, only the first $n - 1$ powers of $A^i$ are linearly independent. Thus

$$y \equiv 0 \quad \Longleftrightarrow \quad Cx_0 = CAx_0 = CA^2x_0 = \cdots = CA^{n-1}x_0$$

or,

$$y = 0 \quad \Longleftrightarrow \quad \begin{bmatrix} C \\ CA \\ \vdots \\ CA^{n-1} \end{bmatrix} x_0 = 0, \quad \Longleftrightarrow \quad \mathcal{O}x_0 = 0$$

and the condition $\mathcal{O}x_0 = 0 \iff x_0 = 0$ is satisfied if and only if $\text{rank}(\mathcal{O}) = n$. $\square$

The matrix $\mathcal{O}$ is called the *observability matrix*. Given these conditions, we can redefine observability of linear time-invariant state space realizations as follows.

**Definition 11.2** *The state space realization (11.1) [or (11.6)] is said to be observable if*

$$rank \left( \begin{bmatrix} C \\ CA \\ \vdots \\ CA^{n-1} \end{bmatrix} \right) = n$$

### 11.1.2   Observer Form

The following result is well known. Consider the linear time-invariant state space realization (11.1), and let the transfer function associated with this state space realization be

$$\widehat{H}(s) = C(sI - A)^{-1}B = \frac{p_{n-1}s^{n-1} + p_{n-2}s^{n-2} + \cdots + p_0}{s^n + q_{n-1}s^{n-1} + \cdots + q_0}$$

Assuming that $(A, C)$ form an observable pair, there exists a nonsingular matrix $T \in \mathbb{R}^{n \times n}$ such that defining new coordinates $\bar{x} = Tx$, the state space realization (11.1) takes the so-called observer form:

$$\dot{\bar{x}} = \begin{bmatrix} 0 & 0 & 0 & \cdots & -q_0 \\ 1 & 0 & 0 & \cdots & -q_1 \\ 0 & 1 & 0 & \cdots & -q_2 \\ \vdots & & \ddots & & \vdots \\ 0 & 0 & & 1 & -q_{n-1} \end{bmatrix} \begin{bmatrix} \bar{x}_1 \\ \bar{x}_2 \\ \vdots \\ \vdots \\ \bar{x}_n \end{bmatrix} + \begin{bmatrix} p_0 \\ p_1 \\ \vdots \\ \vdots \\ p_{n-1} \end{bmatrix} u$$

$$y = \begin{bmatrix} 0 & \cdots & 1 \end{bmatrix} \bar{x}$$

### 11.1.3   Observers for Linear Time-Invariant Systems

Now consider the following observer structure:

$$\dot{\hat{x}} = A\hat{x} + Bu + L(y - C\hat{x}) \tag{11.8}$$

where $L \in \mathbb{R}^{n-1}$ is the so-called *observer gain*. Defining the observer error

$$\tilde{x} \stackrel{def}{=} x - \hat{x} \tag{11.9}$$

we have that

$$\begin{aligned} \dot{\tilde{x}} &= \dot{x} - \dot{\hat{x}} = (Ax + Bu) - (A\hat{x} + Bu + Ly - LC\hat{x}) \\ &= (A - LC)(x - \hat{x}) \end{aligned}$$

or

$$\dot{\tilde{x}} = (A - LC)\tilde{x}. \tag{11.10}$$

Thus, $\tilde{x} \to 0$ as $t \to \infty$, provided the so-called *observer error dynamics* (11.10) is asymptotically (exponentially) stable. This is, of course, the case, provided that the eigenvalues of the matrix $A - LC$ are in the left half of the complex plane. It is a well-known result that, if the observability condition (11.7) is satisfied, then the eigenvalues of $(A - LC)$ can be placed anywhere in the complex plane by suitable selection of the observer gain $L$.

### 11.1.4 Separation Principle

Assume that the system (11.1) is controlled via the following control law:

$$u = K\hat{x} \tag{11.11}$$
$$\dot{\hat{x}} = A\hat{x} + Bu + L(y - C\hat{x}) \tag{11.12}$$

*i.e.* a state feedback law is used in (11.11). However, short of having the true state $x$ available for feedback, an estimate $\hat{x}$ of the true state $x$ was used in (11.11). Equation (11.12) is the observer equation. We have

$$\begin{aligned} \dot{x} &= Ax + BK\hat{x} \\ &= Ax + BKx - BK\tilde{x} \\ &= (A + BK)x - BK\tilde{x}. \end{aligned}$$

Also

$$\dot{\tilde{x}} = (A - LC)\tilde{x}$$

Thus

$$\begin{bmatrix} \dot{x} \\ \dot{\tilde{x}} \end{bmatrix} = \begin{bmatrix} A + BK & -BK \\ 0 & A - LC \end{bmatrix} \begin{bmatrix} x \\ \tilde{x} \end{bmatrix}$$
$$\dot{\bar{x}} = \bar{A}\bar{x}$$

where we have defined

$$\bar{x} \overset{def}{=} \begin{bmatrix} x \\ \tilde{x} \end{bmatrix}$$

The eigenvalues of the matrix $\bar{A}$ are the union of those of $(A + BK)$ and $(A - LC)$. Thus, we conclude that

(i) The eigenvalues of the observer are not affected by the state feedback and vice versa.

(ii) The design of the state feedback and the observer can be carried out independently. This is called the *separation principle*.

## 11.2   Nonlinear Observability

Now consider the system

$$\psi_{nl} \begin{cases} \dot{x} = f(x) + g(x)u & f : \mathbb{R}^n \to \mathbb{R}^n, g : \mathbb{R}^n \to \mathbb{R} \\ y = h(x) & h : \mathbb{R}^n \to \mathbb{R} \end{cases} \qquad (11.13)$$

For simplicity we restrict attention to single-output systems. We also assume that $f(\cdot), g(\cdot)$ are sufficiently smooth and that $h(0) = 0$. Throughout this section we will need the following notation:

- $x_u(t, x_0)$: represents the solution of (11.13) at time $t$ originated by the input $u$ and the initial state $x_0$.

- $y(x_u(t, x_0))$: represents the output $y$ when the state $x$ is $x_u(t, x_0)$.

Clearly

$$y(x_u(t, x_0)) \equiv h(x_u(t, x_0))$$

**Definition 11.3** *A pair of states* $(x_0^1, x_0^2)$ *is said to be* distinguishable *if there exists an input function u such that*

$$y(x_u(t, x_0^1)) \equiv y(x_u(t, x_0^2))$$

**Definition 11.4** *The state space realization* $\psi_{nl}$ *is said to be (locally)* observable *at* $x_0 \in \mathbb{R}^n$ *if there exists a neighborhood $U_0$ of $x_0$ such that every state $x \neq x_0 \in \Omega$ is distinguishable from $x_0$. It is said to be* locally observable *if it is locally observable at each $x_0 \in \mathbb{R}^n$.*

This means that $\psi_{nl}$ is locally observable in a neighborhood $U_0 \subset \mathbb{R}^n$ if there exists an input $u \in \mathbb{R}$ such that

$$y(x_u(t, x_0^1)) \equiv y(x_u(t, x_0^2)) \qquad \forall t \in [0, t] \qquad \Longleftrightarrow \qquad x_0^1 = x_0^2$$

There is no requirement in Definition 11.4 that distinguishability must hold for <u>all</u> functions. There are several subtleties in the observability of nonlinear systems, and, in general, checking observability is much more involved than in the linear case. In the following theorem, we consider an unforced nonlinear system of the form

$$\psi_{nl} \begin{cases} \dot{x} = f(x) & f : \mathbb{R}^n \to \mathbb{R}^n \\ y = h(x) & h : \mathbb{R}^n \to \mathbb{R} \end{cases} \qquad (11.14)$$

and look for observability conditions in a neighborhood of the origin $x = 0$.

**Theorem 11.1** *The state space realization (11.14) is locally observable in a neighborhood $U_0 \subset D$ containing the origin, if*

$$rank\left(\begin{bmatrix} \nabla h \\ \vdots \\ \nabla L_f^{n-1}h \end{bmatrix}\right) = n \quad \forall x \in U_0 \tag{11.15}$$

**Proof:** The proof is omitted. See Reference [52] or [36].

The following example shows that, for linear time-invariant systems, condition (11.15) is equivalent to the observability condition (11.7).

**Example 11.1** *Let*

$$\psi_l \begin{cases} \dot{x} = Ax \\ y = Cx. \end{cases}$$

*Then $h(x) = Cx$ and $f(x) = Ax$, and we have*

$$\nabla h(x) = C$$
$$\nabla L_f h = \nabla(\frac{\partial h}{\partial x}\dot{x}) = \nabla(CAx) = CA$$
$$\vdots$$
$$\nabla L_f^{n-1}h = CA^{n-1}$$

*and therefore $\psi_l$ is observable if and only if $S = \{C, CA, CA^2, \cdots, CA^{n-1}\}$ is linearly independent or, equivalently, if $rank(\mathcal{O}) = n$.*

Roughly speaking, definition 11.4 and Theorem 11.1 state that if the linearization of the state equation (11.13) is observable, then (11.13) is locally observable around the origin. Of course, for nonlinear systems local observability does not, in general, imply global observability.

We saw earlier that for linear time-invariant systems, observability is independent of the input function and the $B$ matrix in the state space realization. This property is a consequence of the fact that the mapping $x_0 \to y$ is linear, as discussed in Section 11.2. Nonlinear systems often exhibit singular inputs that can render the state space realization unobservable. The following example clarifies this point.

**Example 11.2** *Consider the following state space realization:*

$$\psi_{nl} \begin{cases} \dot{x}_1 = x_2(1-u) \\ \dot{x}_2 = x_1 \\ y = x_1 \end{cases}$$

*which is of the form*

$$\begin{cases} \dot{x} = f(x) + g(x)u \\ y = h(x) \end{cases}$$

*with*

$$f(x) = \begin{pmatrix} x_2 \\ x_1 \end{pmatrix}, \quad g(x) = \begin{pmatrix} -x_2 \\ 0 \end{pmatrix}, \quad h(x) = x_1.$$

*If $u = 0$, we have*

$$rank(\{\nabla h, \nabla L_f h\}) = rank(\{[1 \ 0], [0 \ 1]\}) = 2$$

*and thus*

$$\begin{cases} \dot{x} = f(x) \\ y = h(x) \end{cases}$$

*is observable according to Definition 11.4. Now consider the same system but assume that $u = 1$. Substituting this input function, we obtain the following dynamical equations:*

$$\begin{cases} \dot{x}_1 = 0 \\ \dot{x}_2 = x_1 \\ y = x_1. \end{cases}$$

*A glimpse at the new linear time-invariant state space realization shows that observability has been lost.*  □

## 11.2.1  Nonlinear Observers

There are several ways to approach the nonlinear state reconstruction problem, depending on the characteristics of the plant. A complete coverage of the subject is outside the scope of this textbook. In the next two sections we discuss two rather different approaches to nonlinear observer design, each applicable to a particular class of systems.

# 11.3  Observers with Linear Error Dynamics

Motivated by the work on feedback linearization, it is tempting to approach nonlinear state reconstruction using the following three-step procedure:

(i) Find an invertible coordinate transformation that linearizes the state space realization.

(ii) Design an observer for the resulting linear system.

(iii) Recover the original state using the inverse coordinate transformation defined in (i).

More explicitly, suppose that given a system of the form

$$\begin{cases} \dot{x} = f(x) + g(x, u) & x \in \mathbb{R}^n, \ u \in \mathbb{R} \\ y = h(x) & y \in \mathbb{R} \end{cases} \tag{11.16}$$

there exist a diffeomorphism $T(\cdot)$ satisfying

$$z = T(x), \qquad T(0) = 0, \quad z \in \mathbb{R}^n \tag{11.17}$$

and such that, after the coordinate transformation, the new state space realization has the form

$$\begin{cases} \dot{x} = A_0 z + \gamma(y, u) \\ y = C_0 z \end{cases} \qquad y \in \mathbb{R} \tag{11.18}$$

where

$$A_0 = \begin{bmatrix} 0 & 0 & \cdots & 0 \\ 1 & 0 & \cdots & 0 \\ 0 & 1 & \cdots & 0 \\ \vdots & & \ddots & \\ 0 & 0 & 1 & 0 \end{bmatrix}, \quad C_0 = [0 \ 0 \ \cdots \ 0 \ 1], \ \gamma = \begin{bmatrix} \gamma_1(y, u) \\ \gamma_1(y, u) \\ \vdots \\ \vdots \\ \gamma_n(y, u) \end{bmatrix}. \tag{11.19}$$

then, under these conditions, an observer can be constructed according to the following theorem.

**Theorem 11.2** *([52]) If there exist a coordinate transformation mapping (11.16) into (11.18), then, defining*

$$\dot{\hat{z}} = A_0 \hat{z} + \gamma(y, u) - K(y - \hat{z}_n), \qquad \hat{z} \in \mathbb{R}^n \tag{11.20}$$

$$\hat{x} = T^{-1}(z) \tag{11.21}$$

*such that the eigenvalues of $(A_0 + K C_0)$ are in the left half of the complex plane, then $\hat{x} \to x$ as $t \to \infty$.*

**Proof:** Let $\tilde{z} = z - \hat{z}$, and $\tilde{x} = x - \hat{x}$. We have

$$\begin{aligned} \dot{\tilde{z}} &= \dot{z} - \dot{\hat{z}} \\ &= [A_0 z + \gamma(y, u)] - [A_0 \hat{z} + \gamma(y, u) - K(y - \hat{z}_n)] \\ &= (A_0 + K C_0) \tilde{z}. \end{aligned}$$

If the eigenvalues of $(A_0 + K C_0)$ have negative real part, then we have that $\tilde{z} \to 0$ as $t \to \infty$. Using (11.21), we obtain

$$\begin{aligned} \tilde{x} &= x - \hat{x} \\ &= T^{-1}(z) - T^{-1}(z - \tilde{z}) \quad \to 0 \ \text{as} \ t \to \infty \end{aligned}$$

$\square$

**Example 11.3** *Consider the following dynamical system*

$$\begin{cases} \dot{x}_1 = x_2 + 2x_1^2 \\ \dot{x}_2 = x_1 x_2 + x_1^3 u \\ y = x_1 \end{cases} \tag{11.22}$$

*and define the coordinate transformation*

$$\begin{cases} z_1 = x_2 - \frac{1}{2}x_1^2 \\ z_2 = x_1. \end{cases}$$

*In the new coordinates, the system (11.22) takes the form*

$$\begin{cases} \dot{z}_1 = -2y^3 + y^3 u \\ \dot{z}_2 = z_1 + \frac{5}{2}y^2 \\ y = z_2 \end{cases}$$

*which is of the form (11.18) with*

$$A_0 = \begin{bmatrix} 0 & 0 \\ 1 & 0 \end{bmatrix}, \quad C_0 = [0 \ 1], \quad and \ \gamma = \begin{bmatrix} -2y^3 + y^3 u \\ \frac{5}{2}y^2 \end{bmatrix}.$$

*The observer is*

$$\dot{\hat{z}} = A_0 \hat{z} + \gamma(y, u) - K(y - \hat{z}_2)$$

$$\begin{bmatrix} \dot{\hat{z}}_1 \\ \dot{\hat{z}}_2 \end{bmatrix} = \begin{bmatrix} 0 & 0 \\ 1 & 0 \end{bmatrix} \begin{bmatrix} \hat{z}_1 \\ \hat{z}_2 \end{bmatrix} + \begin{bmatrix} -2y^3 + y^3 u \\ 5y^2/2 \end{bmatrix} + \begin{bmatrix} K_1 \\ K_2 \end{bmatrix} (y - \hat{z}_2).$$

*The error dynamics is*

$$\begin{bmatrix} \dot{\tilde{z}}_1 \\ \dot{\tilde{z}}_2 \end{bmatrix} = \begin{bmatrix} 0 & -K_1 \\ 1 & -K_2 \end{bmatrix} \begin{bmatrix} \tilde{z}_1 \\ \tilde{z}_2 \end{bmatrix}$$

*Thus, $\tilde{z} \to 0$ for any $K_1, K_2 > 0$.*                                                                          □

It should come as no surprise that, as in the case of feedback linearization, this approach to observer design is based on cancelation of nonlinearities and therefore assumes "perfect modeling." In general, perfect modeling is never achieved because system parameters cannot be identified with arbitrary precision. Thus, in general, the "expected" cancellations will not take place and the error dynamics will not be linear. The result is that this observer scheme is not robust with respect to parameter uncertainties and that convergence of the observer is not guaranteed in the presence of model uncertainties.

## 11.4 Lipschitz Systems

The nonlinear observer discussed in the previous section is inspired by the work on feedback linearization, and belongs to the category of what can be called the differential geometric approach. In this section we show that observer design can also be studied using a Lyapunov approach. For simplicity, we restrict attention to the case of Lipschitz systems, defined below.

Consider a system of the form:

$$\begin{cases} \dot{x} = Ax + f(x, u) \\ y = Cx \end{cases} \tag{11.23}$$

where $A \in \mathbb{R}^{n \times n}$, $C \in \mathbb{R}^{1 \times n}$, and $f : \mathbb{R}^n \times \mathbb{R} \to \mathbb{R}^n$ is Lipschitz in $x$ on an open set $D \subset \mathbb{R}^n$, i.e., $f$ satisfies the following condition:

$$\|f(x_1, u^*) - f(x_2, u^*)\| \le \gamma \|x_1 - x_2\| \quad \forall x \in D. \tag{11.24}$$

Now consider the following observer structure

$$\dot{\hat{x}} = A\hat{x} + f(\hat{x}, u) + L(y - C\hat{x}) \tag{11.25}$$

where $L \in \mathbb{R}^{n \times 1}$. The following theorem shows that, under these assumptions, the estimation error converges to zero as $t \to \infty$.

**Theorem 11.3** *Given the system (11.23) and the corresponding observer (11.25), if the Lyapunov equation*

$$P(A - LC) + (A - LC)^T P = -Q \tag{11.26}$$

*where $P = P^T > 0$, and $Q = Q^T > 0$, is satisfied with*

$$\gamma < \frac{\lambda_{min}(Q)}{2\lambda_{max}(P)} \tag{11.27}$$

*then the observer error $\tilde{x} = x - \hat{x}$ is asymptotically stable.*

**Proof:**

$$\begin{aligned} \dot{\tilde{x}} &= \dot{x} - \dot{\hat{x}} \\ &= [Ax + f(x, u)] - [A\hat{x} + f(\hat{x}, u) + L(y - C\hat{x})] \\ &= (A - LC)\tilde{x} + f(x, u) - f(\hat{x}, u). \end{aligned}$$

To see that $\dot{\tilde{x}}$ has an asymptotically stable equilibrium point at the origin, consider the Lyapunov function candidate:

$$V(\tilde{x}) \stackrel{def}{=} \tilde{x}^T P \tilde{x}.$$

$$
\begin{aligned}
\dot{V}(\tilde{x}) &= \dot{\tilde{x}} P \tilde{x} + \tilde{x} P \dot{\tilde{x}} \\
&= -\tilde{x}^T Q \tilde{x} + 2\tilde{x}^T P[f(\tilde{x} + \hat{x}, u) - f(\hat{x}, u)]
\end{aligned}
$$

but

$$\|\tilde{x}^T Q \tilde{x}\| \geq \lambda_{min}(Q)\|\tilde{x}\|^2$$

and

$$
\begin{aligned}
\|2x^T P[f[(\tilde{x} + \hat{x}, u) - f(\hat{x}, u)]\| &\leq \|2\tilde{x}^T P\| \, \|f[(\tilde{x} + \hat{x}, u) - f(\hat{x}, u)\| \\
&\leq 2\gamma\lambda_{max}(P)\|\tilde{x}\|^2.
\end{aligned}
$$

Therefore, $\dot{V}$ is negative definite, provided that

$$\lambda_{min}(Q)\|\tilde{x}\|^2 > 2\gamma\lambda_{max}(P)\|\tilde{x}\|^2$$

or, equivalently

$$\gamma < \frac{\lambda_{min}(Q)}{2\lambda_{max}(P)}$$

$\square$

**Example 11.4** *Consider the following system:*

$$
\begin{bmatrix} \dot{x}_1 \\ \dot{x}_2 \end{bmatrix} = \begin{bmatrix} 0 & 1 \\ 1 & 2 \end{bmatrix} \begin{bmatrix} x_1 \\ x_2 \end{bmatrix} + \begin{bmatrix} 0 \\ x_2^2 \end{bmatrix}
$$

$$
y = [1 \ \ 0] \begin{bmatrix} x_1 \\ x_2 \end{bmatrix}
$$

*Setting*

$$
L = \begin{bmatrix} 0 \\ 2 \end{bmatrix}
$$

*we have that*

$$
A - LC = \begin{bmatrix} 0 & 1 \\ -1 & -2 \end{bmatrix}
$$

*Solving the Lyapunov equation*

$$P(A - LC) + (A - LC)^T P = -Q$$

*with $Q = I$, we obtain*

$$P = \begin{bmatrix} 1.5 & -0.5 \\ -0.5 & 0.5 \end{bmatrix}$$

*which is positive definite. The eigenvalues of $P$ are $\lambda_{min}(P) = 0.2929$, and $\lambda_{max}(P) = 1.7071$. We now consider the function $f$. Denoting*

$$x_1 = \begin{bmatrix} \xi_1 \\ \xi_2 \end{bmatrix}, \qquad x_2 = \begin{bmatrix} \mu_1 \\ \mu_2 \end{bmatrix}$$

*we have that*

$$
\begin{aligned}
\|f(x_1) - f(x_2)\|_2 &= \sqrt{(\xi_2^2 - \mu_2^2)^2} \\
&= |\xi_2^2 - \mu_2^2| \\
&= |(\xi_2 + \mu_2)(\xi_2 - \mu_2)| \\
&\leq 2|\xi_2| \, |\xi_2 - \mu_2| = 2k\|\xi_2 - \mu_2| \\
&\leq 2k\|x_1 - x_2\|_2
\end{aligned}
$$

*for all $x$ satisfying $|\xi_2| < k$. Thus, $\gamma = 2k$ and $f$ is Lipschitz $\forall x = [\xi_1 \; \xi_2]^T : |\xi_2| < k$, and we have*

$$\gamma = 2k < \frac{1}{2\lambda_{max}(P)}$$

*or*

$$k < \frac{1}{6.8284}$$

*The parameter $k$ determines the region of the state space where the observer is guaranteed to work. Of course, this region is a function of the matrix $P$, and so a function of the observer gain $L$. How to maximize this region is not trivial (see "Notes and References" at the end of this chapter).*

$\square$

## 11.5  Nonlinear Separation Principle

In Section 11.1.4 we discussed the well-known separation principle for linear time-invariant (LTI) systems. This principle guarantees that output feedback can be approached in two steps:

(i) Design a state feedback law assuming that the state $x$ is available.

(ii) Design an observer, and replace $x$ with the estimate $\hat{x}$ in the control law.

near systems do not enjoy the same properties. Indeed, if the true state
estimate given by an observer, then exponential stability of the observer
., ... general, guarantee closed-loop stability. To see this, consider the following
example.

**Example 11.5** ([47]) Consider the following system:

$$\dot{x} = -x + x^4 + x^2\xi \tag{11.28}$$
$$\dot{\xi} = -k\xi + u \quad k > 0 \tag{11.29}$$

We proceed to design a control law using backstepping. Using $\xi$ as the input in (11.28), we
choose the control law $\phi_1(x) = -x^2$. With this control law, we obtain

$$\dot{x} = -x + x^4 + x^2\phi_1(x) = -x$$

Now define the error state variable

$$z = \xi - \phi_1(x) = \xi + x^2.$$

With the new variable, the system (11.28)–(11.29) becomes

$$\dot{x} = -x + x^2 z \tag{11.30}$$
$$\dot{z} = \dot{\xi} - \dot{\phi}(x) \tag{11.31}$$
$$= -k\xi + u + 2x\dot{x} \tag{11.32}$$
$$= -k\xi + u + 2x(-x + x^4 + x^2\xi) \tag{11.33}$$
$$= -k\xi + u + 2x(-x + x^2 z) \tag{11.34}$$

Letting

$$V(x, \xi) = \frac{1}{2}(x^2 + z^2)$$
$$\Rightarrow \quad \dot{V} = -x^2 + z[x^3 - k\xi + u + 2x(-x + x^2 z)]$$

and taking

$$u = -cz - x^3 + k\xi - 2x(-x + x^2 z), \quad c > 0 \tag{11.35}$$

we have that

$$\dot{V} = -x^2 - cz^2$$

which implies that $x = 0, z = 0$ is a globally asymptotically stable equilibrium point of the
system

$$\dot{x} = -x + x^2 z$$
$$\dot{z} = -cz - x^3$$

*This control law, of course, assumes that both $x$ and $\xi$ are measured. Now assume that only $x$ is measured, and suppose that an observer is used to estimate $\xi$. This is a reduced-order observer, that is, one that estimates only the nonavailable state $\xi$. Let the observer be given by*

$$\dot{\widehat{\xi}} = -k\widehat{\xi} + u.$$

*The estimation error is $\tilde{\xi} = \xi - \widehat{\xi}$. We have*

$$\begin{aligned} \dot{\tilde{\xi}} &= -\dot{\xi} - \dot{\widehat{\xi}} \\ &= -k\xi + u + k\widehat{\xi} - u \\ &= -k\tilde{\xi} \end{aligned}$$

*It follows that*

$$\tilde{\xi}(t) = \tilde{\xi}(0)e^{-kt}$$

*which implies that $\widehat{\xi}$ exponentially converges toward $\xi$. Using the estimated $\xi$ in the control law (11.35), we obtain*

$$\begin{aligned} \dot{x} &= -x + x^2 z + x^2\tilde{\xi} \\ \dot{z} &= -cz - x^3 + 2x^3\tilde{\xi} \\ \dot{\tilde{\xi}} &= -k\tilde{\xi}. \end{aligned}$$

*Even though $\tilde{\xi}$ exponentially converges to zero, the presence of the terms $x^2\xi$ and $2x^3\tilde{\xi}$ leads to finite escape time for certain initial conditions. To see this, assume that the error variable $z \equiv 0$. We have*

$$\begin{aligned} \dot{x} &= -x + x^2\tilde{\xi} \\ \tilde{\xi}(t) &= \tilde{\xi}(0)e^{-kt} \end{aligned}$$

*Solving for $x$, we obtain*

$$x(t) = \frac{x_0(1+k)}{(1 + k - \tilde{\xi}_0 x_0)e^t + \tilde{\xi}_0 x_0 e^{-kt}}$$

*which implies that the state $x$ grows to infinity in finite time for any initial condition $\tilde{\xi}_0 x_0 > 1 + k$.* □

## Notes and References

Despite recent progress, nonlinear observer design is a topic that has no universal solution and one that remains a very active area of research. Observers with linear error dynamics were first studied by Krener and Isidori [45] and Bestle and Zeitz [10] in the single-input–single-output case. Extensions to multivariable systems were obtained by Krener and Respondek [46] and Keller [42]. Section 11.3 is based on Reference [45] as well as References [52] and [36]. Besides robustness issues, one of the main problems with this approach to observer design is that the conditions required to be satisfied for the existence of these observers are very stringent and are satisfied by a rather narrow class of systems.

Observers for Lipschitz systems were first discussed by Thau [80]. Section 11.4 is based on this reference. The main problem with this approach is that choosing the observer gain $L$ so as to satisfy condition (11.27) is not straightforward. See References [61], [60], and [96] for further insight into how to design the observer gain $L$.

There are several other approaches to nonlinear observer design, not covered in this chapter. The list of notable omissions includes adaptive observers [52]-[54], high gain-observers, [4], [5], and observer backstepping, [47].

Output feedback is a topic that has received a great deal of attention in recent years. As mentioned, the separation principle does not hold true for general nonlinear systems. Conditions under which the principle is valid were derived by several authors. See References [90], [82], [83], [8], [4] and [5].

# Appendix A

# Proofs

The purpose of this appendix is to provide a proof of several lemmas and theorems encountered throughout the book. For easy of reference we re-state the lemmas and theorems before each proof.

## A.1  Chapter 3

**Lemma 3.1**: $V : D \to \mathbb{R}$ *is positive definite if and only if there exists class* $\mathcal{K}$ *functions* $\alpha_1$ *and* $\alpha_2$ *such that*

$$\alpha_1(\|x\|) \leq V(x) \leq \alpha_2(\|x\|) \quad \forall x \in B_r \subset D.$$

*Moreover, if* $D = \mathbb{R}^n$ *and* $V(\cdot)$ *is radially unbounded, then* $\alpha_1$ *and* $\alpha_2$ *can be chosen in the class* $\mathcal{K}_\infty$.

**Proof:** Given $V(x) : D \subset \mathbb{R}^n \to \mathbb{R}$, we show that $V(x)$ is positive definite if and only if there exist $\alpha_1 \in \mathcal{K}$ such that $\alpha_1(\|x\|) \leq V(x)$ $\forall x \in D$. It is clear that the existence of $\alpha_1$ is a sufficient condition for the positive definiteness of $V$. To prove necessity, define

$$\chi(y) = \min_{y \leq \|x\| \leq r} V(x); \quad \text{for } 0 \leq y \leq r$$

The function $\chi(\cdot)$ is well defined since $V(\cdot)$ is continuous and $y \leq \|x\| \leq r$ defines a compact set in $\mathbb{R}^n$. Moreover, $\chi(\cdot)$ so defined has the following properties:

(i) $\mathcal{X} \leq V(x)$, $\quad 0 \leq \|x\| \leq r$.

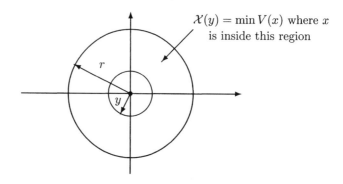

Figure A.1: Asymptotically stable equilibrium point.

(ii) It is continuous.

(iii) It is positive definite [since $V(x) > 0$].

(iv) It satisfies $\chi(0) = 0$.

Therefore, $\chi$ is "almost" in the class $\mathcal{K}$. It is not, in general, in $\mathcal{K}$ because it is not strictly increasing. We also have that

$$\chi(\|x\|) \leq V(x) \quad \text{for } 0 \leq \|x\| \leq r$$

$\chi(\cdot)$ is not, however, in the class $\mathcal{K}$ since, in general, it is not strictly increasing. Let $\alpha_1(y)$ be a class $\mathcal{K}$ function such that $\alpha_1(y) \leq k\chi(y)$ with $0 < k < 1$. Thus

$$\alpha_1(\|x\|) \leq \chi(\|x\|) \leq V(x) \quad \text{for } \|x\| \leq r.$$

The function $\alpha_1$ can be constructed as follows:

$$\alpha_1(s) = \min \left[ \frac{s}{y}\chi(y) \right] \quad s \leq y \leq r$$

this function is strictly increasing. To see this notice that

$$\frac{\alpha_1}{s} = \min \left[ \frac{\chi(y)}{y} \right] \quad s \leq y \leq r \tag{A.1}$$

is positive. It is also nondecreasing since as $s$ increases, the set over which the minimum is computed keeps "shrinking" (Figure A.2). Thus, from (A.1), $\alpha_1$ is strictly increasing.

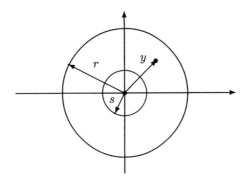

Figure A.2: Asymptotically stable equilibrium point.

This proves that there exists $\alpha_1(\cdot) \in \mathcal{K}$ such that

$$\alpha_1(\|x\|) \le V(x) \quad \text{for } \|x\| \le r$$

The existence of $\alpha_2(\cdot) \in \mathcal{K}$ such that

$$V(x) \le \alpha_2(\|x\|) \quad \text{for } \|x\| \le r$$

can be proved similarly. □

To simplify the proof of the following lemma, we assume that the equilibrium point is the origin, $x_e = 0$.

**Lemma 3.2**: *The equilibrium $x_e$ of the system (3.1) is stable if and only if there exists a class $\mathcal{K}$ function $\alpha(\cdot)$ and a constant $\epsilon$ such that*

$$\|x(0)\| < \delta \quad \Rightarrow \quad \|x(t)\| \le \alpha(\|x(0)\|) \quad \forall t \ge 0. \tag{A.2}$$

**Proof**: Suppose first that (A.2) is satisfied. We must show that this condition implies that for each $\epsilon_1, \exists \delta_1 = \delta_1(\epsilon_1) > 0$:

$$\|x(0)\| < \delta_1 \quad \Rightarrow \quad \|x(t)\| < \epsilon_1 \quad \forall t \ge t_0. \tag{A.3}$$

Given $\epsilon_1$, choose $\delta_1 = \min\{\delta, \alpha^{-1}(\epsilon_1)\}$. Thus, for $\|x(0)\| < \delta_1$, we have

$$\|x(t)\| \le \alpha(\|x(0)\|) < \alpha(\alpha^{-1}(\epsilon_1)) = \epsilon_1.$$

For the converse, assume that (A.3) is satisfied. Given $\epsilon_1$, let $\Delta \subset \mathbb{R}^+$ be the set defined as follows:

$$\Delta = \{\delta_1 \in \mathbb{R}^+ : \|x(0)\| < \delta_1 \;\Rightarrow\; \|x(t)\| < \epsilon_1\}.$$

This set is bounded above and therefore has a supremum. Let $\delta^* = \sup(\Delta)$. Thus,

$$\|x(0)\| < \delta^* \;\Rightarrow\; \|x(t)\| < \epsilon_1 \qquad \forall t \geq t_0.$$

The mapping $\Phi : \epsilon_1 \to \delta^*(\epsilon_1)$ satisfies the following: (i) $\Phi(0) = 0$, (ii) $\Phi(\epsilon_1) > 0$ $\forall \epsilon > 0$, and (iii) it is nondecreasing, but it is not necessarily continuous. We can find a class $\mathcal{K}$ function $\xi(\cdot)$ that satisfies $\xi(r) \leq \Phi(r)$, for each $r \in \mathbb{R}^+$. Since the inverse of a function in the class $\mathcal{K}$ is also in the same class, we now define:

$$\alpha(\cdot) = \xi^{-1}(\cdot) \quad \text{and } \alpha(\cdot) \in \mathcal{K} .$$

Given $\epsilon$, choose any $x(0)$ such that $\alpha(\|x(0)\|) = \epsilon$. Thus $\|x(0)\| < \delta^*$, and we have

$$\|x(t)\| < \epsilon = \alpha(\|x(0)\|) \text{ provided that } \|x(0)\| < \delta^*.$$

$\square$

**Lemma 3.3** *The equilibrium $x_e = 0$ of the system (3.1) is asymptotically stable if and only if there exists a class $\mathcal{KL}$ function $\beta(\cdot, \cdot)$ and a constant $\epsilon$ such that*

$$\|x_0\| < \delta \;\Rightarrow\; \|x(t)\| \leq \beta(\|x_0\|, t) \quad \forall t \geq 0. \tag{A.4}$$

**Proof:** Suppose that A.4 is satisfied. Then

$$\|x(t)\| \leq \beta(\|x_0\|, 0) \quad \text{whenever } \|x_0\| < \delta$$

and the origin is stable by Lemma 3.2. Moreover, $\beta$ is in class $\mathcal{KL}$ which implies that $\beta(\|x_0\|, t) \to 0$ as $t \to \infty$. Thus, $\|x(t)\| \to 0$ as $t \to \infty$ and $x = 0$ is also convergent. It then follows that $x = 0$ is asymptotically stable. The rest of the proof (i.e., the converse argument) requires a tedious constructions of a $\mathcal{KL}$ function $\beta$ and is omitted. The reader can consult Reference [41] or [88] for a complete proof. $\square$

**Theorem 3.5** *A function $g(x)$ is the gradient of a scalar function $V(x)$ if and only if the matrix*

$$g(x) = \begin{bmatrix} \frac{\partial g_1}{\partial x_1} & \frac{\partial g_2}{\partial x_1} & \cdots & \frac{\partial g_n}{\partial x_1} \\ \frac{\partial g_1}{\partial x_2} & \frac{\partial g_2}{\partial x_2} & \cdots & \frac{\partial g_n}{\partial x_2} \\ \vdots & & & \vdots \\ \frac{\partial g_1}{\partial x_n} & \frac{\partial g_2}{\partial x_n} & \cdots & \frac{\partial g_n}{\partial x_n} \end{bmatrix}$$

*is symmetric.*

**Proof**: To prove necessity, we assume that $g(x) = \nabla V(x)$. Thus

$$g(x) = \begin{bmatrix} \frac{\partial V}{\partial x_1} \\ \vdots \\ \frac{\partial V}{\partial x_n} \end{bmatrix} = \begin{bmatrix} g_1 \\ \vdots \\ g_n \end{bmatrix}.$$

Thus

$$T = \begin{bmatrix} \frac{\partial^2 V}{\partial x_1 \partial x_1} & \frac{\partial^2 V}{\partial x_2 \partial x_1} & \cdots & \frac{\partial^2 V}{\partial x_n \partial x_1} \\ \frac{\partial^2 V}{\partial x_1 \partial x_2} & \frac{\partial^2 V}{\partial x_2 \partial x_2} & \cdots & \frac{\partial^2 V}{\partial x_n \partial x_2} \\ \vdots & & & \vdots \\ \frac{\partial^2 V}{\partial x_1 \partial x_n} & \frac{\partial^2 V}{\partial x_2 \partial x_n} & \cdots & \frac{\partial^2 V}{\partial x_n \partial x_n} \end{bmatrix}$$

and the result follows since $\frac{\partial^2 V}{\partial x_i \partial x_j} = \frac{\partial^2 V}{\partial x_j \partial x_i}$. To prove sufficiency, assume that

$$\frac{\partial g_i}{\partial x_j} \frac{\partial g_j}{\partial x_i}$$

and consider the integration as defined in (3.7):

$$
\begin{aligned}
V(x) &= \int_0^x g(x)\, dx = \int_0^{x_1} g_1(s_1, 0, \cdots, 0)\, ds_1 \\
&+ \int_0^{x_2} g_2(x_1, s_2, 0, \cdots, 0)\, ds_2 + \cdots + \int_0^{x_n} g_n(x_1, x_2, \cdots, s_n)\, ds_n \quad (A.5)
\end{aligned}
$$

We now compute the partial derivatives of this function:

$$
\begin{aligned}
\frac{\partial V}{\partial x_1} &= g_1(x_1, 0, \cdots, 0) + \int_0^{x_2} \frac{\partial g_2}{\partial x_1}(x_1, s_2, 0, \cdots, 0)\, ds_2 + \cdots \\
&+ \int_0^{x_n} \frac{\partial g_n}{\partial x_1}(x_1, x_2, \cdots, s_n)\, ds_n \\
&= g_1(x_1, 0, \cdots, 0) + \int_0^{x_2} \frac{\partial g_1}{\partial x_2}(x_1, s_2, 0, \cdots, 0)\, ds_2 + \cdots \\
&+ \int_0^{x_n} \frac{\partial g_1}{\partial x_n}(x_1, x_2, \cdots, s_n)\, ds_n
\end{aligned}
$$

where we have used the fact that $\frac{\partial g_i}{\partial x_j} \frac{\partial g_j}{\partial x_i}$, by assumption. It then follows that

$$\frac{\partial V}{\partial x_1} = g_1(x_1, 0, \cdots, 0) + g_1(x_1, s_2, \cdots, 0)|_0^{x_2} + \cdots$$
$$+ g_1(x_1, x_2, \cdots, s_n)|_0^{x_n}$$
$$= g_1(x)$$

Proceeding in the same fashion, it can be shown that

$$\frac{\partial V}{\partial x_i} = g_i(x) \quad \forall i.$$

$\square$

**Lemma 3.4**: *If the solution $x(t, x_0, t_0)$ of the system (3.1) is bounded for $t > t_0$, then its (positive) limit set $N$ is (i) bounded, (ii) closed, and (iii) nonempty. Moreover, the solution approaches $N$ as $t \to \infty$.*

**Proof:** Clearly $N$ is bounded, since the solution is bounded. Now take an arbitrary infinite sequence $\{t_n\}$ that tends to infinity with $n$. Then $x(t_n, x_0, t_0)$ is a bounded sequence of points, and by properties of bounded sequences, it must contain a convergent subsequence that converges to some point $p \in \mathbb{R}^n$. By Definition 3.12 $p \in N$, and thus $N$ is nonempty. To prove that it is closed, we must show that it contains all its limit points. To show that this is the case, consider the sequence of points $\{q_i\}$ in $N$, which approaches a limit point $q$ of $N$ as $i \to \infty$. Given $\epsilon > 0, \exists q_k \in N$ such that

$$\|q_k - q\| < \frac{\epsilon}{2}$$

Since $q_k$ is in $N$ there is a sequence $\{t_n\}$ such that $t_n \to \infty$ as $n \to \infty$, and

$$\|x(t_n, x_0, t_0) - q_k\| < \frac{\epsilon}{2}.$$

Thus,

$$\|x(t_n, x_0, t_0) - q\| < \epsilon$$

which implies that $q \in N$ and thus $N$ contains its limit points and it is closed. It remains to show that $x(t, x_0, t_0)$ approaches $N$ as $t \to \infty$. Suppose that this is not the case. Then there exists a sequence $\{t_n\}$ such that

$$\|p - x(t_n, x_0, t_0)\| > \epsilon \qquad \text{as } n \to \infty$$

for any $p$ in the closed set $N$. Since the infinite sequence $x(t_n, x_0, t_0)$ is bounded, it contains some subsequence that converges to a point $q \notin N$. But this is not possible since it contradicts Definition 3.12.

$\square$

**Lemma 3.5**: *The positive limit set $N$ of a solution $x(t, x_0, t_0)$ of the autonomous system*

$$\dot{x} = f(x) \tag{A.6}$$

*is invariant with respect to (A.6).*

**Proof**: We need to show that if $x_0 \in N$, then $x(t, x_0, t_0) \in N$ for all $t > t_0$. Let $p \in N$. Then there exists a sequence $\{t_n\}$ such that $t_n \to \infty$ and $x(t_n, x_0, t_0) \to p$ as $n \to \infty$. Since the solution of (A.6) is continuous with respect to initial conditions, we have

$$\lim_{n \to \infty} x[t, x(t_n, x_0, t_0), t_0] = x(t, p, t_0).$$

Also,

$$x[t, x(t_n, x_0, t_0), t_0] = x[t + t_n, x_0, t_0].$$

It follows that

$$\lim_{n \to \infty} x[t + t_n, x_0, t_0] = x(t, p, t_0) \tag{A.7}$$

and since $x(t, x_0, t_0)$ approaches $N$ as $t \to \infty$, (A.7) shows that $x(t, p, t_0)$ is in $N \forall t$.  □

## A.2   Chapter 4

**Theorem 4.6**: *Consider the system $\dot{x} = A(t)x$. The equilibrium state $x = 0$ is exponentially stable if and only if for any given symmetric, positive definite continuous and bounded matrix $Q(t)$, there exist a symmetric, positive definite continuously differentiable and bounded matrix $P(t)$ such that*

$$-Q(t) = P(t)A(t) + A^T(t)P(t) + \dot{P}(t) \tag{A.8}$$

**Proof**: Sufficiency is straightforward. To prove necessity, let

$$P(t) = \int_t^\infty \Phi^T(\tau, t)Q(\tau)\Phi(\tau, t)\, d\tau.$$

Thus,

$$
\begin{aligned}
x^T P x &= \int_t^\infty x^T \Phi^T(\tau, t)Q(\tau)\Phi(\tau, t)x\, d\tau \\
&= \int_t^\infty [\Phi^T(\tau, t)x(t)]^T Q(\tau)[\Phi(\tau, t)x(t)]d\tau \\
&= \int_t^\infty x(\tau)^T Q(\tau)x(\tau)d\tau.
\end{aligned}
$$

Since the equilibrium is globally asymptotically stable, there exist $k_1, k_2$ such that (see, for example, Chen [15] pp. 404)

$$\|x(t)\| \leq k_1 e^{-k_2(t-t_0)}.$$

Thus

$$x^T P x \;=\; \int_t^\infty k_1 e^{-k_2(\tau-t)} Q(\tau) k_1 e^{-k_2(\tau-t)} \, d\tau$$

$$=\; \int_t^\infty k_1^2 Q(\tau) e^{-2k_2(\tau-t)} \, d\tau.$$

The boundedness of $Q(t)$ implies that there exist $M$: $\|Q(t)\| \leq M$. Thus

$$x^T P x = \int_t^\infty k_1^2 M e^{-2k_2(\tau-t)} \, d\tau.$$

or

$$x^T P x \leq \frac{k_1^2 M}{2k_2}.$$

Also, this implies that $x^T P x \leq A_2 \|x\|^2$. On the other hand, since $Q$ is positive definite, $\exists a > 0$ such that $x^T Q x \geq a x^T x$. Also, $A$ bounded implies that $\|A(t)\| < N$, $\forall t \in \mathbb{R}$. Thus

$$\Rightarrow \quad x^T P x \;=\; \int_t^\infty x^T \Phi^T(\tau, t) Q(\tau) \Phi(\tau, t) x \, d\tau$$

$$=\; \int_t^\infty x(\tau)^T Q(\tau) x(\tau) \, d\tau$$

$$\geq\; \int_t^\infty a \frac{\|A(\tau)\|}{N} \|x(\tau)\|^2 \, d\tau$$

$$\geq\; \int_t^\infty \frac{a}{N} \|x(\tau)^T A(\tau) x(\tau)\| \, d\tau$$

$$\geq\; \left\| \int_t^\infty \frac{a}{N} x(\tau)^T \frac{d}{d\tau} x(\tau) \, d\tau \right\| \frac{a}{N} x^T x.$$

Thus,

$$x^T P x \geq \frac{a}{N} x^T x$$

or

$$A_1 \|x\|^2 \leq x^T P x$$

$$\Rightarrow \quad A_1 \|x\|^2 \leq x^T P x \leq A_2 \|x\|^2$$

which shows that $P$ is positive definite and bounded. That $P$ is symmetric follows from the definition. Thus, there remains to show that $P$ satisfies (A.8). We know that (see Chen [15], Chapter 4,

$$\frac{\partial}{\partial t} \Phi(\tau, t) = -\Phi(\tau, t) A(t).$$

Thus,

$$
\begin{aligned}
\dot{P} &= \int_t^\infty \Phi^T(\tau, t) Q(\tau) \frac{\partial}{\partial t} \Phi(\tau, t) \, d\tau + \int_t^\infty \frac{\partial}{\partial t} \Phi^T(\tau, t) Q(\tau) \Phi(\tau, t) \, d\tau - Q(t) \\
&= -\left[ \int_t^\infty \Phi^T(\tau, t) Q(\tau) \Phi(\tau, t) \, d\tau \right] A(t) - A^T(t) \left[ \int_t^\infty \Phi^T(\tau, t) Q(\tau) \Phi(\tau, t) \, d\tau \right] - Q(t)
\end{aligned}
$$

which implies that

$$
\dot{P} = -P(t)A(t) - A^T(t)P(t) - Q(t).
$$

$\square$

## A.3   Chapter 6

**Theorem 6.1** *Consider a function $\widehat{F}(s) \in \mathcal{R}(s)$. Then $\widehat{F}(s) \in \widehat{\mathcal{A}}$ if and only if (i) $\widehat{F}(s)$ is proper, and (ii) all poles of $\widehat{F}(s)$ lie in the left half of the complex plane.*

**Proof:** Assume first that $\widehat{F}(s)$ satisfies (i) and (ii), and let $n(s) \, and \, d(s) \in \mathcal{R}[s]$ respectively, be the numerator and denominator polynomials of $\widehat{F}(s)$. Dividing $n(s)$ by $d(s)$, we can express $\widehat{F}(s)$ as:

$$
\widehat{F}(s) = k + \frac{n(s)}{d(s)} = k + \widehat{G}(s) \tag{A.9}
$$

where $k \in \mathbb{R}$ and $\widehat{G}(s)$ is strictly proper. Expanding $\widehat{G}(s)$ in partial fractions and antitransforming (A.9), we have

$$
f(t) = \mathcal{L}^{-1}\{\widehat{F}(s)\} = k\delta(t) + \sum_{i=1}^n g_i(t)
$$

where $n$ is the degree of $d(s)$ and each $g_i$ has one of the following forms [$u(t)$ represents the unit step function]: (i) $kt^{m-1}e^{\lambda t}u(t)$, if $\lambda < 0$ is a real pole of multiplicity $m$; and (ii) $ke^{(\sigma + \jmath\omega)t}u(t), \sigma < 0$, if $\lambda = (\sigma + \jmath\omega)$ is a complex pole. It follows that $f(t) \in \mathcal{A}$, and so $\widehat{F}(s) \in \widehat{\mathcal{A}}$. To prove the converse, notice that if (i) does not hold, then $f(\cdot)$ contains derivatives of impulses and so does not belong to $\mathcal{A}$. Also, if (ii) does not hold, then, clearly, for some $i$, $g_i \notin \mathcal{L}_1$ and then $f(t) \notin \mathcal{A}$ and $\widehat{F}(s) \notin \widehat{\mathcal{A}}$, which completes the proof. $\square$

**Theorem 6.2** *Consider a linear time-invariant system $H$, and let $h(\cdot)$ represent its impulse response. Then $H$ is $\mathcal{L}_p$ stable if and only if $h(\cdot) = h_0\delta(t) + h_a(t) \in \mathcal{A}$ and moreover, if $H$ is $\mathcal{L}_p$ stable, then $\|Hx\|_p \leq \|h\|_{\mathcal{A}}\|x\|_p$.*

**Proof:** The necessity is obvious. To prove sufficiency, assume that $h(\cdot) \in \mathcal{A}$ and consider the output of $H$ to an input $u \in \mathcal{L}_p$ . We have

$$h(t) * u(t) = h_0 u(t) + \int_0^t h_a(t - \tau) u(\tau) \, d\tau$$

$$= g_1 + g_2$$

We analyze both terms separately. Clearly

$$\|g_{1T}\|_{\mathcal{L}_p} = \|h_0 u(t)\|_{\mathcal{L}_p}$$

$$= |h_0| \, \|u(t)\|_{\mathcal{L}_p}. \tag{A.10}$$

For the second term we have that

$$|g_2(t)| = |\int_0^t h_a(t - \tau) u(\tau) \, d\tau| \le \int_0^t |h_a(t - \tau)| \, |u(\tau)| \, d\tau$$

and choosing $p$ and $q \in \mathbb{R}^+$ such that $1/p + 1/q = 1$, it follows that

$$|g_2(t)| \le \int_0^t |h_a(t - \tau)| \, |u(\tau)| \, d\tau = \int_0^t |h_a(t - \tau)|^{1/p} \, |h_a(t - \tau)|^{1/q} |u(\tau)| \, d\tau. \tag{A.11}$$

Thus, by Hölder's inequality (6.5) with $f(\cdot) = |h_a(t - \tau)|^{1/p} |u(\tau)|$ and $g(\cdot) = |h_a(t - \tau)|^{1/q}$, we have that

$$
\begin{aligned}
|g_2(t)| &\le \left( \int_0^t |h_a(t - \tau)| \, d\tau \right)^{1/q} \left( \int_0^t |h_a(t - \tau)||u(\tau)|^p \, d\tau \right)^{1/p} \\
&\le (\|(h_a)_t\|_{\mathcal{L}_1})^{1/q} \left( \int_0^t |h_a(t - \tau)||u(\tau)|^p \, d\tau \right)^{1/p}.
\end{aligned}
$$

Thus

$$
\begin{aligned}
(\|(g_2)_T\|_{\mathcal{L}_p})^p &= \int_0^T |g_2(s)|^p \, ds \\
&\le \int_0^T (\|(h_a)_T\|_{\mathcal{L}_1})^{p/q} \left( \int_0^t |h_a(s - \tau)||u(\tau)|^p \, d\tau \right) \, ds \\
&= (\|(h_a)_T\|_{\mathcal{L}_1})^{p/q} \int_0^T \left( \int_0^t |h_a(s - \tau)||u(\tau)|^p \, d\tau \right) \, ds
\end{aligned}
$$

and reversing the order of integration, we obtain

$$(\|(g_2)_T\|_{\mathcal{L}_p})^p \le (\|(h_a)_T\|_{\mathcal{L}_1})^{p/q} \int_0^t |u(\tau)|^p \left( \int_0^T |h_a(s - \tau)| \, ds \right) \, d\tau$$

$$\leq \ (\|(h_a)_T\|_{\mathcal{L}_1})^{p/q} \ \|(h_a)_T\|_{\mathcal{L}_1} \ (\|u_T\|_{\mathcal{L}_1})^p$$
$$= \ (\|(h_a)_T\|_{\mathcal{L}_1})^p \ (\|u_T\|_{\mathcal{L}_1})^p.$$

It follows that

$$\|(g_2)_T\|_{\mathcal{L}_p} \ \leq \ \|(h_a)_T\|_{\mathcal{L}_1} \ \|u_T\|_{\mathcal{L}_1}. \tag{A.12}$$

Thus, from (A.12) and (A.10) we conclude that

$$\|(h * u)_T\|_{\mathcal{L}_p} \ \leq \ \|(g_1)_T\|_{\mathcal{L}_p} \ + \ \|(g_2)_T\|_{\mathcal{L}_p}$$
$$\leq \ (|h_0| + \|(h_a)_T\|_{\mathcal{L}_1}) \ \|u_T\|_{\mathcal{L}_p}$$

and since, by assumption, $u \in \mathcal{L}_p$ and $h(\cdot) \in \mathcal{A}$, we can take limits as $T \to \infty$. Thus

$$\|Hu\|_{\mathcal{L}_p} = \|h * u\|_{\mathcal{L}_p} = (|h_0| + \|h_a\|_{\mathcal{L}_1}) \ \|u\|_{\mathcal{L}_p}.$$

$\square$

**Theorem 6.6** Consider the feedback interconnection of the subsystems $H_1$ and $H_2 : \mathcal{L}_{2e} \to \mathcal{L}_{2e}$. Assume $H_2$ is a nonlinearity $\phi$ in the sector $[\alpha, \beta]$, and let $H_1$ be a linear time-invariant system with transfer function $\widehat{G}(s)$ of the form:

$$\widehat{G}(s) = \widehat{g}(s) + \frac{n(s)}{d(s)}$$

Where $\widehat{G}(s)$ satisfies assumptions (i)-(iii) stated in Theorem 6.6. Under these assumptions, if one of the following conditions is satisfied then the system is $\mathcal{L}_2$ stable:

(a) If $0 < \alpha < \beta$: The Nyquist plot of $\widehat{G}(s)$ is bounded away from the critical circle $\mathcal{C}^*$, centered on the real line and passing through the points $(-\alpha^{-1} + \jmath 0)$ and $(-\beta^{-1} + \jmath 0)$, and encircles it $\nu$ times in the counterclockwise direction, where $\nu$ is the number of poles of $\widehat{G}(s)$ in the open right half plane.

(b) If $0 = \alpha < \beta$: $\widehat{G}(s)$ has no poles in the open right half plane and the Nyquist plot of $\widehat{G}(s)$ remains to the right of the vertical line with abscissa $-\beta^{-1}$ for all $\omega \in \mathbb{R}$.

(c) If $\alpha < 0 < \beta$: $\widehat{G}(s)$ has no poles in the closed right half of the complex plane and the Nyquist plot of $\widehat{G}(s)$ is contained entirely within the interior of the circle $\mathcal{C}^*$.

**Proof:** Consider the system $S$ of equations (6.23) and (6.24) (Fig. 6.6) and define the system $S_K$ by applying a loop transformation of the Type I with $K = q$ defined as follows:

$$q \stackrel{def}{=} \frac{(\beta + \alpha)}{2}. \tag{A.13}$$

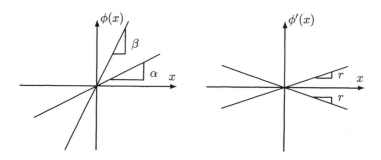

Figure A.3: Characteristics of $H_2$ and $H_2'$.

We have

$$H_1' = H_1[1 + qH_1]^{-1} = \frac{\widehat{G}(s)}{[1 + q\widehat{G}(s)]} \qquad (A.14)$$

$$H_2' = H_2 - q. \qquad (A.15)$$

By Theorem 6.4, the stability of the original system $S$ can be analyzed using the modified system $S_K$. The significance is that the type I loop transformation has a considerable effect on the nonlinearity $\phi$. Namely, if $H_2 = \phi \in [\alpha, \beta]$, it is easy to see that $H_2' = \phi' \in [-r, r]$, where

$$r \stackrel{def}{=} \frac{(\beta - \alpha)}{2} \qquad \text{(see Figure A.3.)}$$

Thus, the gain of $H_2'$ is

$$\gamma(H_2') = r \qquad (A.16)$$

According to the small gain theorem, if the following two conditions are satisfied, then the system is stable[1].

(i) $H_1'$ and $H_2'$ are $\mathcal{L}_2$ stable.

(ii) $\gamma(H_1')\gamma(H_2') < 1$.

**Condition (i):** $H_2'$ is bounded by assumption, and moreover, $\gamma(H_2') = r$. By lemma 6.9.1, $H_1'$ is $\mathcal{L}_2$ stable if and only if the Nyquist plot of $\widehat{G}(s)$ encircles the point $(-q^{-1} + \jmath 0)$ $\nu$ times in counterclockwise direction.

---

[1]Here condition (ii) actually implies condition (i). Separation into two conditions will help clarify the rest of the proof.

**Condition (ii):** $H_1$ and $H_1'$ are linear time-invariant systems; thus, the $\mathcal{L}_2$ gain of $H_1$ is

$$\gamma(H_1) = \|\widehat{G}\|_\infty = \sup_\omega |\widehat{G}(\jmath\omega)|.$$

Thus

$$\gamma(H_1)\gamma(H_2) < 1 \tag{A.17}$$

if and only if

$$r\left[\sup_\omega \left|\frac{\widehat{G}(\jmath\omega)}{[1+q\widehat{G}(\jmath\omega)]}\right| < 1\right. \tag{A.18}$$

or

$$r|\widehat{G}(\jmath\omega)| < |1+q\widehat{G}(\jmath\omega)| \qquad \forall\omega \in \mathbb{R}. \tag{A.19}$$

Let $\widehat{G}(\jmath\omega) = x + \jmath y$. Then equation (A.19) can be written in the form

$$r^2(x^2+y^2) < (1+qx)^2 + q^2y^2$$
$$\Rightarrow x^2(q^2-r^2) + y^2(q^2-r^2) + 2qx + 1 > 0$$
$$\Rightarrow \alpha\beta(x^2+y^2) + (\alpha+\beta)x + 1 > 0. \tag{A.20}$$

We now analyze conditions (a)–(c) separately:

(a) $0 < \alpha < \beta$: In this case we can divide equation (A.20) by $\alpha\beta$ to obtain

$$x^2 + y^2 + \frac{(\alpha+\beta)}{\alpha\beta}x + \frac{1}{\alpha\beta} > 0 \tag{A.21}$$

which can be expressed in the following form

$$(x+a)^2 + y^2 > R^2 \tag{A.22}$$

where

$$a = \frac{\alpha+\beta}{2\alpha\beta} \qquad R = \frac{(\beta-\alpha)}{2\alpha\beta} \tag{A.23}$$

Inequality (A.22) divides the complex plane into two regions separated by the circle $\mathcal{C}$ of center $(-a+\jmath 0)$ and radius $R$. Notice that if $y = 0$, then $\mathcal{C}$ satisfies $(x+a)^2 = R^2$, which has the following solutions: $x_1 = (-\beta^{-1} + \jmath 0)$ and $x_2 = (-\alpha^{-1} + \jmath 0)$. Thus $\mathcal{C} = \mathcal{C}^*$, *i.e.* $\mathcal{C}$ is equal to the critical circle $\mathcal{C}^*$ defined in Theorem 6.6. It is easy to see that inequality (A.22) is satisfied by points *outside* the critical circle. Therefore, the gain condition $\gamma(H_1')\gamma(H_2') < 1$ is satisfied if the Nyquist plot of $\widehat{G}(s)$ lies outside the circle $\mathcal{C}^*$. To satisfy the stability condition (i), the Nyquist plot of $\widehat{G}(s)$ must encircle the point $(-q^{-1} + \jmath 0)$, $\nu$ times in counterclockwise direction. Since the critical point $(-q^{-1} + \jmath 0)$ is located *inside* the circle $\mathcal{C}^*$, it follows that the Nyquist plot of $\widehat{G}(s)$ must encircle *the entire critical circle* $\mathcal{C}^*$ $\nu$ times in the counterclockwise direction, and part (a) is proved.

(b) $0 = \alpha < \beta$: In this case inequality (A.22) reduces to $x > \beta^{-1}$, or $\Re[\widehat{G}(\jmath\omega)] > -\beta^{-1}$. Thus, condition (ii) is satisfied if the Nyquist plot of $\widehat{G}(s)$ lies entirely to the right of the vertical line passing through the point $(-\beta^{-1} + \jmath 0)$. But $-q^{-1} = -2(\alpha + \beta)^{-1}$, and if $\alpha = 0, -q^{-1} = -2\beta^{-1}$. Thus, the critical point is inside the forbidden region and therefore $\widehat{G}(s)$ must have no poles in the open right half plane.

(c) $\alpha < 0 < \beta$: In this case dividing (A.22) by $\alpha\beta$, we obtain

$$x^2 + y^2 + \frac{(\alpha + \beta)}{\alpha\beta}x + \frac{1}{\alpha\beta} < 0 \qquad (A.24)$$

where we notice the change of the inequality sign with respect to case (a). Equation (A.24) can be written in the form

$$(x + a)^2 + y^2 < R^2 \qquad (A.25)$$

where $a$ and $R^2$ are given by equation (A.23). As in case (a) this circle coincides with the critical circle $C^*$, however, because of the change in sign in the inequality, (A.24) is satisfied by points inside $C^*$. It follows that condition (ii) is satisfied if and only if the Nyquist plot of $\widehat{G}(s)$ is contained entirely within the interior of the circle $C^*$. To satisfy condition (i) the Nyquist plot of $\widehat{G}(s)$ must encircle the point $(-q^{-1} + \jmath 0)$ $\nu$ times in the counterclockwise direction. However, in this case $(-q^{-1} + \jmath 0)$ lies outside the circle $C^*$. It follows that $\widehat{G}(s)$ cannot have poles in the open right half plane. Poles on the imaginary axis are not permitted since in this case the Nyquist diagram tends to infinity as $\omega$ tends to zero, violating condition (A.24). This completes the proof of Theorem 6.6.                                                                              □

## A.4   Chapter 7

**Theorem 7.4**: *Consider the system (7.1). Assume that the origin is an asymptotically stable equilibrium point for the autonomous system $\dot{x} = f(x, 0)$, and that the function $f(x, u)$ is continuously differentiable. Under these conditions (7.1) is locally input-to-state-stable.*

**Proof of theorem 7.4**: We will only sketch the proof of the theorem. The reader should consult Reference [74] for a detailed discussion of the ISS property and its relationship to other forms of stability. Given that $x = 0$ is asymptotically stable, the converse Lyapunov theorem guarantees existence of a Lyapunov function $V$ satisfying

$$\alpha_1(\|x\|) \leq V(x(t)) \leq \alpha_2(\|x\|) \qquad \forall x \in D \qquad (A.26)$$

$$\frac{\partial V(x)}{\partial x}f(x, 0) \leq -\alpha_3(\|x\|) \qquad \forall x \in D \qquad (A.27)$$

and

$$\left\|\frac{\partial V}{\partial x}\right\| \leq k_1 \|x\|, \qquad k_1 > 0 \qquad\qquad (A.28)$$

We need to show that, under the assumptions of the theorem, there exist $\alpha \in \mathcal{K}$ such that

$$\frac{\partial V}{\partial x}f(x,u) \leq -\alpha(\|x\|), \qquad \forall x : \|x\| > \sigma(\|u\|).$$

To this end we reason as follows. By the assumptions of the theorem, $f(\cdot, \cdot)$ satisfies

$$\|f(x,u) - f(x,0)\| \leq L \|u\|, \qquad L > 0 \qquad\qquad (A.29)$$

the set of points defined by $u : u(t) \in \mathbb{R}^m$ and $\|u\| \leq \delta$ form a compact set. Inequality (A.29) implies that in this set

$$\|f(x,u)\| \leq \|f(x,0)\| + \epsilon, \qquad \epsilon > 0.$$

Now consider a function $\alpha \in \mathcal{K}$ with the property that

$$\alpha(\|x\|) \geq \alpha_3(\|x\|) + \epsilon, \qquad \forall x : \|x\| \geq \mathcal{X}(\|u\|).$$

With $\alpha$ so defined, we have that,

$$\frac{\partial V}{\partial x}f(x,u) \leq -\alpha(\|x\|), \qquad \forall x : \|x\| \geq \mathcal{X}(\|u\|)$$

and the result follows. □

**Theorem 7.5**: *Consider the system (7.1). Assume that the origin is an exponentially stable equilibrium point for the autonomous system $\dot{x} = f(x,0)$, and that the function $f(x,u)$ is continuously differentiable and globally Lipschitz in $(x,u)$. Under these conditions (7.1) is input-to-state stable.*

**Proof of Theorem 7.5**: The proof follows the same lines as that of Theorem 7.4 and is omitted. □

**Proof of theorem 7.7**: Assume that $(\alpha, \sigma)$ is a supply pair with corresponding ISS Lyapunov function $V$, i.e.,

$$\nabla V(x) \cdot f(x,u) \leq -\alpha(\|x\|) + \sigma(\|u\|) \qquad \forall x \in \mathbb{R}^n, u \in \mathbb{R}^m. \qquad (A.30)$$

We now show that given $\tilde{\sigma} \in \mathcal{K}_\infty, \exists \tilde{\alpha}$

$$\nabla V(x) \cdot f(x,u) \leq -\tilde{\alpha}(\|x\|) + \tilde{\sigma}(\|u\|) \qquad \forall x \in \mathbb{R}^n, u \in \mathbb{R}^m. \qquad (A.31)$$

To this end we consider a new ISS Lyapunov function candidate $W$, defined as follows

$$W = \rho \circ V = \rho(V(x)), \quad \text{where} \tag{A.32}$$

$$\rho(s) = \int_0^s q(t)\, dt \tag{A.33}$$

where $q(\cdot) : \mathbb{R}^+ \to \mathbb{R}$ is a positive definite, smooth, nondecreasing function.

We now look for an inequality of the form (A.30) in the new ISS Lyapunov function $W$. Using (A.32) we have that

$$\begin{aligned}
\dot{W} &= \frac{\partial W}{\partial x} \cdot f(x,u) = \rho'(V(x))\dot{V}(x,u) \\
&= q[V(x)]\,[\sigma(\|u\|) - \alpha(\|x\|)].
\end{aligned} \tag{A.34}$$

Now define

$$\theta = \overline{\alpha} \circ \alpha{-}1 \circ (2\sigma)$$

that is,

$$\theta(s) = \overline{\alpha}(\alpha^{-1}(2\sigma)). \tag{A.35}$$

We now show that the right hand-side of (A.34) is bounded by

$$q[\theta(\|u\|)]\sigma(\|u\|) - \frac{1}{2}q[V(x)]\alpha(\|x\|). \tag{A.36}$$

To see this we consider two separate cases:

(i)  $\sigma(\|u\|) \le \frac{1}{2}\alpha(\|x\|)$: In this case, the right hand-side of (A.34) is bounded by

$$-\frac{1}{2}q[V(x)]\alpha(\|x\|)$$

(ii)  $\frac{1}{2}\alpha(\|x\|) \le \sigma(\|u\|)$: In this case, we have that $V(x) \le \overline{\alpha}(\|x\|) \le \theta(\|u\|)$.

From (i) and (ii), the the right-hand side of (A.34) is bounded by (A.36).

The rest of the proof can be easily completed if we can show that there exist $\tilde{\alpha} \in \mathcal{K}_\infty$ such that

$$q[\theta(r)]\sigma(r) - \frac{1}{2}q[\underline{\alpha}(s)]\alpha(s) \le \tilde{\sigma}(r) - \tilde{\alpha}(s) \qquad \forall r, s \ge 0. \tag{A.37}$$

To this end, notice that for any $\beta$, $\tilde{\beta} \in \mathcal{K}_\infty$ the following properties hold:

**Property 1**: if $\beta = O(\tilde{\beta}(r))$ as $r \to \infty$, then there exist a positive definite, smooth, and nondecreasing function $q$:

$$q(r)\beta(r) \le \tilde{\beta}(r) \qquad \forall r \in [0,\infty).$$

□With this property in mind, consider $\sigma \in \mathcal{K}_\infty$ [so that $\theta$, defined in (A.35) is $\mathcal{K}_\infty$], and define

$$\beta(r) = \sigma(\theta^{-1}(r)), \quad \tilde{\beta}(r) = \tilde{\sigma}(\theta^{-1}(r)).$$

With these definitions, $\beta(\cdot)$ and $\tilde{\beta}(\cdot) \in \mathcal{K}_\infty$. By property 1, there exist a positive definite, smooth, and nondecreasing function $q(\cdot)$:

$$q(r)\beta(r) \le \tilde{\beta}(r) \qquad \forall r \in [0, \infty).$$

Thus

$$q[\theta(r)]\sigma(r) \le \tilde{\sigma}. \tag{A.38}$$

Finally, defining

$$\tilde{\alpha}(s) = \frac{1}{2}q(\underline{\alpha}(s))\alpha(s) \tag{A.39}$$

we have that

$$\begin{cases} q[\underline{\alpha}(s)]\alpha(s) & \ge & 2\tilde{\alpha}(s) \\ q[\theta(s)]\sigma(s) & \le & \tilde{\sigma}(r) \end{cases} \tag{A.40}$$

and the theorem is proved substituting (A.40) into (A.37). □

**Proof of Theorem 7.8**: As in the case of Theorem 7.8 we need the following property

**Property 2**: if $\tilde{\beta} = O(\beta(r))$ as $r \to 0^+$, then there exist a positive definite, smooth, and nondecreasing function $q$:

$$\tilde{\beta}(s) \le q(s)\beta(s) \qquad \forall s \in [0, \infty).$$

□Defining $\beta(r) = \frac{1}{2}\alpha[\theta^{-1}(s)]$ we obtain $\tilde{\beta}(r) = \tilde{\alpha}[\theta^{-1}(s)]$. By property 2, there exist $q$:

$$\tilde{\beta} \le q(s)\beta(s) \qquad \forall s \in [0, \infty).$$

Thus

$$-\frac{1}{2}q[\theta(s)]\alpha(s) \le -\tilde{\alpha}(s). \tag{A.41}$$

Finally, defining

$$\tilde{\sigma}(r) = q[\theta(r)]\sigma(s) \tag{A.42}$$

we have that (A.40) is satisfied, which implies (A.37). □

## A.5    Chapter 8

**Theorem 8.3**: *Let $H : \mathcal{X}_e \to \mathcal{X}_e$, and assume that $(I+H)$ is invertible in $\mathcal{X}_e$, i.e., assume that $(I + H)^{-1} : \mathcal{X}_e \to \mathcal{X}_e$. Define the function $S : \mathcal{X}_e \to \mathcal{X}_e$:*

$$S = (H - I)(I + H)^{-1}. \tag{A.43}$$

*We have*

*(a)  $H$ is passive if and only if the gain of $S$ is at most 1, that is, $S$ is such that*

$$\|(Sx)_T\|_{\mathcal{X}} \leq \|x_T\|_{\mathcal{X}} \qquad \forall x \in \mathcal{X}_e, \forall T \in \mathcal{X}_e \tag{A.44}$$

*(b)  $H$ is strictly passive and has finite gain if and only if the gain of $S$ is less than 1.*

(a) : By assumption, $(I + H)$ is invertible, so we can define

$$\begin{align}
x \;&\overset{def}{=}\; (I + H)^{-1}y \tag{A.45}\\
(I + H)x \;&=\; y \tag{A.46}\\
Hx \;&=\; y - x \tag{A.47}\\
\Rightarrow Sy \;&=\; (H - I)(I + H)^{-1}y \;=\; (H - I)x \tag{A.48}
\end{align}$$

$$\begin{align}
\|Sy\|_T^2 \;&=\; \langle (H - I)x, (H - I)x >_T \\
\;&=\; \langle Hx - x, Hx - x > T \\
\;&=\; \|Hx\|_T^2 + \|x\|_T^2 - 2\langle x, Hx >_T \tag{A.49}
\end{align}$$

$$\begin{align}
\|y\|_T^2 \;&=\; \langle y, y >_T \\
\;&=\; \langle (I + H)x, (I + H)x >_T \\
\;&=\; \|Hx\|_T^2 + \|x\|_T^2 + 2\langle x, Hx >_T \tag{A.50}
\end{align}$$

Thus, subtracting (A.50) from (A.49), we obtain

$$\|Sy\|_T^2 = \|y\|_T^2 - 4\langle x, Hx >_T. \tag{A.51}$$

Now assume that $H$ is passive. In this case, $\langle x, Hx >_T \geq 0$, and (A.51) implies that

$$\|Sy\|_T^2 \leq \|y\|_T^2 \quad \Rightarrow \quad \|Sy\|_T \leq \|y\|_T \tag{A.52}$$

which implies that (A.44) is satisfied. On the other hand, if the gain of $S$ is less than or equal to 1, then (A.51) implies that

$$4\langle x, Hx >_T = \|y\|_T^2 - \|Sy\|_T^2 \geq 0$$

and thus $H$ is passive. This completes the proof of part (a).

(b) : Assume first that $H$ is strictly passive and has finite gain. The finite gain of $H$ implies that

$$\|Hx\|_T \leq \gamma(H)\|x\|_T.$$

Substituting (A.47) in the last equation, we have

$$
\begin{aligned}
\|y - x\|_T &\leq \gamma(H)\|x\|_T \\
\mid \|y\|_T - \|x\|_T \mid &\leq \gamma(H)\|x\|_T \\
\|y\|_T &\leq (1 + \gamma(H))\|x\|_T
\end{aligned}
$$

or

$$(1 + \gamma(H))^{-1}\|y\|_T \leq \|x\|_T. \tag{A.53}$$

Also, if $H$ is strictly passive, then

$$
\begin{aligned}
\langle x, Hx \rangle_T &\geq \delta\|x\|_T^2 \\
\Rightarrow \quad 4\langle x, Hx \rangle_T &\geq 4\delta\|x\|_T^2. \tag{A.54}
\end{aligned}
$$

Substituting (A.53) in (A.54), we obtain

$$
\begin{aligned}
4\langle x, Hx \rangle_T &\geq 4\delta(1 + \gamma(H))^{-2}\|y\|_T^2 \\
\text{or} \quad 4\langle x, Hx \rangle_T &\geq \delta'\|y\|_T^2 \tag{A.55}
\end{aligned}
$$

where

$$0 < \delta' < \min[1, 4\delta(1 + \gamma(H))^{-2}].$$

Thus, substituting (A.55) in (A.51), we see that

$$
\begin{aligned}
\|Sy\|_T^2 + \delta'\|y\|_T^2 &\leq \|y\|_T^2 \\
\Rightarrow \quad \|Sy\|_T &\leq (1 - \delta')^{1/2}\|y\|_T \tag{A.56}
\end{aligned}
$$

and since $0 < \delta' < 1$, we have that $0 < (1 - \delta')^{1/2} < 1$, which implies that $S$ has gain less than 1.

For the converse, assume that $S$ has gain less than 1. We can write

$$\|Sy\|_T^2 \leq (1 - \delta')\|y\|_T^2 \qquad 0 < \delta' < 1. \tag{A.57}$$

Substituting $\|Sy\|_T^2$ by its equivalent expression in (A.51), we have

$$
\begin{aligned}
\|y\|_T^2 - 4\langle x, Hx \rangle_T &\leq \|y\|_T^2 - \delta'\|y\|_T^2 \\
\Rightarrow \quad 4\langle x, Hx \rangle_T &\geq \delta'\|y\|_T^2
\end{aligned}
$$

and substituting $\|y\|_T^2$ using (A.50), we obtain

$$
\begin{aligned}
4\langle x, Hx \rangle_T &\geq \delta'(\|Hx\|_T^2 + \|x\|_T^2 + 2\langle x, Hx \rangle_T) \\
(4 - 2\delta')\langle x, Hx \rangle_T &\geq \delta'(\|Hx\|_T^2 + \|x\|_T^2) \geq \delta'\|x\|_T^2
\end{aligned}
$$

and since $0 < \delta' < 1$, we have

$$
\langle x, Hx \rangle_T \geq \frac{\delta'}{(4 - 2\delta')}\|x\|_T^2
$$

so that $H$ is strictly passive. To see that $H$ has also finite gain, substitute $\|Sy\|_T^2$ and $\|y\|_T^2$ in (A.56) by their equivalent expressions in (A.49) and (A.50) to obtain

$$
\begin{aligned}
\delta'\|Hx\|_T^2 &\leq (4 - 2\delta')\langle x, Hx \rangle_T - \delta'\|x\|_T^2 \\
&\leq (4 - 2\delta')\|Hx\|_T\|x\|_T - \delta'\|x\|_T^2 \quad \text{by Schwartz' inequality}
\end{aligned}
$$

and since $\delta'\|x\|_T^2 \geq 0$,

$$
\delta'\|Hx\|_T^2 \leq (4 - 2\delta')\|Hx\|_T\|x\|_T
$$

so that

$$
\|Hx\|_T \leq \frac{(4 - 2\delta')}{\delta'}\|x\|_T
$$

from where it follows that $H$ has finite gain. This completes the proof. $\qquad\square$

**Theorem 8.10** *Consider the feedback interconnection of Figure 7.5, and assume that*

*(i) $H_1$ is linear time-invariant, strictly proper, and SPR.*

*(ii) $H_2$ is passive (and possibly nonlinear).*

*Under these assumptions, the feedback interconnection is input-output-stable.*

**Proof**: We need to show that $y \in \mathcal{X}$ whenever $u \in \mathcal{X}$. To see this, we start by applying a loop transformation of the type I, with $K = -\epsilon$, $\epsilon > 0$, as shown in Figure A.4.

We know that the feedback system $S$ of Figure 7.5 is input-output stable if and only if the system $S_\epsilon$ of Figure A.4 is input-output stable. First consider the subsystem $H_2' = H_2 + \epsilon I$. We have,

$$
\langle x_T, (H_2 + \epsilon I)x_T \rangle = \langle x_T, H_2 x_T \rangle + \epsilon\langle x_T, x_T \rangle \geq \epsilon\langle x_T, x_T \rangle .
$$

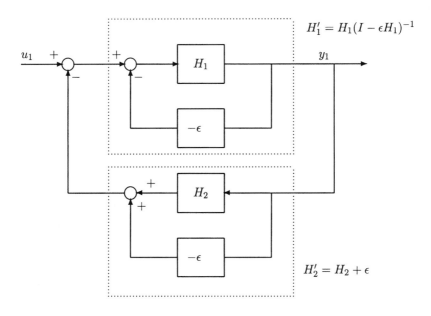

Figure A.4: The feedback system $S_\epsilon$.

It follows that $H_2'$ is strictly passive for any $\epsilon > 0$. We now argue that $H_1'$ is SPR for sufficiently small $\epsilon$. To see this, notice that, denoting $\gamma(H_1)$, then by the small gain theorem, $H_1'$ is stable for any $\epsilon < \gamma$. Also, since $H_1$ is SPR, there exist positive definite matrices $P$ and $L$, a real matrix $Q$ and $\mu$ sufficiently small such that (8.29)–(8.30) are satisfied. But then it follows that, given $P, L, Q, \mu$,

$$P(A + \epsilon BC) + (A + \epsilon BC)^T P = -QQ^T - \mu L + 2\epsilon C^T C$$

and provided that $0 \leq \epsilon < \epsilon^* = \mu\lambda_{min}[L]/(2\lambda_{max}[C^T C])$, the matrix $\mu L - 2\epsilon C^T C$ is positive definite. We conclude that for all $\epsilon \in (0, \min(\epsilon^*, \gamma^{-1}))$, $H_1'$ is strong SPR, hence passive, and $H_2'$ is strictly passive. Thus, the result follows from Theorem 8.4. $\qquad\square$

## A.6   Chapter 9

**Theorem 9.2**: *The nonlinear system $\psi$ given by (9.13) is QSR-dissipative [i.e., dissipative with supply rate given by (9.6)] if there exists a differentiable function $\phi : \mathbb{R}^n \to \mathbb{R}$ and function $L : \mathbb{R}^n \to \mathbb{R}^q$ and $W : \mathbb{R}^n \to \mathbb{R}^{q \times m}$ satisfying*

$$\phi(x) \;\geq\; 0, \quad \phi(0) = 0 \tag{A.58}$$

$$\frac{\partial\phi}{\partial x} f(x) \;=\; h^T(x)Qh(x) - L^T(x)L(x) \tag{A.59}$$

$$\frac{1}{2}g^T \left(\frac{\partial\phi}{\partial x}\right)^T \;=\; S^T h(x) - W^T L(x) \tag{A.60}$$

$$R \;=\; W^T W. \tag{A.61}$$

**Proof**: Sufficiency was proved in Chapter 9. To prove necessity we show that the available storage $\phi_a(\cdot)$ defined in Definition 9.5 is a solution of equations (A.61)(A.61) for appropriate functions $L$ and $W$.

Notice first that for any state $x_0$ there exist a control input $u \in \mathcal{U}$ that takes the state from any initial state $x(t_{-1})$ at time $t = t_{-1}$ to $x_0$ at $t = 0$. Since the system is dissipative, we have that

$$\int_{t_0}^{t_1} \omega(t) \, \mathrm{d}t = \phi(x(t_1)) - \phi(x(t_0)) \;\geq\; 0.$$

In particular

$$\int_{t_{-1}}^{T} \omega(t) \, \mathrm{d}t = \int_{t_{-1}}^{0} \omega(t) \, \mathrm{d}t + \int_{0}^{T} \omega(t) \, \mathrm{d}t = \phi(x(T)) - \phi(x(t_{-1})) \;\geq\; 0$$

that is

$$\int_0^T \omega(t) \, dt \geq - \int_{t-1}^0 \omega(t) \, dt$$

The right-hand side of the last inequality depends only on $x_0$, whereas $u$ can be chosen arbitrarily on $[0, T]$. Hence there exists a bounded function $C : \mathbb{R}^n \to \mathbb{R}$ such that

$$\int_0^T \omega(t) \, dt \geq C(x_0) > -\infty$$

which implies that $\phi_a$ is bounded. By Theorem 9.1 the available storage is itself a storage function, i.e.,

$$\int_0^t \omega(s) \, ds \geq \phi_a(x(t)) - \phi_a(x_0) \quad \forall t \geq 0$$

which, since $\phi_a$ is differentiable by the assumptions of the theorem, implies that

$$d(x, u) \overset{def}{=} \omega - \frac{d\phi_a(x)}{dt} \geq 0.$$

Substituting (9.13), we have that

$$\begin{aligned}
d(x, u) &= -\frac{d\phi_a(x)}{dt} + \omega(u, y) \\
&= -\frac{\partial \phi_a(x)}{\partial x}[f(x) + g(x)u] + y^T Q y + 2 y^T S u + u^T R u \\
&= -\frac{\partial \phi_a(x)}{\partial x} f(x) - \frac{\partial \phi_a(x)}{\partial x} g(x) u + h^T Q h + 2 h^T S u + u^T R u
\end{aligned}$$

we notice that $d(x, u)$, so defined, has the following properties: (i) $d(x, u) \geq 0$ for all $x, u$; and (ii) it is quadratic in $u$. It then follows that $d(x, u)$ can be factored as

$$\begin{aligned}
d(x, u) &= [L(x) + W u]^T [L(x) + W u] \\
&= L^T L + 2 L^T W u + u^T W^T W u
\end{aligned}$$

which implies that

$$\begin{aligned}
R &= W^T W \\
\frac{\partial S_a}{\partial x} f &= h^T(x) Q h(x) - L^T(x) L(x) \\
\frac{1}{2} g^T \frac{\partial S_a}{\partial x} &= S^T h(x) - W^T L(x).
\end{aligned}$$

This completes the proof. $\qquad\qquad\qquad\qquad\qquad\qquad\qquad\qquad\qquad\qquad\qquad\square$

## A.7   Chapter 10

**Theorem 10.2**: *The system (10.28) is input–state linearizable on $D_0 \subset D$ if and only if the following conditions are satisfied:*

(i) *The vector fields $\{g(x),\ ad_f g(x), \cdots,\ ad_f^{n-1} g(x)\}$ are linearly independent in $D_0$. Equivalently, the matrix*

$$C = [g(x),\ ad_f g(x), \cdots,\ ad_f^{n-1} g(x)]_{n \times n}$$

*has rank $n$ for all $x \in D_0$.*

(ii) *The distribution $\Delta = span\{g,\ ad_f g, \cdots,\ ad_f^{n-2} g\}$ is involutive in $D_0$.*

**Proof**: Assume first that the system (10.28) is input–state linearizable. Then there exists a coordinate transformation $z = T(x)$ that transforms (10.28) into a system of the form $\dot{z} = Az + Bv$, with $A = A_c$ and $B = B_c$, as defined in Section 10.2. From (10.17) and (10.18) we know that $T$ is such that

$$
\begin{aligned}
\frac{\partial T_1}{\partial x} f(x) &= T_2(x) \\
\frac{\partial T_2}{\partial x} f(x) &= T_3(x) \\
&\ \ \vdots \\
\frac{\partial T_{n-1}}{\partial x} f(x) &= T_n(x) \\
\frac{\partial T_n}{\partial x} f(x) &\neq 0
\end{aligned}
\tag{A.62}
$$

and

$$
\begin{aligned}
\frac{\partial T_1}{\partial x} g(x) &= 0 \\
\frac{\partial T_2}{\partial x} g(x) &= 0 \\
&\ \ \vdots \\
\frac{\partial T_{n-1}}{\partial x} g(x) &= 0 \\
\frac{\partial T_n}{\partial x} g(x) &\neq 0.
\end{aligned}
\tag{A.63}
$$

Equations (A.62) and (A.63) can be rewritten as follows:

$$L_f T_i = L_f T_{i+1}, \qquad i = 1, 2, \cdots, n-1. \tag{A.64}$$

$$L_g T_1 = L_g T_2 = \cdots = L_g T_{n-1} = 0 \quad L_g T_n \neq 0 \tag{A.65}$$

By the Jabobi identity we have that

$$\begin{aligned} \nabla T_1 [f, g] &= \nabla(L_g T_1) f - \nabla(L_f T_1) g \\ &= 0 - L_g T_2 = 0 \end{aligned}$$

or

$$\nabla T_1 \, ad_f g = 0.$$

Similarly

$$\nabla T_1 \, ad_f^k g = 0 \quad k = 0, 1, \cdots, n-2 \tag{A.66}$$

$$\nabla T_1 \, ad_f^{n-1} g \neq 0. \tag{A.67}$$

We now claim that (A.66)-(A.67) imply that the vector fields $g$, $ad_f g, \cdots, ad_f^{n-1} g$ are linearly independent. To see this, we use a contradiction argument. Assume that (A.66) (A.67) are satisfied but the $g$, $ad_f g, \cdots, ad_f^{n-1} g$ are not all linearly independent. Then, for some $i \leq n-1$, there exist scalar functions $\lambda_1(x)$, $\lambda_2(x)$, $\cdots$, $\lambda_{i-1}(x)$ such that

$$ad_f^i g = \sum_{k=0}^{i-1} \lambda_k \, ad_f^k g$$

and then

$$ad_f^{n-1} g = \sum_{k=n-i-1}^{n-2} \lambda_k \, ad_f^k g$$

and taking account of (A.66). we conclude that

$$\nabla T_1 \, ad_f^{n-1} g = \sum_{k=n-i-1}^{n-2} \lambda_k \, \nabla T_1 \, ad_f^k g = 0$$

which contradicts (A.67). This proves that (i) is satisfied. To prove that the second property is satisfied notice that (A.66) can be written as follows

$$\nabla T_1 [g(x), \, ad_f g(x), \cdots, ad_f^{n-2} g(x)] = 0 \tag{A.68}$$

that is, there exist $T_1$ whose partial derivatives satisfy (A.68). Hence $\Delta$ is completely integrable and must be involutive by the Frobenius theorem.

Assume now that conditions (i) and (ii) of Theorem 10.2 are satisfied. By the Frobenius theorem, there exist $T_1(x)$ satisfying

$$L_g T_1(x) = L_{ad_f g} T_1 = \cdots = L_{ad_f^{n-2} g} T_1 = 0$$

and taking into account the Jacobi identity, this implies that

$$L_g T_1(x) = L_g L_f T_1(x) = \cdots = L_g L_f^{n-2} T_1(x) = 0$$

but then we have that

$$\nabla T_1(x)\mathcal{C} = \nabla T_1(x)[g, \, ad_f g(x), \, \cdots, \, ad_f^{n-1} g(x)] = [0, \, \cdots, \, 0, \, L ad_f^{n-1} g T_1(x)].$$

The columns of the matrix $[g, \, ad_f g(x), \, \cdots, \, ad_f^{n-1} g(x)]$ are linearly independent on $D_0$ and so rank$(\mathcal{C}) = n$, by condition (i) of Theorem 10.2, and since $\nabla T_1(x) \neq 0$, it must be true that

$$ad_f^{n-1} g T_1(x) \neq 0$$

which implies, by the Jacobi identity, that

$$L_g L_f^{n-1} T_1(x) \neq 0.$$

$\square$

**Proof of theorem 10.3**:

The proof of Theorem 10.3 requires some preliminary results. In the following lemmas we consider a system of the form

$$\begin{cases} \dot{x} = f(x) + g(x)u, \\ y = h(x), \end{cases} \qquad \begin{aligned} f, g : D \subset \mathbb{R}^n \to \mathbb{R}^n \\ h : D \subset \mathbb{R}^n \to \mathbb{R} \end{aligned} \qquad (A.69)$$

and assume that it has a relative degree $r < n$.

**Lemma A.1** *If the system (A.69) has relative degree $r < n$ in $\Omega$, then*

$$L_{ad_f^j g} L_f^k h(x) = \begin{cases} 0 & 0 \leq j+k < r-1 \\ (-1)^j L_g L_f^{r-1} h(x) \neq 0 & j+k = r-1 \end{cases} \qquad (A.70)$$

$\forall x \in \Omega, \, \forall j \leq r-1, \, k > 0.$

**Proof**: We use the induction algorithm on $j$.

(i) $\underline{j = 0}$: For $j = 0$ condition (A.70) becomes

$$L_g L_f^k h(x) = \begin{cases} 0 & 0 \leq k < r - 1 \\ L_g L_f^{r-1} h(x) & k = r - 1 \end{cases}$$

which is satisfied by the definition of relative degree.

(ii) $\underline{j = i}$: Continuing with the induction algorithm, we assume that

$$L_{ad_f^i g} L_f^k h(x) = \begin{cases} 0 & 0 \leq i + k < r - 1 \\ (-1)^i L_g L_f^{r-1} h(x) & i + k = r - 1 \end{cases} \tag{A.71}$$

is satisfied, and show that this implies that it must be satisfied for $j = i + 1$.
From the Jacobi identity we have that

$$L_{ad_{fg}\beta} \lambda = L_f L_\beta \lambda - L_\beta L_f \lambda$$

for any smooth function $\lambda(x)$ and any smooth vector fields $f(x)$ and $\beta(x)$. Defining

$$\begin{aligned} \lambda &= L_f^k h(x) \\ \beta &= ad_f^i g \end{aligned}$$

we have that

$$L_{ad_f^{i+1} g} L_f^k h(x) = L_f L_{ad_f^i g} L_f^k h(x) - L_{ad_f^i g} L_f^{k+1} h(x) \tag{A.72}$$

Now consider any integer $k$ that satisfies $i + 1 + k \leq r - 1$. The first summand in the right-hand side of (A.72) vanishes, by (A.71). The second term on the right-hand side is

$$-L_{ad_f^i g} L_f^{k+1} h(x) = \begin{cases} 0 & 0 \leq i + k + 1 < r - 1 \\ (-1)^i L_g L_f^{r-1} h(x) & i + k + 1 = r - 1 \end{cases}$$

and thus the lemma is proved. □

**Lemma A.2** *If the relative degree of the system (A.69) is $r$ in $\Omega$, then $\nabla h(x)$, $\nabla L_f h(x)$, $\cdots$, $\nabla L_f^{r-1} h(x)$ are linearly independent in $\Omega$.*

**Proof:** Assume the contrary, specifically, assume that $\nabla h(x)$, $\nabla L_f h(x)$, $\cdots$, $\nabla L_f^{r-1} h(x)$ are not linearly independent in $\Omega$. Then there exist smooth functions $\alpha_1(x)$, $\cdots$, $\alpha_r(x)$ such that

$$\alpha_1 \nabla h(x) + \alpha_2 \nabla L_f h(x) + \cdots + \alpha_r \nabla L_f^{r-1} h(x) = 0. \tag{A.73}$$

Multiplying (A.73) by $ad_f^0 g = g$ we obtain

$$\alpha_1 L_g h(x) + \alpha_2 L_g L_f h(x) + \cdots + \alpha_r L_g L_f^{r-1} h(x) = 0. \qquad (A.74)$$

By assumption, the system has relative degree $r$. Thus

$$L_g L_f^i h(x) = 0 \quad \text{for } 0 \le i \le r - 1$$
$$L_g L_f^{r-1} h(x) \ne 0.$$

Thus (A.74) becomes

$$\alpha_r L_g L_f^{r-1} h(x) = 0$$

and since $L_g L_f^{r-1} \ne 0$ we conclude that $\alpha_r$ must be identically zero in $\Omega$.

Next, we multiply (A.73) by $ad_f g$ and obtain

$$\alpha_1 ad_f gh(x) + \alpha_2 ad_f g L_f h(x) + \cdots + \alpha_r ad_f g L_f^{r-1} h(x) = -\alpha_{r-1} ad_f g L_f^{r-1} h(x) = 0$$

where lemma A.1 was used. Thus, $\alpha_{r-1} = 0$. Continuing with this process (multiplying each time by $ad_f^2 g, \cdots, ad_f^{r-1} g$ we conclude that $\alpha_1, \cdots, \alpha_r$ must be identically zero on $\Omega$, and the lemma is proven. $\qquad \square$

**Theorem 10.3**: *Consider the system (10.41) and assume that it has relative degree $r <$ $n$ $\forall x \in D_0 \subset D$. Then, for every $x_0 \in D_0$ there exist a neighborhood $\Omega$ of $x_0$ and smooth function $\mu_1, \cdots, \mu_{n-r}$ such that*

*(i)* $L_g \mu_i(x) = 0, \quad$ *for $1 \le i \le n - r, \quad \forall x \in \Omega$, and*

*(ii)*

$$T(x) = \begin{bmatrix} \mu_1(x) \\ \vdots \\ \mu_{n-r}(x) \\ --- \\ \psi_1 \\ \vdots \\ \psi_r \end{bmatrix}$$

$$\psi_1 = h(x)$$
$$\psi_2 = L_f h(x)$$
$$\vdots$$
$$\psi_r = L_f^{r-1} h(x)$$

*is a diffeomorphism on $\Omega$.* $\qquad \square$

**Proof of Theorem 10.3**: To prove the theorem, we proceed as follows.

The single vector $g$ is clearly involutive and thus, the Frobenius theorem guarantees that for each $x_0 \in D$ there exist a neighborhood $\Omega$ of $x_0$ and $n-1$ linearly independent smooth functions $\mu_1, \cdots, \mu_{n-1}(x)$ such that

$$L_g \mu_i(x) = 0 \qquad \text{for } 1 \le i \le n-1 \ \forall x \in \Omega$$

also, by (A.73) in Lemma A.2, $\nabla h(x), \cdots, \nabla L_f^{n-2} h(x)$ are linearly independent. Thus, defining

$$T(x) = \begin{bmatrix} \mu_1(x) \\ \vdots \\ \mu_{n-r}(x) \\ --- \\ h(x) \\ \vdots \\ L_f^{r-1} h(x) \end{bmatrix}$$

i.e., with $h(x), \cdots, L_f^{r-1} h(x)$ in the last $r$ rows of $T$, we have that

$$\nabla L_f^{r-1} h(x_0) \notin \text{span}\{\nabla \mu_1, \cdots, \nabla \mu_{n-1}\}$$

$$\Rightarrow \text{rank}\left(\left[\frac{\partial T}{\partial x}(x_0)\right]\right) = n$$

which implies that $\frac{\partial T}{\partial x}(x_0) \neq 0$. Thus, $T$ is a diffeomorphism in a neighborhood of $x_0$, and the theorem is proved. $\qquad\square$

# Bibliography

[1] B. D. O. Anderson and S. Vongpanitlerd, Network Analysis and Synthesis: A Modern Systems Theory Approach, Prentice-Hall, Englewood Cliffs, NJ, 1973.

[2] P. J. Antsaklis and A. N. Michel, Linear Systems, McGraw-Hill, New York, 1977.

[3] V. I. Arnold, Ordinary Differential Equations, MIT Press, Cambridge, MA, 1973.

[4] A. Atassi, and H. Khalil, "A Separation Principle for the Stabilization of a Class of Nonlinear Systems," *IEEE Trans. Automat. Control,*" Vol. 44, No. 9, pp. 1672-1687, 1999.

[5] A. Atassi, and H. Khalil, "A Separation Principle for the Control of a Class of Nonlinear Systems," *IEEE Trans. Automat. Control,*" Vol. 46, No. 5, pp. 742–746, 2001.

[6] J. A. Ball, J. W. Helton, and M. Walker, "$H_\infty$ Control for Nonlinear Systems via Output Feedback," *IEEE Trans. on Automat. Control*, Vol. AC-38, pp. 546–559, 1993.

[7] R. G. Bartle, The Elements of Real Analysis, 2nd ed., Wiley, New York, 1976.

[8] , S. Battilotti, "Global Output Regulation and Disturbance ttenuation with Global Stability via Measurement Feedback for a Class of Nonlinear Systems," *IEEE Trans. Automat. Control*, Vol. AC-41, pp. 315–327, 1996.

[9] W. T. Baumann and W. J. Rugh, "Feedback Control of Nonlinear Systems by Extended Linearization," *IEEE Trans. Automat. Control*, Vol. AC-31, pp. 40–46, Jan. 1986.

[10] D. Bestle and M. Zeitz, "Canonical Form Observer Design for Nonlinear Time-Variable Systems," *Int. J. Control*, Vol. 38, pp. 419–431, 1983.

[11] W. A. Boothby, An Introduction to Differential Manifolds and Riemaniann Geometry, Academic Press, New York, 1975.

[12] R. W. Brockett, "Feedback Invariants for Non-Linear Systems," *Proc. IFAC World Congress*, Vol. 6, pp. 1115–1120, 1978.

[13] W. L. Brogan, Modern Control Theory, 3rd ed., Prentice Hall, ENglewood Cliffs, NJ, 1991.

[14] C. I. Byrnes and A. Isidori, "A Frequency Domain Philosophy for Nonlinear Systems," *Proc. 23rd IEEE Conf. Decision and Control*, pp. 1569–1573, 1984.

[15] C. T. Chen, Linear Systems Theory and Design, 2nd ed., Holt, Rinehart and Winston, New York, 1984.

[16] M. A. Dahleh and J. Pearson, "$\ell^1$-Optimal Controllers for Discrete-Time Systems," Proc. American Control Conf., Seattle, WA, pp. 1964–1968, 1986.

[17] M. A. Dahleh and J. Pearson, "$\ell^1$-Optimal Controllers for MIMO Discrete-Time Systems," *IEEE Trans. on Automatic Control*, Vol. 32, No. 4, 1987.

[18] M. A. Dahleh and J. Pearson, "$\mathcal{L}^1$-Optimal Compensators for Continuous-Time Systems," *IEEE Trans. on Automatic Control*, Vol. 32, No. 10, pp. 889–895, 1987.

[19] M. A. Dahleh and J. Pearson, "Optimal Rejection of Persistent Disturbances, Robust Stability and Mixed Sensitivity Minimization," *IEEE Trans. on Automatic Control*, Vol. 33, No. 8, pp. 722–731, 1988.

[20] M. A. Dahleh and I. J. Diaz-Bobillo, Control of Uncertain Systems: A Linear Programming Approach, Prentice-Hall, Englewood Cliffs, NJ, 1995.

[21] C. A. Desoer and M. Vidyasagar, "Feedback Systems: Input-Output Properties", Academic Press, New York, 1975.

[22] R. L. Devaney, An Introduction to Chaotic Dynamical Systems, 2nd ed.," Addison-Wesley, Reading, MA, 1989.

[23] J. C. Doyle, K. Glover, P. P. Khargonekar and B. A. Francis, "State Space Solutions to Standard $H_2$ and $H_\infty$ Control Probles," *IEEE Trans. on Automatic Control*, Vol. 38, No. 8, pp. 831–847, August 1989.

[24] I. Fantoni and R. Lozano, Non-Linear Control for Underactuated Mechanizal Systems, Springer-Verlag, New York, 2002.

[25] B. Francis, A Course in $H_\infty$ Control Theory, Springer-Verlag, New York, 1987.

[26] J. Guckenheimer and P. Holmes, Nonlinear Oscillations, Dynamical Systems, and Bifurcations of Vector Fields, Springer-Verlag, New York, 1983.

[27] W. Hahn, Stability of Motion, Springer-Verlag, Berlin, 1967.

[28] J. Hauser, S. Sastry, and P. Kokotović, "Nonlinear Control via Approximate Input-Output Linearization: The Ball and Beam Example," *IEEE Trans. on Automat. Control*, Vol 37, pp. 392–398, March 1992.

[29] D. J. Hill and P. J. Moylan, "Dissipative Dynamical Systems: Basic State and Input-Output Properties," *the J. Franklin Inst.*, Vol. 309, No. 1 pp. 327–357, Jan. 1980.

[30] D. J. Hill and P. J. Moylan, "Stability of Nonlinear Dissipative Systems," *IEEE Trans. Automat. Control*, Vol. AC-21, pp. 708–711, Oct. 1976.

[31] D. J. Hill and P. J. Moylan, "Stability Results for Nonlinear Feedback Systems," *Automatica*, Vol. 13, pp. 377–382, July 1977.

[32] M. W. Hirsch and S. Smale, Differential Equations, Dynamical Systems and Linear Algebra, Academic Press, New York, 1974.

[33] P. R. Halmos, Finite Dimensional Vector Spaces, Springer-Verlag, New York, 1974, reprinted 1987.

[34] L. R. Hunt, R. Su, and G. Meyer, "Design for Multi-Input Nonlinear Systems," in *Differential Geometric Control Theory*, R. W. Brockett, R. S. Millman, and H. Sussmann, eds., Birhauser, Boston, pp. 268-298, 1983.

[35] P. Ioannou and G. Tao, "Frequency Domain Conditions for Strictly Positive Real Functions," *IEEE Trans. Automat. Control*, Vol 32, pp. 53–54, Jan. 1987.

[36] A. Isidori, Nonlinear Control Systems, 3rd ed., Springer-Verlag, New York, 1995.

[37] A. Isidori, Nonlinear Control Systems, Springer-Verlag, New York, 1999.

[38] A. Isidori and A. Astolfi, "Disturbance Attenuation and $H_\infty$ Control via Measurement Feedback in Nonlinear Systems," *IEEE Trans. Automat. Control*, Vol. AC-37, pp. 1283–1293, 1992.

[39] B. Jakubczyk and W. Respondek, "On Linearization of Control Systems," *Bull. Acad. Polonaise Sci. Ser. Sci. Math.*, Vol. 28, pp. 1-33, 1980.

[40] T. Kailath, Linear Systems, Prentice-Hall, Englewood Cliffs, NJ, 1980.

[41] H. K. Khalil, Nonlinear Systems, 2nd ed., Prentice-Hall, Englewood Cliffs, NJ, 1996.

[42] H. Keller, "Non-Linear Observer Design by Transformation into a Generalized Observer Canonical Form," *Int. J. Control*,", Vol. 46, No. 6, pp. 1915–1930, 1987.

[43] P. Kokotović, "The Joy of Feedback: Nonlinear and Adaptive," *IEEE Control Sys. Mag.*, pp. 7–17, June 1992.

[44] N. N. Krasovskii, Stability of Motion, Stanford University Press, Stanford, CA, 1963.

[45] A. J. Krener and A. Isidori, "Linearization by Output Injection and Nonlinear Observers," *Syst. Control Lett.*, Vol. 3, pp. 47–52, 1983.

[46] A. J. Krener and W. Respondek, "Nonlinear Observers with Linearizable Error Dynamics," *SIAM J. Control Optimization*, Vol. 23, pp. 197–216, 1985.

[47] M. Krstić, I. Kanellakopoulos, and P. Kokotović, Nonlinear and Adaptive Control Design, Wiley, New York, 1995.

[48] J. P. LaSalle, and S. Lefschetz, Stability by Lyapunov's Direct Method, Academic Press, New York, 1961.

[49] J. P. LaSalle, "Some Extension of Lyapunov's Second Method," *IRE Trans. of Circuit Theory*, Vol. 7, No. 4, pp. 520–527, Dec. 1960.

[50] E. N. Lorenz, "Deterministic Nonperiodic Flow," *J. Atmos. Sci.*, Vol. 20, pp. 130–141, 1963.

[51] I. J. Maddox, Elements of Functional Analysis, 2nd Edition, Cambridge University Press, Cambridge, UK, 1988.

[52] R. Marino and P. Tomei, Nonlinear Control Design: Geometric, Adaptive and Robust, Prentice-Hall, Englelwood Cliffs, NJ, 1995.

[53] R. Marino and P. Tomei, "Adaptive Observers for a Class of Multi-Output Nonlinear Systems," *Int. J. Adaptive Control Signal Process.*,", Vol. 6, No. 6, pp. 353–365, 1992.

[54] R. Marino and P. Tomei, "Adaptive Observers with Arbitrary Exponential Rate of Convergence for Nonlinear Systems," *IEEE Trans. Automat. Control*, Vol. 40, No. 7, pp. 1300–1304, 1995.

[55] R. K. Miller and A. N. Michel, Ordinary Differential Equations, Academic Press, New York, 1982.

[56] K. Narendra and A. M. Anaswami, Stable Adaptive Systems, Prentice-Hall, Englewood Cliffs, NJ. 1989.

[57] H. Nijmeijer and A. van der Schaft, Nonlinear Dynamical Control Systems, Springer Verlag, Berlin, 1990.

[58] E. Ott, Chaos in Dynamical Systems, Cambridge University Press, Cambridge, UK, 1993.

[59] L. Perko, Differential Equations and Dynamical Systems, Springer-Verlag, New York, 1991.

[60] S. Raghavan and J. K. Hedrick, "Observer Design for a Class of Nonlinear Systems," Int. J. Control, Vol. 59, No. 2, pp. 515–528, 1994.

[61] R. Rajamani, "Observers for Lipschitz Nonlinear Systems," IEEE Trans. Automat. Control, Vol. AC-43, pp. 397–401, March 1998.

[62] W. Rudin, Principles of Mathematical Analysis, 3rd ed., McGraw-Hill, New York, 1976.

[63] W. J. Rugh, "Analytical Framework for Gain Scheduling", IEEE Control Systems Magazine, Vol. 11, No. 1, pp. 79–84, Jan. 1991.

[64] W. J. Rugh, Linear System Theory, 2nd ed., Prentice Hall, Englewood Cliffs, NJ, 1996.

[65] I. W. Sandberg, "On the $\mathcal{L}_2$-Boundedness of Solutions of Nonlinear Functional Equations," Bell Sys. Tech. J., Vol.43, pp. 1581–1599, 1964.

[66] I. W. Sandberg, "An observation Concerning the Application of the Contraction Mapping Fixed-Point Theorem and a Result Concerning the Norm-Boundedness of Solutions of Nonlinear Functional Equations," Bell Sys. Tech J., Vol.44, pp. 1809–1812, 1965.

[67] I. W. Sandberg, "Some Results Concerning the Theory of Physical Systems Governed by Nonlinear Functional Equations," Bell Sys. Tech J., Vol.44, pp. 871–898, 1965.

[68] J.-J. E. Slotine and W. Li, Applied Nonlinear Control, Prentice-Hall, Englewood Cliffs, NJ, 1991.

[69] E. D. Sontag, "Smooth Stabilization Implies Coprime Factorization," IEEE Trans. Automat. Control, vol. 34, pp. 435–443, 1989.

[70] E. D. Sontag, "State Space and I/O Stability for Nonlinear Systems," Feedback Control, Nonlinear Systems and Complexity, B. Francis and A. R. Tannenbaum, eds., Lecture Notes in Control and Information Sciences, Springer-Verlag, Berlin, pp. 215–235, 1995.

[71] E. D. Sontag, "Further Facts about Input-to-State Stabilization," IEEE Trans. Automat. Control, Vol.35, No.4, pp. 473–476, 1990.

[72] E. D. Sontag, "On the Input-to-State Stability Property," *European Journal of Control,* pp. 24-36, 1995.

[73] E. D. Sontag and Y. Wang, "On Characterizations of the Input-to-State Stability Property," *Syst. Control Lett.*, Vol. 24, pp. 351-359, 1995.

[74] E. D. Sontag and Y. Wang, "New Characterizations of Input-to-State Stability," *IEEE Trans. Automat. Control*, Vol. 41, pp. 1283–1294, 1996.

[75] E. D. Sontag and A. Teel, "Changing Supply Functions in Input/State Stable Systems," *IEEE Trans. Automat. Control*, Vol. 40, pp. 1476–1478, 1995.

[76] S. H. Strogatz, Nonlinear Dynamics and Chaos, Addison-Wesley, Reading, MA, 1994.

[77] R. Su, "On the Linear Equivalents of Nonlinear Systems, *Syst. Control Lett.*, Vol. 2, pp. 48–52, 1982.

[78] G. Tao and P. Ioannou, "Strictly Positive Real Matrices and the Lefschetz-Kalman-Yakubovich Lemma," *IEEE Trans. Automat. Control*, Vol 33, pp. 1183–1185, Dec. 1988.

[79] J. H. Taylor, "Strictly Positive Real Functions and the Lefschetz-Kalman-Yakubovich (LKY) Lemma," *IEEE Trans. Circuits Sys.*, pp. 310–311, March 1974.

[80] F. E. Thau, "Observing the State of Nonlinear Dynamic Systems," *Int. J. Control*, Vol. 17, No. 3, pp. 471–479, 1973.

[81] D. L. Trumper, S. M. Olson and P. K. Subrahmanyan, "Linearizing Control of Magnetic Suspension Systems," *IEEE Trans. Control Sys. Technol.*, Vol. 5, No. 4, pp. 427–438, July 1997.

[82] J. Tsinias, "A Generalization of Vidyasagar's theorem on Stability Using State Detection," *Syst. Control Lett.*, Vol. 17, pp. 37–42, 1991.

[83] J. Tsinias, "Sontag's Input-to-State Stability Condition and Global Stabilization Using State Detection," *Syst. Control Lett.*, Vol. 20, pp. 219–226, 1993.

[84] B. Van der Pol, *Radio Review*, Vol. 1, pp. 704–754, 1920.

[85] A. van der Schaft, $\mathcal{L}_2$-Gain and Passivity Techniques in Nonlinear Control, Springer Verlag, London, UK, 1999.

[86] A. van der Schaft, "On a State Space Approach to Nonlinear $H_\infty$ Control," *Syst. Control Lett.*, Vol. 16, pp. 1–8, 1991.

[87] A. van der Schaft, "$\mathcal{L}_2$-Gain Analysis of Nonlinear Systems and Nonlinear State Feedback," *IEEE Trans. Automat. Control*, Vol. 37, pp. 770–784, 1992.

[88] M. Vidyasagar, Nonlinear Systems Analysis, 2nd ed., Prentice Hall, Englewood Cliffs, NJ, 1993.

[89] M. Vidyasagar, "Optimal Rejection of Persistent Bounded Disturbances," *IEEE Trans. Automat. Control*, Vol. 31, pp. 527–535, 1986.

[90] M. Vidyasagar, "On the Stabilization of Nonlinear Systems Using State Detection," *IEEE Trans. Automat. Control*, Vol. AC-25, No. 3, pp. 504–509, 1980.

[91] J. C. Willems, "Dissipative Dynamical Systems. Part I: General Theory," *Arch. Rational Mech. Anal.*, Vol. 45, pp. 321–351, 1972.

[92] J. C. Willems, The Analysis of Feedback Systems, MIT Press, Cambridge, MA, 1971.

[93] J. C. Willems, "The Generation of Lyapunov Functions for Input-Output Stable Systems," *SIAM J. Control*, Vol. 9, pp. 105–133, Feb. 1971.

[94] J. C. Willems, "Mechanisms for the Stability and Instability in Feedback Systems," *IEEE Proc.*, Vol. 64, pp. 24–35, 1976.

[95] J. L. Willems, Stability Theory of Dynamical Systems, Wiley, New York, 1970.

[96] S. H. Zak, "On the Stabilization and Observation of Nonlinear/Uncertain Dynamic Systems," *IEEE Trans. Automat. Control*, Vol. AC-35, No. 5. pp. 604–607, May 1990.

[97] G. Zames, "Functional Analysis Applied to Nonlinear Feedback Systems," *IEEE Trans. Circuit Theory*, Vol. CT-10, pp. 392–404, Sept. 1963.

[98] G. Zames, "On the Input/Output Stability of Nonlinear Time-Varying Feedback Systems", Parts I and II , *IEEE Trans. Automat. Control*, Vol. AC-11, pp. 228–238, and 465–477, 1966.

[99] G. Zames, "Feedback and Optimal Sensitivity: Model Reference Transformations, Multiplicative Seminorms and Approximate Inverses," *IEEE Trans. Automat. Control*, Vol. AC-26, pp. 301–320, 1981.

[100] K. Zhou, J. C. Glover, and J. C. Doyle, Robust and Optimal Control, Prentice-Hall, Englewood Cliffs, NJ, 1996.

# List of Figures

# Index